Comparative Cognition

James R. Anderson • Hika Kuroshima
Editors

Comparative Cognition

Commonalities and Diversity

Springer

Editors
James R. Anderson
Department of Psychology,
Graduate School of Letters
Kyoto University
Sakyo-ku, Kyoto, Japan

Hika Kuroshima
Department of Psychology,
Graduate School of Letters
Kyoto University
Sakyo-ku, Kyoto, Japan

ISBN 978-981-16-2027-0 ISBN 978-981-16-2028-7 (eBook)
https://doi.org/10.1007/978-981-16-2028-7

This Springer imprint is published by the registered company Springer Nature Singapore Pte Ltd.
The registered company address is: 152 Beach Road, #21-01/04 Gateway East, Singapore 189721, Singapore

Foreword

Professor Kazuo Fujita is a leading researcher in the field of comparative cognitive science. Comparative cognitive science is a branch of behavioral science that aims to understand the phylogeny of cognitive functions by comparing the cognitive functions of various animals, including humans (Fujita 1998, 2004). These cognitive functions include everything we believe to be the work of the mind, including, but not limited to, reasoning, emotion, and volition. In other words, the objective of comparative cognitive science is to elucidate how the mind evolves (Fujita 2004). In the past few decades, the number of researchers in comparative cognitive science has significantly increased all over the world. In Japan, one of the most important developments has been the adoption of the term "comparative cognitive science" in the examination section of the Ministry of Education, Culture, Sports, Science and Technology's Science Research Grants. This change is owing, in part, to Professor Fujita's remarkable research achievements in this area. Since Professor Fujita set up a laboratory in the Faculty of Letters at Kyoto University in 1996, he has been studying a wide variety of animals, including macaques, capuchin and squirrel monkeys, tree shrews, rats, hamsters, degus, pigeons, bantams, cockatiels, and bearded dragons, as well as domesticated animals: dogs, cats, and horses. In particular, his research on dogs and cats has attracted considerable international attention, with discoveries reported in international print media, on the internet, and on radio and TV programs both in Japan and further afield.

Traditionally, research in comparative psychology had focused on a limited number of species of nonhuman animals such as pigeons, rats, and mice. However, for decades Professor Fujita has been challenging this narrow traditional focus, strongly advocating broadening the scope of comparisons. The Fujita Lab continues to practice what he preached, by trying to analyze and map the diversity of the mind, through often highly innovative research on a variety of comparison species.

I have had the honor of collaborating with Prof. Fujita on two major projects, the twenty-first Century COE at Kyoto University and the Grant-in-Aid for Scientific Research (S), a large-scale scientific research fund at the Ministry of Education, Culture, Sports, Science and Technology. The Grant-in-Aid for Scientific Research

(S) consists of two key research themes. The first is a multifaceted study of the developmental processes in the active access to the inner parts of one's mind (i.e., consciousness and introspection), and the understanding of the mental states of others (i.e., mind-reading). This project involves an extensive number of species and developmental comparisons, and to date it has produced 444 research papers. The second theme aims to provide empirical insights into the developmental processes in the ability to freely manipulate mental representations—the self-reliance of the mind—again, through extensive species and developmental comparisons. At the time of writing (February 2021), this project is ongoing, involving over 180 researchers. I am also privileged to have co-edited two books in English with Professor Fujita. The first is Fujita, K. & Itakura, S. (2006). *Diversity of Cognition: Evolution, Development, Domestication and Pathology. Kyoto University Press.* This volume aims to shed light on the diversity of cognition in a wide variety of species, with 17 world-leading researchers from seven countries describing diverse modes of cognition in insects, humans of different ages, and various organisms in different pathological states, inspiring scientists and students to understand cognition from a genetic and adaptive perspective. The second book is Itakura, S. & Fujita, K. (2008). *Origins of the Social Mind: Evolutionary and Developmental Views. Springer.* It is a collection of empirical studies arising from conceptual insights from cognitive development, primate and avian cognition, written by some of the world's leading researchers. This book, with its comprehensive and novel research, is intended to appeal to anyone who is interested in how the social mind has developed and evolved.

The present volume consists of 18 chapters written by international researchers and students who are either closely related to Prof. Fujita or whose research has in some ways been influenced by him. The chapters cover a wide range of topics and species and are grouped under sections entitled: Visual Perception and Illusions, Face and Body Perception, Social Cognition, Social Emotions, Metacognition, and Companion Animals. The book fittingly reflects Prof. Fujita's philosophy concerning the diversity of research, as well as the evolution of his own particular interests over the past four decades. I would like to close this foreword by mentioning three of Kazuo Fujita's strong beliefs about comparative cognitive science. First, comparative cognitive science does not care about whether or not there is a mind in a particular life form. The answer to the question of what a mind is varies by definition, and the question itself is futile. The notion of the mind is perhaps better left undefined. Second, comparative cognitive science studies the diversity of the mind. The work of the mind varies depending on the organism's way of life. There are many species of animals living diverse lives on earth, and so the work of the mind must be diverse. Humans are just one species in our planet's diverse ecology. We often falsely believe humans to be unique and the best-adapted species, not because it is so, but because we ourselves are human beings, with an anthropocentric perspective. Third, comparative cognitive science cares about the evolution of mind, not only in humans but also in other animal species. This is because the minds of all species, including humans, have evolved together over the same evolutionary time in a highly diverse manner. In order to understand this diversity, one must understand all minds, not just our own. Indeed, we cannot fully understand the human mind without knowing about nonhuman minds. I sincerely hope and strongly believe that comparative cognitive science, in which Professor Kazuo Fujita has played a leading role, will continue to flourish for many years to come.

References

1. Fujita K (1998) Hikakunintikagaku heno shotai. [Introduction to comparative cognitive science.] Nakanishiya, Kyoto
2. Fujita K (2004) Hikakunintikagaku. [Comparative cognitive science.] In: Hatano G, Ohotsu Y (Eds)Nintikagaku Heno Shotai [Introduction to cognitive science] Kenkyusha, Tokyo, pp. 122–140
3. Fujita K, Itakura S (2006) Diversity of cognition: evolution, development, domestication and pathology. Kyoto University Press, Kyoto
4. Itakura S, Fujita K (2008) Origins of the social mind: evolutionary and developmental views. Springer, Tokyo

Center for Baby Science
Doshisha University
Kizugawa, Kyoto, Japan

Shoji Itakura

Preface and Acknowledgments

We express our sincere thanks to everyone who has contributed to bringing this volume to completion. In early 2019, as the official retirement of Kazuo Fujita from his position as Professor at Kyoto University approached, we started to think about organizing a collection of chapters that would give readers an accessible overview of some current "hot" topics in comparative cognitive science while celebrating Fujita's four decades of contributions to teaching and research in this field. One of us (Anderson) has worked in collaboration with Fujita for more than two of those decades, since first receiving the invitation to join him in Kyoto as Visiting Professor, in 1998. That visit started an almost unbroken run of twice-yearly visits to Fujita's lab, each of which gave rise to new studies on various topics in primate physical and social cognition. The other (Kuroshima) was then a master's student, working on several projects in Fujita's recently established animal laboratory. Under his supervision, she learned not only how to conduct noninvasive psychological research on a variety of animal species but also how to care for them and how to mentor younger students, all of which she continues to do today.

The origins of the book lie in our recognition of the important role of Fujita's research and his friendship in own research careers, as well as his positive influence on many other students and colleagues in Japan and elsewhere. To get things started, we wrote down a long list of potential contributors and then set about whittling it down to a shortlist. In the end, we contacted more than 20 people, including several former students of Fujita who were still active in academia, and other comparative cognitive psychologists whom we knew to be familiar with aspects of Fujita's research. The rate at which we received agreements to participate in the project was truly gratifying, as is seeing the volume finally completed and in print, particularly given various delays caused directly or indirectly by the COVID-19 pandemic.

In the pages that follow, the influence of Fujita's skillful and engaging teaching is reflected in the chapters written by former students of his laboratory at Kyoto University. The range of questions posed, the examples of experimental creativity, and attention to details all reflect, at least in part, his guidance and positive supervision. Indeed, the influence of his overall approach to research can be seen in the diversity of species and topics covered across all of the chapters, along with various conceptual and procedural innovations described therein. Fujita has not only inspired students, he has also stimulated research ideas and new directions for numerous other renowned researchers. We are grateful to those who made the book possible by generously contributing descriptions of their own studies and their thoughts about diverse topics in comparative cognition.

It is also our pleasure to acknowledge colleagues who carefully reviewed and offered thoughtful comments on earlier drafts of chapters. There is no doubt that the final versions benefited greatly from the support and encouragement of our reviewers, some of whom willingly worked on more than one chapter. For reviewing, we thank Elsa Addessi, Christian Agrillo, Sarah Brosnan, Jie Gao, Kazuo Goto, Katie Leighty, Toymoi Matsuno, Bill McGrew, Annika Paukner, Alan Silberberg, Sarah Vick, Kristyn Vitale, and Sota Watanabe. We also extend our thanks to staff at Springer for their support and patience, in particular, Fumiko Yamaguchi and Suraj Kumar. We are sure that Fujita-sensei would also like to see a statement of thanks to all the animals that have helped in their own ways to deepen our knowledge of other species' minds, and so we are happy to acknowledge this. Finally, we are grateful to Professor Shoji Itakura for providing the foreword at relatively short notice; it nicely captures the spirit of what the volume is intended to be.

Sakyo-ku, Kyoto, Japan
February 19, 2021

Jim Anderson
Hika Kuroshima

Contents

List of Contributors

James R. Anderson Department of Psychology, Kyoto University Graduate School of Letters, Kyoto, Japan

Michael J. Beran Language Research Center, Georgia State University, Atlanta, GA, USA

Lauren Brubaker Department of Animal and Rangeland Sciences, Oregon State University, Corvallis, OR, USA

Josep Call School of Psychology and Neuroscience, University of St Andrews, St Andrews, UK

Hitomi Chijiiwa Department of Psychology, Kyoto University Graduate School of Letters, Kyoto, Japan

Fabrice Damon Centre des Sciences du Gout et de l'Alimentation, Université Bourgogne Franche-Comté, Dijon, France

Kun Guo School of Psychology, University of Lincoln, Lincoln, UK

Yuko Hattori Primate Research Institute, Kyoto University, Inuyama, Japan

Chihiro Hiramatsu Department of Human Science, Faculty of Design, Kyushu University, Fukuoka, Japan

Sumie Iwasaki Institute of Human and Social Sciences, Kanazawa University, Kanazawa, Japan

Reiki Kishimoto Graduate School of Arts and Sciences, University of Tokyo, Tokyo, Japan

Hika Kuroshima Department of Psychology, Kyoto University Graduate School of Letters, Kyoto, Japan

Toyomi Matsuno Faculty of Economics, Hosei University, Tokyo, Japan

David Méary Laboratoire de Psychologie et Neurocognition, Université Grenoble Alpes, Grenoble, France

Hiromitsu Miyata Faculty of Letters, Arts and Sciences, Waseda University, Tokyo, Japan

Eduardo B. Ottoni Department of Experimental Psychology, University of São Paulo, São Paulo, Brazil

Audrey E. Parrish Department of Psychology, The Citadel, Charleston, SC, USA

Olivier Pascalis Laboratoire de Psychologie et Neurocognition, Université Grenoble Alpes, Grenoble, France

Giovanna Rosenlicht Department of Animal and Rangeland Sciences, Oregon State University, Corvallis, OR, USA

Manon K. Schweinfurth School of Psychology and Neuroscience, University of St Andrews, St Andrews, UK

Saho Takagi Japan Society for the Promotion of Science, Tokyo, Japan
Azabu University, Kanagawa, Japan

Ayaka Takimoto-Inose Department of Behavioral Science, Graduate School of Humanities and Human Sciences, Hokkaido University, Sapporo, Japan

Lauren E. Thielke Department of Animal and Rangeland Sciences, Oregon State University, Corvallis, OR, USA

Monique A. R. Udell Department of Animal and Rangeland Sciences, Oregon State University, Corvallis, OR, USA

Tomokazu Ushitani Graduate School of Humanities, Chiba University, Chiba, Japan

Kristyn R. Vitale Department of Animal and Rangeland Sciences, Oregon State University, Corvallis, OR, USA

Shelby S. H. Wanser Department of Animal and Rangeland Sciences, Oregon State University, Corvallis, OR, USA

Sota Watanabe Division of General Education, Osaka Kyoiku University, Osaka, Japan

Part I
Visual Perception and Illusions

Chapter 1
Amodal Completion, and Recognizing the Meaning of Cognitive Diversity

Tomokazu Ushitani

Abstract Amodal completion is the visual recognition of occluded portions of objects that are partly occluded by other objects. In this chapter, I review comparative cognitive studies on amodal completion, with emphasis on contributions by Kazuo Fujita and his collaborators. This topic has been studied in many animal species, and diversity in the underlying visual processing mechanisms has been revealed in several cases. The importance of diversity in these and other comparative cognitive studies is discussed.

Keywords Amodal completion · Visual perception · Cognitive diversity · Negative findings · Kazuo Fujita

Introduction: What Is Amodal Completion?

> "For there is nothing covered, that shall not be revealed; neither hid, that shall not be known."—Luke, 12: 22

Because we live in a three-dimensional world, many objects that we look at are partly occluded by other objects just in front of the former. An object that you are observing—for example, a painting on the wall of your room—may be partly occluded by the computer monitor on which you are reading this chapter. Even though you have no retinal images of the occluded portion of the painting, your visual system, if it is healthy, fills the absent information so that you recognize the whole percept of the painting. One may think that recognizing the whole percept of a partly occluded object might require some knowledge about the complete object, but such a filling-in process can occur even with unfamiliar objects. This suggests that the process occurs early in the visual recognition process, i.e., the perceptual stage; therefore, the process is referred to as perceptual completion.

T. Ushitani (✉)
Graduate School of Humanities, Chiba University, Chiba, Japan
e-mail: ushitani@faculty.chiba-u.jp

J. R. Anderson, H. Kuroshima (eds.), *Comparative Cognition*,
https://doi.org/10.1007/978-981-16-2028-7_1

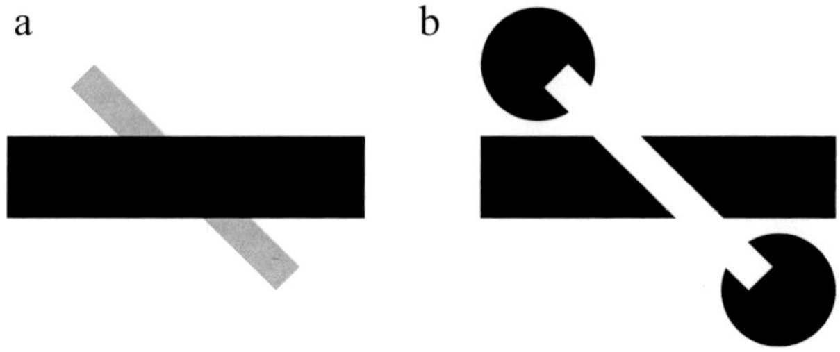

Fig. 1.1 (**a**) An example of amodal completion; (**b**) an example of modal completion

Perceptual completion is also called "amodal completion," in contrast to modal completion. Figure 1.1a is an example of amodal completion; we can see a gray rod located "behind" the black rectangle, though we can also recognize that these are two-dimensional figures with, in reality, no depth relationship between the elements. Although the retinal images are of two fragments of the gray rod at the top and bottom of the rectangle, human observers perceive the two fragments as continuous as a result of the completion of the occluded, central portion of the gray rod. A related phenomenon is modal completion. Figure 1.1b is an example of the Kanizsa illusion, a well-known modal completion phenomenon, in which you can see a white rod in front of the black rectangle. Modal completion is a phenomenon in which observers perceive visible qualities such as clear contours and a brighter surface shown in this example, even though there are no physical contours or surface in reality – such that the Kanizsa illusion is frequently called "subjective contour," whereas amodal completion provides us with no such percepts but just the recognition of existence of hidden contours or surfaces.

The view that amodal completion occurs at the perceptual level rather than knowledge level receives support from two types of empirical facts. First, experimental studies have shown that human infants as young as 4 months perceptually completed the occluded portion of a moving rod when it moved behind a block (Kellman and Spelke 1983). Four-month-old infants saw a rod moving behind a block so that the central portion of the former was occluded by the latter. The infants who lost attention in this event (habituation) regained it when they saw the rod with a gap at its central portion but not when they saw it with no gap, suggesting that the habituated stimulus had been perceived as the intact rod (i.e. the occluded portion had been completed). Second, Sugita (1999) found that cells in primary visual cortex (therefore at a very early stage in the visual pathway) in Japanese macaques responded both to a unitary rod and the same partly occluded rod, but not to a truncated rod even though in the latter two rods (partly occluded, and truncated) the visible parts of the rod were the same (Other physiological evidence had come from a study of capuchin monkey brain by Júnior et al. 1992).

Although studies go back a long way, amodal completion has attracted relatively little research attention probably because it is ubiquitous and common place in our daily life. However, amodal completion is a critical feature of in object recognition in humans, as illustrated by some cases of visual agnosia. Patients with integrative agnosia, a deficit in object recognition caused by a brain lesion, often experience

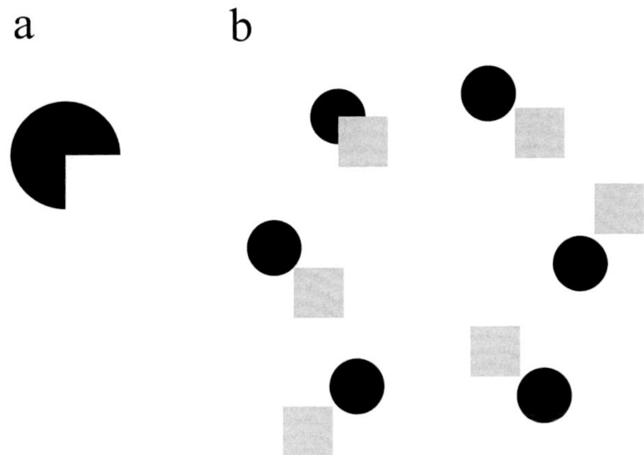

Fig. 1.2 Schematic illustration of stimuli used by Rauschenberger and Yantis (2001). (**a**) a target; (**b**) an example of search display

particular difficulty recognizing partly occluded objects (see Behrmann 2003, for a review). They cannot name visually presented objects, despite still having normal visual acuity and knowledge about the objects themselves, suggesting that amodal completion is important for accurately recognizing things that we see.

One reason why people pay little attention to amodal completion may be that we do it in an automatic and unconscious way. Rauschenberger and Yantis (2001) required participants to find partly truncated disks (i.e., notched circles, or more colloquially, pacman-like shapes; Fig. 1.2a) among multiple intact disks (distractors) in a standard visual-search task. In a condition in which a square was placed adjacent to the target (truncated disk) so as to fill the truncated part of the latter (Fig. 1.2b), the participants' accuracy and latency both deteriorated, because they completed the truncated part of the disk, meaning that the target now looked like an intact disk and easily confused with the distractors.

All of the studies described above suggest that for humans, amodal completion is a fundamental and probably essential visual function. When and why amodal completion has evolved is an important question if we want to better understand not only amodal completion itself but also the nature of the human visual system. In the rest of this chapter, I review comparative cognitive studies on amodal completion, with particular emphasis on how Kazuo Fujita and collaborators have contributed not only to the study of amodal completion, but also to comparative cognition more generally.

Comparative Research on Amodal Completion

The first attempt to find amodal completion in nonhuman animals was conducted by Cerella (1980). Individually tested pigeons were presented with a triangle (Fig. 1.3b) or non-triangle shapes (e.g., circles, stars, etc.; Fig. 1.3a) and trained to peck at the triangle but not at the other shapes. On test trials, the pigeons were presented with a

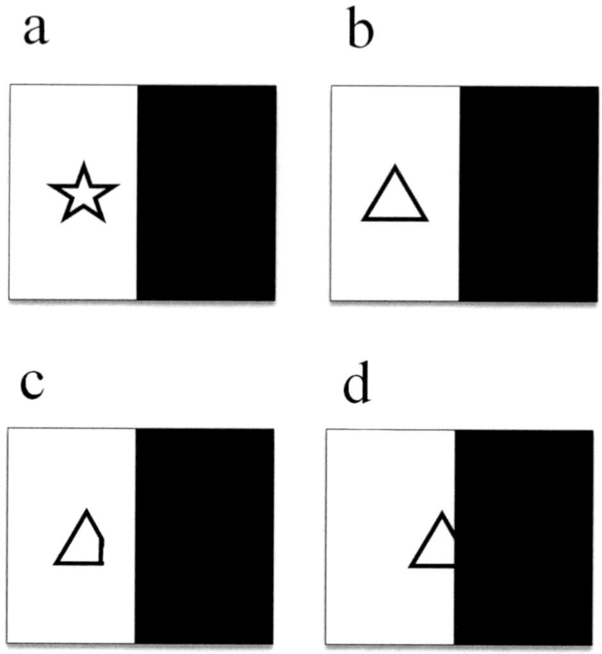

Fig. 1.3 Schematic illustration of stimuli used in Cerella (1980). Examples of (**a**) S-. (**b**) S+, (**c**) test stimulus (truncated condition), (**d**) test stimulus (occlusion condition)

partly truncated triangle (Fig. 1.3c) or the same triangle but with its truncated side abutting on a larger rectangle (in which due to amodal completion we see an intact triangle; Fig. 1.3d). If the pigeons perceptually completed the truncated part of the truncated triangle, they should respond to the triangle placed adjacent to the rectangle (Fig. 1.3d) but not (or, at least, less frequently) to the truncated triangle placed apart from the rectangle (Fig. 1.3c). However, the results were in the opposite direction: the pigeons responded more frequently to the latter stimulus than the former, thus showing no evidence of amodal completion.

More than a decade later, Sekuler et al. (1996) trained pigeons to discriminate between an intact circle and a pacman-like shape in a "yes/no" procedure. More specifically, a discriminative stimulus, either the intact circle or the pacman-like shape, appeared and the birds were rewarded for pecking at the corresponding key out of the two keys located either to the left or right of the discriminative stimulus. On test trials in which a pacman-like shape was presented with an adjacent rectangle filling its "truncated" part (to humans the pacman-like shape looked like an intact circle), the pigeons responded to the key corresponding to the pacman-like shape. This result suggested that pigeons do not perceptually complete the occluded portion.

Following up on these studies of pigeons, Fujita and his colleagues set out to investigate whether nonhuman primates (chimpanzee and rhesus macaques) would perceptually complete occluded portions of objects.

They trained a chimpanzee, named Ai, to match either one unitary rod (Fig. 1.4a) or a non-unitary rod (a set of two aligned rods; Fig. 1.4b)—the latter was identical to

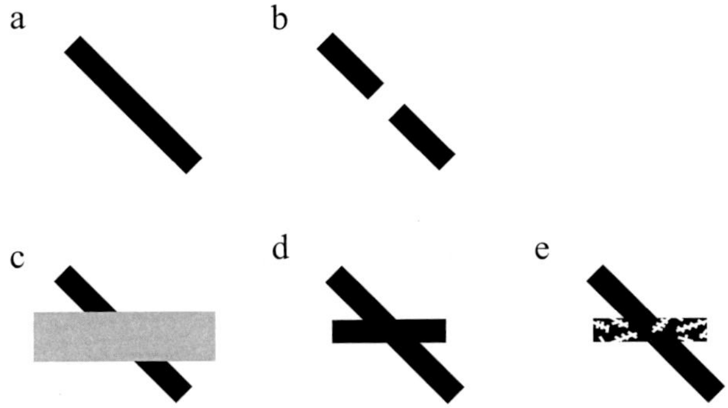

Fig. 1.4 Schematic illustration of stimuli used in Fujita and Giersch (2005), Sato et al. (1997), Ushitani et al. (2001)

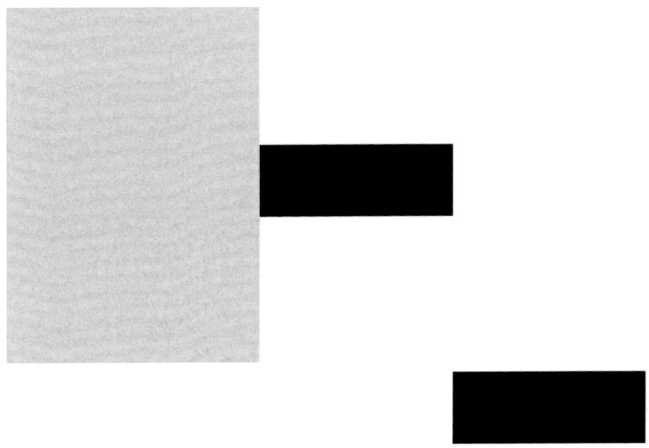

Fig. 1.5 Schematic illustration of stimuli used in Fujita (2001). The length of the black rod just adjacent to the gray rectangle appears longer than the identical one placed apart from the rectangle

the former except for a gap at its center—to the identical sample stimulus. On test trials, a rectangle covered the center of the rod (Fig. 1.4c) such that the rod appeared to be the unitary rod if the observer perceptually completed the occluded portion of the rod, but appeared to be the non-unitary rods if there was no perceptual completion. Ai matched the unitary rod to the test stimulus, suggesting that she experienced amodal completion (Sato et al. 1997). Subsequent studies have confirmed the existence of amodal completion in other chimpanzees (Tomonaga and Imura 2010; Ushitani et al. 2010).

Fujita (2001) approached amodal completion in rhesus macaques in a different way from the chimpanzee study. When a rod is placed adjacent to a larger figure, the length of the former tends to be perceived as longer than it really is (Fig. 1.5). According to Kanizsa (1979), amodal completion occurs at the contour conjunction between the rod and the rectangle in Fig. 1.5, such that the rectangle is perceived to partly occlude the end of the rod and thus our perceptual system "guesses" the rod's length as longer than its actual length. Would nonhuman animals experience this illusion? Fujita first trained two rhesus macaques to respond to the "correct" stimulus depending on the length of a bar placed apart from a large rectangle, presented above the two choice stimuli. When the rod was one of the longer (shorter) rods, touching the left (right) one of the two choice stimuli resulted in a reward. The percentages of "long" reports by both monkeys in the test condition in which the rod was just adjacent to the large rectangle exceeded that in the baseline condition (the rod apart from the rectangle), suggesting that the Kanizsa length illusion (and thus amodal completion) also occurred in rhesus macaques.

Fujita also trained three pigeons to discriminate rod length, using the same procedure as with the rhesus monkeys. However, the birds' performances in test trials (rod adjacent to the large rectangle) were the same as in baseline trials; therefore there was no evidence of amodal completion in pigeons. Several years later, Fujita and I investigated amodal completion in pigeons using a different procedure, but we replicated the above result (Fujita and Ushitani 2005; Ushitani and Fujita 2005; Ushitani et al. 2001). I will return to these studies in a later section.

In parallel with the pigeon projects, Fujita and colleagues adapted the procedure that had been used with the chimpanzee (Sato et al. 1997) for capuchin monkeys. More specifically, as discriminative stimuli, two types of unitary rod were added that combined with another shape (either a straight rectangle or a rectangle with dendritic incisions (Fig. 1.4d, e)). Three male capuchin monkeys that were competent at four-alternative matching-to-sample tasks chose the unitary rod as the correct comparison on test probe trials in which the center of the sample rod was covered by another figure (humans complete the occluded portion and perceive the unitary rod) (Fujita and Giersch 2005). These results not only clearly demonstrated amodal completion in capuchin monkeys, but also revealed details of their percept—how they perceive occluded portions. One of the underlying principles in Fujita's works is that demonstrations should be only the first step, and that further investigations to discover details and mechanisms of a target cognitive process must be conducted. In other words, the final purpose of comparative cognitive research is not to demonstrate odd, isolated abilities in particular species, but to draw a broader picture of how various species perceive and recognizing their world.

Since the studies described above, comparative cognitive research on amodal completion has been thriving. Table 1.1 presents a comprehensive list of studies on amodal completion in nonhumans. Fujita's original and systematic studies covered four species, in a major contribution to elucidating the perceptual worlds of other species. Careful readers may notice that the results for pigeons in Table 1.1 contain a mixture of positive and negative. The following section is dedicated to this issue.

Pigeon Problem

Fujita was my mentor during my undergraduate years and graduate studies. We investigated various visual processes in pigeons, including amodal completion. My first work (Ushitani et al. 2001), involved testing pigeons using a procedure similar to that used for chimpanzee (Sato et al. 1997). Pigeons were trained to match a unitary rod or a non-unitary rod (see above for details) to the sample stimulus in an identical matching-to-sample task. On probe test trials, pigeons were presented with

Table 1.1 Studies of amodal completion in nonhuman animals

Species	Authors	Results
Chimpanzee	Sato et al. (1997)	P
	Tomonaga and Imura (2010)	P
	Ushitani et al. (2001)	P
Rhesus macaque	Fujita (2001)	P
Baboon	Deruelle et al. (2000)	N/P(*1)
	Fagot et al. (2006)	P
Tufted capuchin monkey	Fujita and Giersch (2005)	P
Mouse	Kanizsa et al. (1993)	P? (*2)
Reef fish	Darmaillacq et al. (2011)	P
Redtail splitfin	Sovrano and Bisazza (2008)	P
Chicken	Regolin and Vallortigara (1995)	P
	Lea et al. (1996)	P
	Forkman (1998)	P
	Nakamura et al. (2010)	N
	Nakamura et al. (2011)	N
African gray parrot	Pepperberg and Nakayama (2016)	P
Bengalese finch	Takahasi and Okanoya (2013)	P
Pigeon	Cerella (1980)	N
	Sekuler et al. (1996)	N
	Watanabe and Furuya (1997)	N
	Shimizu (1998)	N
	Fujita (2001)	N
	Ushitani et al. (2001)	N
	DiPietro et al. (2002)	N/P (*3)
	Fujita and Ushitani (2005)	N
	Ushitani and Fujita (2005)	N
	Aust and Huber (2006)	N
	Nagasaka et al. (2007)	P
	Nagasaka and Wasserman (2008)	P

Note: The list does not include electrophysiological or neurophysiological studies. P = positive evidence of amodal completion, N = negative reports of amodal completion; *1 positive, but only with 3D presentation of stimuli; *2: positive, but with no control condition; *3 positive, but only after special training

a rod with the center portion was occluded by a large rectangle. The pigeons matched the non-unitary rod, thus showing no evidence of amodal completion.

Fujita and Ushitani (2005) further investigated pigeons in a paradigm similar to that of Rauschenberger and Yantis (Rauschenberger and Yantis 2001; See Sect. 1), but using diamond shapes instead of disks. We first trained pigeons (and instructed human participants) to detect a diamond shape partly truncated on one side among intact diamond shapes. In contrast to humans, pigeons showed no deterioration in performance when a square was placed adjacent to the truncated diamond part, again suggesting an absence of amodal completion in pigeons. We also conducted a study using more ecologically-meaningful stimuli for pigeons (Ushitani and Fujita 2005) than those typically used in previous studies. More specifically, pigeons were exposed to either a photograph of food (such as a kernel of corn or a safflower seed) or a non-food object (such as a screw nut or a paper clip) and required to peck at the former but not the latter. Before the test phase, the pigeons were exposed to search displays consisting of multiple food and non-food photos; they successfully pecked only at food photos. On test trials, one food photo was partly occluded by an object (a photo of pigeon feather in Experiments 1 and 2 or a photo of a piece of paper with saw-tooth-like edges in Experiment 3), to simulate the daily foraging situation for the pigeons in their home cage. We compared the priority of the choice (pecking or not, and sequential order of pecking between stimuli) for the partly occluded and partly truncated photos of food. The results showed more frequently and faster response sequence choices for the truncated food photos than the occluded food photos. This was the third set of results from my studies suggesting no evidence of amodal completion in pigeons. In our paper (Ushitani and Fujita 2005), we refer to the negative results for pigeons as the "pigeon problem," borrowing from the term "gorilla problem" used in the social cognition domain.

Members of Fujita's laboratory have also investigated whether bantams (a strain of domestic chicken) perceptually complete occluded portions of objects (Nakamura et al. 2010, 2011). Their results have revealed no evidence of amodal completion in bantams, thus showing that the pigeon problem is not limited to pigeons. Considering that other studies have shown evidence of amodal completion in domestic chickens (Forkman 1998; Lea et al. 1996; Regolin and Vallortigara 1995) the mixed results should perhaps be referred to as the "bird problem" (see Table 1.1). The topic of amodal completion in bird species has in fact given rise to controversy, with some researchers presenting evidence of amodal completion in pigeons (Nagasaka et al. 2007; Nagasaka and Wasserman 2008). In reply, Nakamura et al. (2010) have shown that alternative accounts for the apparent positive results in pigeons are plausible. Although we should currently refrain from drawing a definitive conclusion amodal completion in pigeons, we may note that their ability to complete occluded portions of objects, if it exists, may be limited to special types of stimuli or situations, or indeed after special types of training. In the next section, I share some of the lessons that I learned through the study of pigeon perception in the collaboration with Fujita, and suggest how we should promote comparative cognition in the future.

Concluding Comments: Diversity in Cognition

As mentioned earlier, I repeatedly obtained "negative" results regarding amodal completion in pigeons (Fujita and Ushitani 2005; Ushitani and Fujita 2005; Ushitani et al. 2001), which is to say that we found no evidence that pigeons perceptually complete occluded portions of objects. Instead, the results of Sekuler et al. (1996) and Fujita and Ushitani (2005) suggest that the pigeons perceive the shape of the figure as it is—a pacman-like shape is perceived as pacman-like shape, not as an intact circle as seen by humans. Given several studies presenting positive evidence for amodal completion in pigeons, a longer list of studies with negative results, including direct species comparisons using the same similar methods (e.g., Fujita and Giersch 2005; Sato et al. 1997; Ushitani et al. 2001) underlines the need for a conservative interpretation: pigeons rarely complete occluded portions of objects.

"Negative" results—in terms of suggesting different processes from humans—sometimes draw negative responses from researchers. When I presented negative results concerning amodal completion in pigeons—later published in Ushitani et al. (2001) and Ushitani and Fujita (2005)—at academic conferences, some researchers commented that my methods were inadequate, or that negative results were meaningless and that drawing any conclusion from them was a mistake, and likely to lead to fruitless future studies. Although it is clear that caution is required when interpreting both positive and negative results and to drawing conclusions, after the extensive debate surrounding replicability in psychology in the 2010s (e.g., Stanley et al. 2018), researchers have become increasingly open to reports of negative results. In retrospect, those criticisms of my negative findings about amodal completion in completion were inappropriate. Even two decades ago, Fujita saw the importance of negative findings, and encouraged us to continue exploring and presenting our findings on amodal completion in pigeons.

Fujita also imparted to his students the importance of an evolutionary perspective on visual information processing, with or without amodal completion. In one of our papers (Ushitani and Fujita 2005), we raised two evolutionary hypotheses regarding pigeons' lack of perceptual completion of occluded portions of objects. First, as grain feeders, pigeons do not naturally forage on small, moving animals such as bugs or worms that might move behind objects such as stones or leaves. Actually, Plowright et al. (1998) reported a species difference in performance on an invisible displacement task between pigeons and a predatory species, the hill myna. Second, birds' aerial lifestyle may require fast processing: They may prioritize local information such as a fragmented image of an object as it is rather than processing more global information, including completing images of multiple fragmented images. However, given that domestic chickens—omnivorous birds that also feeding on small animals and fly less, repeatedly did not complete occluded portions (Nakamura et al. 2010, 2011) may challenge these hypotheses. At the very least, amodal completion processes require a relatively long time (100–250 ms; Rauschenberger and Yantis 2001), so lighter processes that demand less time and fewer brain resources would favor local processing of fragments over global processing including amodal completion. Systematic species comparisons will continue to clarify the evolution of amodal completion.

To discuss evolutionary processes that lead to the emergence of a cognitive function, we need to understand the diversity of that function, as exemplified by studies of amodal completion. Whereas many mammals may complete occluded portions of objects, the reports on avian species are mixed. The small amount of positive results along with a more a substantial accumulation of negative results in pigeons indicates that amodal completion in this species occurs, at most, in strictly limited conditions. The notion of a dichotomy appears less scientific and less amenable to good research than a perspective that accepts the possibility of a continuum of abilities.

References

Aust U, Huber L (2006) Does the use of natural stimuli facilitate amodal completion in pigeons? Perception 35:333–349

Behrmann M (2003) Neuropsychological approaches to perceptual organization: evidence from visual agnosia. In: Peterson MA, Rhodes G (eds) Perception of faces, objects, and scenes: analytic and holistic processes. Oxford University Press, New York, pp 295–334

Cerella J (1980) The pigeon's analysis of pictures. Pattern Recogn 12:1–6

Darmaillacq A-S, Dickel L, Rahmani N, Shashar N (2011) Do reef fish, *Variola louti* and *Scarus niger*, perform amodal completion? Evidence from a field study. J Comp Psychol 125:273–277

Deruelle C, Barbet I, Dépy D, Fagot J (2000) Perception of partly occluded figures by baboons (*Papio papio*). Perception 29:1483–1497

DiPietro NT, Wasserman EA, Young ME (2002) Effects of occlusion on pigeons' visual object recognition. Perception 31:1299–1312

Fagot J, Barbet I, Parron C, Deruelle C (2006) Amodal completion by baboons (*Papio papio*): contribution of background depth cues. Primates 47:145–150

Forkman B (1998) Hens use occlusion to judge depth in a two dimensional picture. Perception 27:861–867

Fujita K (2001) Perceptual completion in rhesus monkeys (*Macaca mulatta*) and pigeons (*Columba livia*). Percept Psychophys 63:115–125

Fujita K, Giersch A (2005) What perceptual rules do capuchin monkeys (*Cebus apella*) follow in completing partly occluded figures? J Exp Psychol Anim Behav Process 31:387–398

Fujita K, Ushitani T (2005) Better living by not completing: a wonderful peculiarity of pigeon vision? Behav Processes 69:59–66

Júnior MF, Rosa MG, Gattass R, Rocha-Miranda CE (1992) Dynamic surrounds of receptive fields in primate striate cortex: a physiological basis for perceptual completion? Proc Natl Acad Sci 89:8547–8551

Kanizsa G (1979) Organization in vision: essays on gestalt perception. Praeger, New York

Kanizsa G, Renzi P, Conte S, Compostela C, Guerani L (1993) Amodal completion in mouse vision. Perception 22:713–721

Kellman PJ, Spelke ES (1983) Perception of partly occluded objects in infancy. Cogn Psychol 15:483–524

Lea SEG, Slater AM, Ryan CME (1996) Perception of object unity in chicks: a comparison with the human infant. Infant Behav Dev 19:501–504

Nagasaka Y, Lazareva OF, Wasserman EA (2007) Prior experience affects amodal completion in pigeons. Percept Psychophys 69:596–605

Nagasaka Y, Wasserman EA (2008) Amodal completion of moving objects by pigeons. Perception 37:557–570

Nakamura N, Watanabe S, Betsuyaku T, Fujita K (2010) Do bantams (*Gallus gallus domesticus*) experience amodal completion? An analysis of visual search performance. J Comp Psychol 124:331–335

Nakamura N, Watanabe S, Betsuyaku T, Fujita K (2011) Do bantams (*Gallus gallus domesticus*) amodally complete partly occluded lines? An analysis of line classification performance. J Comp Psychol 125:411–419

Pepperberg IM, Nakayama K (2016) Robust representation of shape in a Grey parrot (*Psittacus erithacus*). Cognition 153:146–160

Plowright CMS, Reid S, Kilian T (1998) Finding hidden food: behavior on visible displacement tasks by mynahs (*Gracula religiosa*) and pigeons (*Columba livia*). J Comp Psychol 112:13–25
Rauschenberger R, Yantis S (2001) Masking unveils pre-amodal completion representation in visual search. Nature 410:369–372
Regolin L, Vallortigara G (1995) Perception of partly occluded objects by young chicks. Percept Psychophys 57:971–976
Sato A, Kanazawa S, Fujita K (1997) Perception of object unity in a chimpanzee (*Pan troglodytes*). Jpn Psychol Res 39:191–199
Sekuler AB, Lee JAJ, Shettleworth SJ (1996) Pigeons do not complete partly occluded figures. Perception 25:1109–1120
Shimizu T (1998) Conspecific recognition in pigeons (*Columba livia*) using dynamic video images. Behaviour 135:43–53
Sovrano VA, Bisazza A (2008) Recognition of partly occluded objects by fish. Anim Cogn 11:161–166
Stanley TD, Carter EC, Doucouliagos H (2018) What meta-analyses reveal about the replicability of psychological research. Psychol Bull 144:1325–1346
Sugita Y (1999) Grouping of image fragments in primary visual cortex. Nature 401:269–272
Takahasi M, Okanoya K (2013) An invisible sign stimulus: completion of occluded visual images in the Bengalese finch in an ecological context. Neuroreport 24:370–374
Tomonaga M, Imura T (2010) Pacman in the sky with shadows: the effect of cast shadows on the perceptual completion of occluded figures by chimpanzees and humans. Behav Brain Funct 6:38
Ushitani T, Fujita K (2005) Pigeons do not perceptually complete partly occluded photos of food: an ecological approach to the "pigeon problem". Behav Processes 69:67–78
Ushitani T, Fujita K, Yamanaka R (2001) Do pigeons (*Columba livia*) perceive object unity? Anim Cogn 4:153–161
Ushitani T, Imura T, Tomonaga M (2010) Object-based attention in chimpanzees (*Pan troglodytes*). Vision Res 50:577–584
Watanabe S, Furuya I (1997) Video display for study of avian visual cognition: from psychophysics to sign language. Int J Comp Psychol 10:111–127

Chapter 2
Visual Illusions: Insights from Comparative Cognition

Audrey E. Parrish

Abstract Visual illusions emerge as a byproduct of an otherwise highly functional perceptual system, leading to dissociations of the physical world and the perceptual interpretation of environmental stimuli. Insights from comparative cognition research are discussed in this chapter, including the role of species' unique evolutionary pressures, perceptual mechanisms, and the impact of experimental paradigms on the emergence of illusory phenomena, including visual illusions and perceptual completion.

Keywords Perception · Visual illusions · Perceptual completion · Pigeons · Primates

Introduction

The comparative study of animal perception sheds light on the neural and cognitive mechanisms underlying perceptual experiences, as well as the role of ontogenetic and environmental influences. This endeavor—the study of how nonhuman animals (hereafter animals) interpret their physical worlds via diverse sensory systems—has a long history in psychology and related fields. The famed ethologist, von Uexküll (1934/1957), emphasized the need to evaluate an animal's unique perceptual experience as a function of the environment in which it evolved. In particular, the concept of the *Umwelt* or the animal's "self-world" emerged, in which each species' sensory system affords it a unique lens through which it perceives and subsequently interacts with its external world.

A comparative approach to the study of perception contributes to our understanding of the mechanisms underlying both the accurate interpretation of environmental stimuli and misperceptions of the external world. These misperceptions emerge as visual illusions, which are largely considered to be byproducts of an otherwise

A. E. Parrish (✉)
Department of Psychology, The Citadel, Charleston, SC, USA

J. R. Anderson, H. Kuroshima (eds.), *Comparative Cognition*,
https://doi.org/10.1007/978-981-16-2028-7_2

efficient perceptual system that quickly processes vast amounts of incoming information. The manner in which animals experience visual illusions has attracted the interest of psychologists for the better part of the last century, with the earliest comparative studies emerging in the 1920s (Révész 1924; Warden and Baar 1929). A variety of illusory arrays have since been presented to diverse taxa across the animal kingdom including insects, fish, reptiles, and mammals (for recent reviews, see Agrillo et al. 2020; Feng et al. 2017; Kelley and Kelley 2014), with an increased focus on the mechanisms that underlie visual illusions. Similarities in the processing of illusory stimuli are thought to reflect commonalities in the neural mechanisms of visual processing that extend beyond a species-unique interpretation of the external world.

Kazuo Fujita and his colleagues championed the systematic study of visual illusions as perceived by pigeons and primates, conducting influential work on the Ponzo, Müller-Lyer, and Ebbinghaus illusions, as well as perceptual completion. Fujita's and others' application of the psychophysical approach to the study of illusions and perceptual processing has opened a window into the inner workings of animal minds (see Matsuno and Fujita 2009 for a review), an approach that many comparative psychologists continue to apply in their investigations of visual illusions and perceptual phenomena. Moreover, Fujita's work highlights the importance of systematically varying the parameters of illusory figures to determine the mechanisms that lead to their emergence (or not), as well as the invaluable role of directly comparing species using nearly identical methods. In this chapter, I present a review of comparative research on visual illusions with an emphasis on the illusions (and perceptual mechanisms) most heavily studied by Fujita and his colleagues.

The Psychophysical Study of Geometric Illusions

Geometric illusions emerge when the physical dimensions of a stimulus (e.g., area, length, height, orientation) are misperceived as a function of the context in which it is presented. Geometric illusions are comprised of multiple elements, including a to-be-judged target stimulus and an illusory-inducing context (e.g., see the Ponzo illusion in Fig. 2.1a, in which a horizontal, target line placed closer to the apex of two converging lines is misperceived as longer than an identical line placed farther from the apex). Geometric illusions often are discussed in light of a species' ability to organize the elements within an illusory array, such that perceptual grouping of the elements leads to greater sensitivity to illusions, whereas an inability to or delay in grouping may disrupt their emergence. Thus, perceptual mechanisms that underlie perceptual grouping, in particular those that vary across species, have been a large focus within comparative investigations.

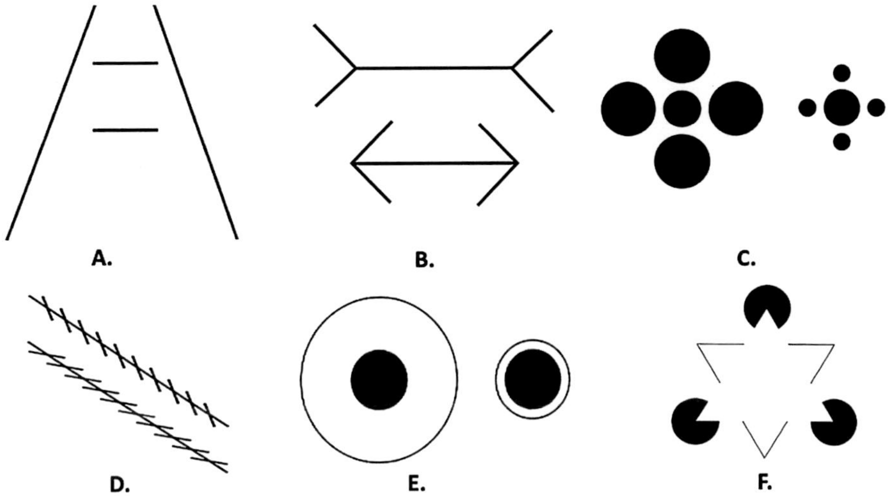

Fig. 2.1 Visual illusions including the (**a**) Ponzo illusion, (**b**) Müller-Lyer illusion, (**c**) Ebbinghaus illusion, (**d**) Zöllner illusion, (**e**) Delboeuf illusion, and (**f**) Kanizsa triangle (perceptual completion)

The Ponzo Illusion

In one of their most well-known extensions of the psychophysical approach to illusory phenomena, Fujita and colleagues investigated perception of the Ponzo illusion by pigeons (*Columba livia*), rhesus macaques (*Macaca mulatta*), and chimpanzees (*Pan troglodytes*) as compared to humans (*Homo sapiens*) in a series of studies (Fig. 2.1a; see Fujita et al. 2012 for a review). As aforementioned, in the Ponzo illusion, human subjects typically perceive a horizontal line as longer when positioned closer to the apex of two converging lines forming an inverted V-shape than when it is positioned farther from the apex. In their first investigation of the Ponzo illusion, Fujita et al. (1991) simultaneously presented pigeons with two lines of differing lengths. In this relative discrimination task, pigeons judged lines placed closer to the apex of the converging lines as longer than lines placed farther from the apex. Thus, the context created by the converging lines created an illusory effect for judging line length as compared to a control condition when the surrounding lines were parallel for the pigeons.

As an extension of this finding, an absolute classification task was presented to pigeons (Fujita et al. 1991) and primates (rhesus monkeys and chimpanzees; Fujita 1997). Subjects first were trained to classify a series of horizontal lines of various lengths as "short" or "long" relative to a pre-determined length. After training on this initial discrimination was completed, converging lines were placed at a standard location (middle context) and, in test trials, horizontal lines were positioned closer to the apex (low context) or farther from the apex (high context). If subjects perceived the Ponzo illusion, the prediction was that they would be more likely to classify the horizontal line as "long" in the low context and "short" in the high context. All species perceived the illusion in a human-like direction, with greater "long" responses for horizontal lines positioned closer to the apex of the converging lines.

An important species-difference emerged from this work; pigeons demonstrated a higher sensitivity to the low context condition (line near apex) versus the high context condition (line farther from apex) relative to primates. The authors attributed this to differential sensitivity to contrast and assimilation effects, which are responsible for the emergence of the Ponzo illusion and other size illusions (Fujita et al. 1991; Fujita 1997). The assimilation-contrast theory posits that contrast effects lead to greater perceived differences in stimuli that are positioned farther apart. Assimilation effects lead to perceptual integration of nearby stimuli and subsequent underestimation of perceived differences (see Goto et al. 2007; Robinson 1998). In this case, pigeons readily assimilated the horizontal line with the nearby converging lines, leading to overestimates in line length, subsequently strengthening the effect of the illusion (Fujita et al. 1991; Fujita 1997; Pressey 1971).

Fujita and colleagues also assessed the perspective theory according to which subjects perceive apparent depth within two-dimensional figures (Gregory 1963). For the Ponzo illusion, perspective cues would lead subjects to perceive the line nearest the apex as more distant and thus longer to maintain size constancy. To test this possibility, pigeons and primates were presented with two conditions that varied in terms of the orientation of the illusory-inducing lines (converging versus parallel). All species perceived the Ponzo illusion in the predicted direction but to similar degrees across conditions (perceptive and non-perspective, Fujita et al. 1991; Fujita 1997), suggesting that perspective cues did not readily impact illusion emergence. Similarly, additional illusory-inducing lines surrounding the horizontal target line (predicted to increase the strength of the illusion by adding perspective cues) failed to enhance the illusion in pigeons (Fujita et al. 1991, 1993).

In a separate assessment of the perspective hypothesis, Fujita (1996) added a real image of a highway scene behind the Ponzo array, which was positioned upright (predicted to increase the strength of the illusion via depth cues) or inverted. The strength of the illusion was amplified among human subjects in the presence of the upright highway scene (even when the converging lines of the Ponzo array were removed). However, this effect was reversed among rhesus monkeys who were more sensitive to the illusion when presented with the traditional line-drawn array versus the real highway scene. Fujita (1997) further isolated features of the illusory-inducing context for the Ponzo, presenting a horizontal line with small dots on either end or short lines rather than the longer converging lines in the traditional illusion. Interestingly, results varied across primates: humans demonstrated a stronger illusion with these truncated formats versus the traditional lines, whereas the strength of the illusion was equivalent for chimpanzees across formats and no illusion emerged for rhesus monkeys using the dot and shortened line elements. Thus, although all species tested perceived the Ponzo illusion in a similar direction, the mechanisms underlying its emergence may not be ubiquitous.

The Ponzo illusion and the related corridor illusion, in which objects are depicted on a corridor background with seemingly more distant objects appearing as larger, have been documented in various species including chimpanzees (Imura et al. 2008), baboons (*Papio papio*: Barbet and Fagot 2002), and horses (*Equus ferus caballus*: Timney and Keil 1996), suggesting continuity in mammalian sensitivity to pictorial depth cues. Size constancy is important in estimating the true size of objects presented both close by and at a distance via depth cues. In a recent study by Hataji et al. (2020), pigeons were shown to be sensitive to the corridor illusion (overestimating target size for objects located higher or presumably farther away), and pigeons prioritized pictorial cues over motion parallax depth cues for size constancy similar to humans. These results, along with other studies on avian object perception (e.g., Cavoto and Cook 2006; Cook and Katz 1999; Reid and Spetch 1998), suggest that size constancy manifests similarly across vertebrate species tested to date including humans, despite variance in visual processing systems.

The Müller-Lyer, Ebbinghaus, and Zöllner Illusions

To further explore the differential perception of geometric illusions among primates and pigeons, Fujita and colleagues investigated the well-known Müller-Lyer illusion, in which the length of a horizontal line is misperceived as a function of brackets in the shape of arrowheads attached to either end of the line (see Fig. 2.1b). When the brackets are outward-facing in the typical shape of an arrow, the horizontal line is underestimated in length relative to when the arrows are inverted, which leads to overestimation of line length. Using an absolute classification task similar to the Ponzo procedure, Nakamura et al. (2006, 2009) conducted several studies of the Müller-Lyer illusion with pigeon and human subjects. Ultimately, both species were biased to classify horizontal lines with inverted brackets as "long" as compared to lines with outward-facing brackets; however, the species differed in the extent to which they perceived the illusion. Specifically, pigeons did not perceive the reversed illusion when the distance between the horizontal line and the brackets was increased. This reversal is common in humans, emerging as a byproduct of contrast effects between the line and the surrounding context (Nakamura et al. 2006, 2009). Consistent with pigeons' perception of the Ponzo illusion, these results suggest that pigeons are differentially sensitive to assimilation and contrast effects relative to primates.

Several other animal species have been presented with the traditional Müller-Lyer stimuli, and most perceive the illusion in a human-like direction, including capuchin monkeys (*Sapajus apella*: Suganuma et al. 2007), rhesus monkeys (Tudusciuc and Nieder 2010), horses (Cappellato et al. 2020), an African grey parrot (*Psittacus erithacus*: Pepperberg et al. 2008), ring doves (*Turtur risorius*: Warden and Baar 1929), redtail splitfin fish (*Xenotoca eiseni:* Sovrano et al. 2016), guppies (*Poecilia reticulata*: Santacà and Agrillo 2020), and garden ants (*Lasius niger*: Sakiyama and Gunji 2013), suggesting that the mechanisms underlying size illusions are phylogenetically widespread. However, some discrepancies across species have emerged, with bamboo sharks (*Chiloscyllium griseum*: Fuss et al. 2014) and goldfish

(*Carassius auratus*: Wyzisk and Neumeyer 2007) failing to perceive the Müller-Lyer illusion. In a recent study that used a relative discrimination task, domestic dogs (*Canis familiaris*) were found to perceive the Müller-Lyer illusion in a human-like direction (Keep et al. 2018). However, follow-up experiments revealed that dogs were using overall array size instead of responding on the basis of the target line alone, underscoring the need for careful control conditions in illusion studies.

The role of assimilation and contrast mechanisms in the emergence of geometric illusions was studied further using the well-known Ebbinghaus illusion (see Fig. 2.1c; Nakamura et al. 2008, 2014). This illusion emerges when a central target dot is misperceived as a function of surrounding inducer dots, such that a central dot encircled by larger dots appears smaller (contrast effects) than when surrounded by smaller dots (assimilation effects). Employing an absolute classification test, Nakamura et al. (2008, 2014) trained pigeons and chickens (*Gallus gallus domesticus*) to classify a black circle (the central target dot) as large or small. Inducer circles of a constant size were gradually faded in from light to dark to habituate subjects to their presence. Illusion test trials with large or small inducer dots revealed that both pigeons and chickens perceived the Ebbinghaus illusion in the reversed direction from human subjects: they overestimated central dots surrounded by large inducers and underestimated central dots surrounded by small inducers. Nakamura et al. (2008, 2014) ruled out the possibility that subjects erroneously responded on the basis of total surface area (target dot and inducers combined) by introducing targets both with and without inducers. Furthermore, it did not appear that the birds were responding on the basis of inducer size alone, as the percentage of "large" responses varied as a function of central dot size within each condition. Thus, for pigeons and bantams, a reversed Ebbinghaus illusion may emerge due, at least in part, to an increased susceptibility to assimilation mechanisms that would lead to the integration of the target dot with both small and large inducers.

Similarly, pigeons and bantams perceived a reversed Zöllner illusion (see Fig. 2.1d; Watanabe et al. 2011, 2013), in which, for humans, two parallel lines appear to converge when oblique crosshatches are superimposed on the lines. In control trials, the birds were trained to peck towards one end of a pair of converging lines that appeared narrower (or wider—counterbalanced across subjects). In critical illusion trials, birds of both species perceived the illusory lines as nonparallel, but in the opposite direction from what is observed in humans (i.e., selection of the end that appears to deviate for humans as "narrower"). These avian results stand in contrast with outcomes of primate work, in which baboons (Benhar and Samuel 1982) and rhesus monkeys perceived the Zöllner illusion in a human-like direction (Agrillo et al. 2014). The mechanisms underlying this illusion relate to its global perception (Parlangeli and Roncato 1995) and the misperception of acute angles that are formed when the crosshatches intersect with the main lines; these angles are perceptually enlarged leading one to misperceive the lines as converging (White 1972). The species' differences observed in the Zöllner likely may again reflect differential sensitivity to contrast and assimilation mechanisms across primates and avian species (see Beran et al. 2020 for a review). Assimilation would lead pigeons and chickens to underestimate the acute angles between the main lines and the cross-hatches giving rise to the reversed effect, whereas primates are likely sensitive to both contrast and assimilation, necessary for the emergence of the Zöllner effect.

Perceptual Organization and Visual Illusions

Differential sensitivity to contrast and assimilation mechanisms is directly tied to a concept that has been touched upon only briefly in this chapter thus far. Specifically, perceptual organization describes the process by which the perceptual system groups individual elements within an array into cohesive units, reflecting objects and figures within a visual scene. The readiness with which perceptual systems organize stimuli in their environment varies across species, such that some species display a global-to-local precedence (i.e., the global figure is perceived prior to the local elements) or a local-to-global precedence (i.e., the individual elements are perceived prior to the global figure). It is established that humans display a global precedence when presented with hierarchical stimuli in which the global figure is comprised of the same or different local elements (e.g., Navon 1977; a letter H comprised of small Hs or Ss). Varying the stimulus design can shift humans towards a local precedence when the individual elements of an array are visually emphasized (see Kimchi 1992, for a review).

Much of the comparative work surrounding perceptual organization has demonstrated that pigeons demonstrate a robust local precedence (e.g., Cavoto and Cook 2001), while primates show mixed results, with chimpanzees sometimes demonstrating a human-like global precedence (e.g., Fujita and Matsuzawa 1990; Hopkins and Washburn 2002) and monkeys displaying a local precedence or none at all (e.g., baboons: Deruelle and Fagot 1998; Fagot and Deruelle 1997; rhesus macaques: Hopkins and Washburn 2002; capuchin monkeys: De Lillo et al. 2005; Spinozzi et al. 2006, tamarins: *Saguinus oedipus*, Neiworth et al. 2006). However, it is important to note that, as with humans, methodological design can impact processing mode in nonhuman primates, particularly if the inter-stimulus distance and overall array size are decreased, which favors perceptual grouping and global processing (e.g., Matsuno and Tomonaga 2007; Neiworth et al. 2006; Tanaka and Fujita 2000). As discussed by Matsuno and Fujita (2009), nonhuman primates do not appear to lack the ability for perceptual organization despite these performance differences across monkeys, apes, and humans, but rather the difference lies within the degree to which primates perceptually group the individual elements within a visual array and attend to larger spatial areas.

Critical to this discussion, perceptual grouping supports the emergence of illusions (e.g., Roberts et al. 2005). For example, perception of the Ebbinghaus illusion (Fig. 2.1c) is dependent upon perceiving the central dot as a component of the larger set of inducer circles. A cross-cultural study underscored this link by demonstrating that Himba tribesman, a locally oriented human population, perceived a weakened Ebbinghaus illusion (de Fockert et al. 2007). Additionally, a weakened Ebbinghaus illusion has been documented among individuals diagnosed with autism spectrum disorder, who also adopt a more locally oriented processing style (e.g., Dakin and Frith 2005; Happé 1996). Comparative results further support this link between perceptual organization and illusory perception of the Ebbinghaus array, as one example. Species with a global precedence perceive the Ebbinghaus in a human-like direction (e.g., redtail splitfin fish (Sovrano et al. 2015) and dolphins, *Tursiops truncatus* (Murayama et al. 2012)). However, a reversed or weakened Ebbinghaus has been documented among species characterized by a local precedence (e.g., bamboo sharks (Fuss and Schluessel 2017), baboons (Parron and Fagot 2007), bantams and pigeons (Nakamura et al. 2008, 2014)).

It is important to note that processing mode is not solely predictive of illusory perception; rather, methodology plays a critical role in this relationship. Experimental methods supporting perceptual grouping of the target stimulus and illusory context can favor global processing and, in turn, illusory perception. For example, human perception of the Ebbinghaus illusion is weakened by disrupting perceptual grouping via isolating the central target dot and the inducers (e.g., increasing the inter-stimulus distance or brightness/shape of the stimuli, erasing the outer edges of the inducer circles (e.g., Choplin and Medin 1999; Coren and Enns 1993; Roberts et al. 2005)). Similarly, comparative studies often employ training procedures that visually isolate the target stimulus and illusory context, a necessary component in training non-verbal species on the discrimination rule. However, these procedures may inadvertently weaken perceptual grouping and subsequent illusory emergence, particularly among locally oriented species (e.g., gradually fading in inducer circles (Nakamura et al. 2008, 2014); differently colored inducer and target dots (Parron and Fagot 2007)).

The necessity of isolating the illusory target as the discriminative stimulus without disrupting perceptual grouping or habituating subjects to the illusory context has proven challenging in comparative studies. For example, in Parrish et al. (2015), rhesus macaques and capuchin monkeys initially perceived a reversed Delboeuf illusion in a relative discrimination task (Fig. 2.1e), in which a target dot typically appears to be larger when encircled by a small ring (assimilation effect) and smaller when encircled by a large ring (contrast effect). However, an absolute classification task revealed that rhesus and capuchin monkeys were, in fact, sensitive to the Delboeuf illusion consistent with human subjects, classifying the central dot as increasingly larger as ring size decreased. Critically, in the relative discrimination task, the monkeys also responded to the inducer rings rather than discriminating based solely on the target dots, a confound that was more easily isolated using the absolute discrimination task. In a recent study by Qadri and Cook (2019), starlings presented with the Ebbinghaus illusion demonstrated a similar response, appearing to perceive a reversed illusion before follow-up experiments revealed that they were inadvertently incorporating the inducer circles into the size discrimination, an important confound to control for when assessing sensitivity to the illusion (see also Keep et al. (2018) for a similar finding in dogs with the Müller-Lyer illusion).

Beyond stimulus design, the angle from which visual arrays are viewed can impact perceptual grouping and subsequent emergence of visual illusions in humans (e.g., Oyama 1960; Weintraub 1979). Subsequently, the manner in which animals interact with visual stimuli may lead to closer or more distant perspectives, the latter of which promotes perceptual grouping. For example, although domestic chickens have shown evidence of a reversed Ebbinghaus illusion when a pecking response is required, which necessitates a closer perspective (Nakamura et al. 2014), positive evidence of the illusion was documented in 4-day-old chicks of the same species required to approach the Ebbinghaus array from a farther distance to recover food rewards (Rosa Salva et al. 2013). It is proposed that the more distant viewing angle in the chick study, along with a decrease in the distance between inducer and target circles, facilitated perceptual grouping and subsequent illusory perception. The authors also discussed the role of the pecking response as linked to the central visual field, which is associated with localized rather than global or holistic processing (see Rosa Salva et al. 2013; Goodale 1983; Lamb and Robertson 1988). Future studies are needed to parse out the role of development in the emergence of the Ebbinghaus and other visual illusions, as a critical variable between these studies was age (4-day-old chicks vs. 6-month-old chickens). These contrasting findings illustrate how the methodologies by which we assess visual illusions interact directly with perceptual processing mode as well as contrast and assimilation mechanisms and subsequent illusory perceptions.

Perceptual Completion

Organisms routinely are faced with incomplete visual information from the environment, in which an object is partially occluded behind another object (e.g., a predator partially hidden behind a tree). However, these objects are figurally completed via subjective (illusory) contours and thus perceived as whole instead of fragmented (e.g., the predator is perceived as a complete organism; Kanizsa 1979; Michotte et al. 1991; Nieder 2002). Perceptual completion represents a disconnect between one's perceptual experience and the physical stimuli under scrutiny akin to the visual illusions explored above. Specifically, modal completion occurs when one perceives object borders that are otherwise absent and amodal completion occurs when one visually completes an occluded stimulus. For example, in the famous Kanizsa triangle (Fig. 2.1f), human participants mentally complete and perceive a white triangle despite there being no connective contours. The ability to integrate multiple features of a visual scene as a unified object despite incomplete information underlies form perception, and this ability has been studied systematically by Fujita and colleagues, including research with both primates and pigeons (see Fujita 2001a, 2004, 2012 for reviews).

Several nonhuman species demonstrate human-like perception of illusory contours when presented with amodal completion tasks, including but not limited to chimpanzees (e.g., Fagot and Tomonaga 2001; Sato et al. 1997), baboons (e.g., Fagot et al. 2006), macaque monkeys (e.g., De Weerd et al. 1996; Fujita 2001b; Sugita 1999), capuchin monkeys (e.g., Fujita and Giersch 2005), squirrel monkeys

(*Saimiri*: Nagasaka and Osada 2000), dogs (Byosiere et al. 2019), cats (*Felis catus*: Bravo et al. 1988), rodents (Kanizsa et al. 1993), owls (*Tyto alba*: Nieder and Wagner 1999), chickens (e.g., Forkman and Vallortigara 1999; Lea et al. 1996; Regolin and Vallortigara 1995; Regolin et al. 2004), cuttlefish (*Sepia spp*: Lin and Chiao 2017; Zylinski et al. 2012), bamboo sharks (Fuss et al. 2014), redtail splitfin fish (Sovrano and Bisazza 2009), and goldfish (Wyzisk and Neumeyer 2007). This work also has been extended to more naturalistic stimuli (e.g., partially occluded predators or conspecifics), underscoring the adaptive value of perceptual completion (e.g., Darmaillacq et al. 2011; Okanoya and Takahashi 1999; Tvardíková and Fuchs 2010).

Notably, there appears to be a discrepancy in pigeons' ability to perceptually complete fragmented visual input (see Fujita 2004 for a review). For example, when presented with a target bar overlapped by a larger rectangle, pigeons did not overestimate the length of the bar touching the rectangle as did rhesus monkeys, which indicated that monkeys but not pigeons perceived the bar as continuing behind the rectangle (Fujita 2001b). For similar negative results for amodal completion by pigeons, reference the following (e.g., Cerella 1980; Fujita and Ushitani 2005; Sekuler et al. 1996; Shimizu 1998; Ushitani et al. 2001; Watanabe and Furuya 1997). As raised by Fujita (2004), these discrepancies may reflect potential differences in the adaptation of the pigeon's visual system to environmental demands, such that as grain eaters, pigeons may have less need for perceptual completion than other species. Relatedly, this "pigeon problem" (Fujita 2004, 2012) also may reflect important differences in grouping mechanisms across locally oriented species such as pigeons (see Bruce et al. 2003). Furthermore, positive evidence of perceptual completion by pigeons following extended training (DiPietro et al. 2002), in light of prior experience (Nagasaka et al. 2007), and for moving stimuli (e.g., Nagasaka and Wasserman 2008), suggests context-specificity in this ability.

Continuity and discrepancies in amodal completion and the perception of illusory contours reveal important patterns in the evolution of visual capabilities of species with variant neural mechanisms (see Nieder (2002) for a review). Grouping abilities, depth cues, motion processing, and mechanisms for delineating the boundaries between objects and their backgrounds are critical to perceptual completion. The completion of subjective contours reveals an adaptive response of visual systems to processing environmental stimuli under poor conditions of visibility and often incomplete information. Consistency in responding to such stimuli across fish, birds, and mammals suggests amodal completion is relatively widespread among vertebrates, and perhaps a result of homologous evolutionary pressures (see Rosa Salva et al. (2014) for a review). Also important, however, are the cases in which discrepancies emerge, underscoring the need to situate visual demands in an evolutionary context with an eye towards selective pressures on perception.

Summary

As demonstrated in the works of Fujita and his colleagues alongside the many contributors to comparative perception science, susceptibility to visual illusions reflects the relationship between the visual system, perceptual mechanisms, methodological approach, and species' unique sensory worlds. Furthermore, the performance of animals within these illusory tasks, including performance in control trials for discriminating quantity and size, also depends on concomitant cognitive skills. For instance, visual illusions typically emerge at an individual's discrimination threshold (i.e., the point at which there are just noticeable differences of size, quantity, etc.). These illusory experiences often are quite subtle, generating small subjective differences among stimuli; therefore, an animal must be able to make fine discriminations or have sufficient numerical acuity to perceive certain illusions (e.g., the Solitaire illusion: Miletto Petrazzini et al. 2018). Comparative research is particularly useful in revealing the role of these mechanisms, given the variability in neurobiology, discrimination abilities, and perceptual processing mode across visually oriented species. Furthermore, researchers have begun to explore the roots of inter- and intra-species variance in the perception of illusions to understand better the role of attention, quantitative discrimination abilities, and motivation. There also is a growing interest in the role of experience and development in the perception of visual illusions, which can be systematically modeled within a variety of animal species (e.g., Rosa Salva et al. 2013; Vallortigara 2012). Through this endeavor of investigating the perceptual world of animals, comparative psychologists continue to shed light on the *Umwelt* of both closely and distantly related species.

References

Agrillo C, Parrish AE, Beran MJ (2014) Do rhesus monkeys (*Macaca mulatta*) perceive the Zöllner illusion? Psychon Bull Rev 21:986–994

Agrillo C, Santacà M, Pecunioso A, Petrazzini MEM (2020) Everything is subjective under water surface, too: visual illusions in fish. Anim Cogn 2:1–14

Barbet I, Fagot J (2002) Perception of the corridor illusion by baboons (*Papio papio*). Behav Brain Res 132:111–115

Benhar E, Samuel D (1982) Visual illusions in the baboon (*Papio anubis*). Anim Learn Behav 10:115–118

Beran MJ, Parrish AE, Agrillo C (2020) Zöllner illusion. In: Vonk J, Shackelford TK (eds) Encyclopedia of animal cognition and behavior. Springer, Cham

Bravo M, Blake R, Morrison S (1988) Cats see subjective contours. Vision Res 28:861–865

Bruce V, Green PR, Georgeson MA (2003) Visual perception: physiology, psychology, & ecology. Psychology Press, Hove

Byosiere SE, Chouinard PA, Howell TJ, Bennett PC (2019) Illusory contour perception in domestic dogs. Psychon Bull Rev 26:1641–1649

Cappellato A, Miletto Petrazzini ME, Bisazza A, Dadda M, Agrillo C (2020) Susceptibility to size visual illusions in a non-primate mammal (*Equus caballus*). Animals 10:1673

Cavoto BR, Cook RG (2006) The contribution of monocular depth cues to scene perception by pigeons. Psychol Sci 17:628–634

Cavoto KK, Cook RG (2001) Cognitive precedence for local information in hierarchical stimulus processing by pigeons. J Exp Psychol Anim Behav Process 27:3–16

Cerella J (1980) The pigeon's analysis of pictures. Pattern Recogn 12:1–6

Choplin JM, Medin DL (1999) Similarity of the perimeters in the Ebbinghaus illusion. Percept Psychophys 61:3–12

Cook RG, Katz JS (1999) Dynamic object perception by pigeons. J Exp Psychol Anim Behav Process 25:194–210

Coren S, Enns JT (1993) Size contrast as a function of conceptual similarity between test and inducers. Percept Psychophys 54:579–588

Dakin S, Frith U (2005) Vagaries of visual perception in autism. Neuron 48:497–507

Darmaillacq AS, Dickel L, Rahmani N, Shashar N (2011) Do reef fish, *Variola louti* and *Scarus niger*, perform amodal completion? Evidence from a field study. J Comp Psychol 125:273–277

de Fockert J, Davidoff J, Fagot J, Parron C, Goldstein J (2007) More accurate size contrast judgments in the Ebbinghaus illusion by a remote culture. J Exp Psychol Hum Percept Perform 33:738–742

De Lillo C, Spinozzi G, Truppa V, Naylor DM (2005) A comparative analysis of global and local processing of hierarchical visual stimuli in young children (*Homo sapiens*) and monkeys (*Cebus apella*). J Comp Psychol 119:155–165

De Weerd P, Desimone R, Ungerleider LG (1996) Cue-dependent deficits in grating orientation discrimination after V4 lesions in macaques. Vis Neurosci 13:529–538

Deruelle C, Fagot J (1998) Visual search for global/local stimulus features in humans and baboons. Psychon Bull Rev 5:476–481

DiPietro NT, Wasserman EA, Young ME (2002) Effects of occlusion on pigeons' visual object recognition. Perception 31:1299–1312

Fagot J, Barbet I, Parron C, Deruelle C (2006) Amodal completion by baboons (*Papio papio*): contribution of background depth cues. Primates 47:145–150

Fagot J, Deruelle C (1997) Processing of global and local visual information and hemispheric specialization in humans (*Homo sapiens*) and baboons (*Papio papio*). J Exp Psychol Hum Percept Perform 23:429–442

Fagot J, Tomonaga M (2001) Effects of element separation on perceptual grouping by humans (*Homo sapiens*) and chimpanzees (*Pan troglodytes*): perception of Kanizsa illusory figures. Anim Cogn 4:171–177

Feng LC, Chouinard PA, Howell TJ, Bennett PC (2017) Why do animals differ in their susceptibility to geometrical illusions? Psychon Bull Rev 24:262–276

Forkman B, Vallortigara G (1999) Minimization of modal contours: an essential cross-species strategy in disambiguating relative depth. Anim Cogn 2:181–185

Fujita K (1996) Linear perspective and the Ponzo illusion: a comparison between rhesus monkeys and humans. Jpn Psychol Res 38:136–145

Fujita K (1997) Perception of the Ponzo illusion by rhesus monkeys, chimpanzees, and humans: similarity and difference in the three primate species. Percept Psychophys 59:284–292

Fujita K (2001a) What you see is different from what I see: species differences in visual perception. In: Matsuzawa T (ed) Primate origins of human cognition and behavior. Springer, Berlin, pp 29–54

Fujita K (2001b) Perceptual completion in rhesus monkeys (*Macaca mulatta*) and pigeons (*Columba livia*). Percept Psychophys 63:115–125

Fujita K (2004) How do nonhuman animals perceptually integrate figural fragments? Jpn Psychol Res 46:154–169

Fujita K (2012) Seeing what is not there: illusion, completion, and spatiotemporal boundary formation in comparative perspective. In: Zentall TR, Wasserman EA (eds) The Oxford handbook of comparative cognition. Oxford University Press, New York, pp 25–47

Fujita K, Blough DS, Blough PM (1991) Pigeons see the Ponzo illusion. Anim Learn Behav 19:283–293
Fujita K, Blough DS, Blough PM (1993) Effects of the inclination of context lines on perception of the Ponzo illusion by pigeons. Anim Learn Behav 21:29–34
Fujita K, Giersch A (2005) What perceptual rules do capuchin monkeys (*Cebus apella*) follow in completing partly occluded figures? J Exp Psychol Anim Behav Process 31:387–398
Fujita K, Matsuzawa T (1990) Delayed figure reconstruction by a chimpanzee (*Pan troglodytes*) and humans (*Homo sapiens*). J Comp Psychol 104:345–351
Fujita K, Nakamura N, Sakai A, Watanabe S, Ushitani T (2012) Amodal completion and illusory perception in birds and primates. In: Lazareva OF, Shimizu T, Wasserman EA (eds) How animals see the world: Comparative behavior, biology, and evolution of vision. Oxford University Press, New York, pp 101–116
Fujita K, Ushitani T (2005) Better living by not completing: A wonderful peculiarity of pigeon vision? Behav Processes 69:59–66
Fuss T, Bleckmann H, Schluessel V (2014) Visual discrimination abilities in the gray bamboo shark (*Chiloscyllium griseum*). Fortschr Zool 117:104–111
Fuss T, Schluessel V (2017) The Ebbinghaus illusion in the gray bamboo shark (*Chiloscyllium griseum*) in comparison to the teleost damselfish (*Chromis chromis*). Fortschr Zool 123:16–29
Goodale MA (1983) Visually guided pecking in the pigeon (Columba livia). Brain Behav Evol 22:22–41
Goto T, Uchiyama I, Imai A, Takahashi SY, Hanari T, Nakamura S, Kobari H (2007) Assimilation and contrast in optical illusions. Jpn Psychol Res 49:33–44
Gregory RL (1963) Distortion of visual space as inappropriate constancy scaling. Nature 199:678–680
Happé FG (1996) Studying weak central coherence at low levels: children with autism do not succumb to visual illusions: a research note. J Child Psychol Psychiatry 37:873–877
Hataji Y, Kuroshima H, Fujita K (2020) Dynamic corridor illusion in pigeons: Humanlikepictorial cue precedence over motion parallax cue in size perception. Perception 11:1–13
Hopkins WD, Washburn DA (2002) Matching visual stimuli on the basis of global and local features by chimpanzees (*Pan troglodytes*) and rhesus monkeys (*Macaca mulatta*). Anim Cogn 5:27–31
Imura T, Tomonaga M, Yagi A (2008) The effects of linear perspective on relative size discrimination in chimpanzees (*Pan troglodytes*) and humans (*Homo sapiens*). Behav Processes 77:306–312
Kanizsa G (1979) Organization in vision: Essays on Gestalt perception. Praeger, New York
Kanizsa G, Renzi P, Conte S, Compostela C, Guerani L (1993) Amodal completion in mouse vision. Perception 22:713–721
Keep B, Zulch HE, Wilkinson A (2018) Truth is in the eye of the beholder: Perception of the Müller-Lyer illusion in dogs. Learn Behav 46:501–512
Kelley LA, Kelley JL (2014) Animal visual illusion and confusion: the importance of a perceptual perspective. Behav Ecol 25:450–463
Kimchi R (1992) Primacy of wholistic processing and global/local paradigm: a critical review. Psychol Bull 112:24–38
Lamb MR, Robertson LC (1988) The processing of hierarchical stimuli: Effects of retinal locus, locational uncertainty, and stimulus identity. Percept Psychophys 44:172–181
Lea SE, Slater AM, Ryan CM (1996) Perception of object unity in chicks: A comparison with the human infant. Infant Behav Dev 19:501–504
Lin I, Chiao CC (2017) Visual equivalence and amodal completion in cuttlefish. Front Physiol 8:40
Matsuno T, Fujita K (2009) A comparative psychophysical approach to visual perception in primates. Primates 50:121–130
Matsuno T, Tomonaga M (2007) Global and local visual processing by chimpanzees (*Pan troglodytes*). Jpn J Psychol Sci 25:281–282

Michotte A, Thinès G, Crabbé G (1991) Amodal completion of perceptual structures. In: Thines G, Costall A, Butterworth G (eds) Michotte's experimental phenomenology of perception. Erlbaum, New York, pp 140–167. (Original work published 1964)

Miletto Petrazzini ME, Parrish AE, Beran MJ, Agrillo C (2018) Exploring the solitaire illusion in guppies (*Poecilia reticulata*). J Comp Psychol 132:48–57

Murayama T, Usui A, Takeda E, Kato K, Maejima K (2012) Relative size discrimination and perception of the Ebbinghaus illusion in a bottlenose dolphin (*Tursiops truncatus*). Aquatic Mammals 38:333–342

Nagasaka Y, Lazareva OF, Wasserman EA (2007) Prior experience affects amodal completion in pigeons. Percept Psychophys 69:596–605

Nagasaka Y, Osada Y (2000) Subjective contours, amodal completion, and transparency in animals. Jpn J Anim Psychol 50:61–73

Nagasaka Y, Wasserman EA (2008) Amodal completion of moving objects by pigeons. Perception 37:557–570

Nakamura N, Fujita K, Ushitani T, Miyata H (2006) Perception of the standard and the reversed Müller-Lyer figures in pigeons (*Columba livia*) and humans (*Homo sapiens*). J Comp Psychol 120:252–261

Nakamura N, Watanabe S, Fujita K (2008) Pigeons perceive the Ebbinghaus-Titchener circles as an assimilation illusion. J Exp Psychol Anim Behav Process 34:375–387

Nakamura N, Watanabe S, Fujita K (2009) Further analysis of perception of the standard Müller-Lyer figures in pigeons (*Columba livia*) and humans (*Homo sapiens*): Effects of length of brackets. J Comp Psychol 123:287–294

Nakamura N, Watanabe S, Fujita K (2014) A reversed Ebbinghaus–Titchener illusion in bantams (*Gallus gallus domesticus*). Anim Cogn 17:471–481

Navon D (1977) Forest before trees: The precedence of global features in visual perception. Cogn Psychol 9:353–383

Neiworth JJ, Gleichman AJ, Olinick AS, Lamp KE (2006) Global and local processing in adult humans (*Homo sapiens*), 5-year-old children (*Homo sapiens*), and adult cotton-top tamarins (*Saguinus oedipus*). J Comp Psychol 120:323–330

Nieder A (2002) Seeing more than meets the eye: Processing of illusory contours in animals. J Comp Physiol 188:249–260

Nieder A, Wagner H (1999) Perception and neuronal coding of subjective contours in the owl. Nat Neurosci 2:660–663

Okanoya K, Takahashi M (1999) Shikaku-teki hokan e no seitaigaku-teki apuroochi [Ecological approach to visual completion]. Kokoro no hattatsu: ninchi-teki seicho no kikoo 5:34–41

Oyama T (1960) Japanese studies on the so-called geometrical optical illusions. Psychol Forsch 3:7–20

Parlangeli O, Roncato S (1995) The global figural characteristics in the Zöllner illusion. Perception 24:501–512

Parrish AE, Brosnan SF, Beran MJ (2015) Do you see what I see? A comparative investigation of the Delboeuf illusion in humans (*Homo sapiens*), rhesus monkeys (*Macaca mulatta*), and capuchin monkeys (*Cebus apella*). J Experiment Psychol 41:395–405

Parron C, Fagot J (2007) Comparison of grouping abilities in humans (*Homo sapiens*) and baboons (*Papio papio*) with the Ebbinghaus illusion. J Comp Psychol 121:405–411

Pepperberg IM, Vicinay J, Cavanagh P (2008) Processing of the Müller-Lyer illusion by a Grey parrot (*Psittacus erithacus*). Perception 37:765–781

Pressey AW (1971) An extension of assimilation theory to illusions of size, area, and direction. Percept Psychophys 9:172–176

Qadri MA, Cook RG (2019) Perception of Ebbinghaus–Titchener stimuli in starlings (*Sturnus vulgaris*). Anim Cogn 22:973–989

Regolin L, Marconato F, Vallortigara G (2004) Hemispheric differences in the recognition of partly occluded objects by newly hatched domestic chicks (*Gallus gallus*). Anim Cogn 7:162–170

Regolin L, Vallortigara G (1995) Perception of partly occluded objects by young chicks. Percept Psychophys 57:971–976
Reid SL, Spetch ML (1998) Perception of pictorial depth cues by pigeons. Psychon Bull Rev 5:698–704
Révész G (1924) Experiments on animal space perception. Br J Psychol 14:387–414
Roberts B, Harris MG, Yates TA (2005) The roles of inducer size and distance in the Ebbinghaus illusion (Titchener circles). Perception 34:847–856
Robinson JO (1998) The psychology of visual illusion. Dover, Mineola
Rosa Salva O, Rugani R, Cavazzana A, Regolin L, Vallortigara G (2013) Perception of the Ebbinghaus illusion in four-day-old domestic chicks (*Gallus gallus*). Anim Cogn 16:895–906
Rosa Salva O, Sovrano VA, Vallortigara G (2014) What can fish brains tell us about visual perception? Front Neural Circuit 8:119
Sakiyama T, Gunji YP (2013) The Müller-Lyer illusion in ant foraging. PLoS One 8:e81714
Santacà M, Agrillo C (2020) Perception of the Müller–Lyer illusion in guppies. Curr Zool 66:205–213
Sato A, Kanazawa S, Fujita K (1997) Perception of object unity in a chimpanzee (*Pan troglodytes*). Jpn Psychol Res 39:191–199
Sekuler AB, Lee JA, Shettleworth SJ (1996) Pigeons do not complete partly occluded figures. Perception 25:1109–1120
Shimizu T (1998) Conspecific recognition in pigeons (*Columba livia*) using dynamic video mages. Behaviour 135:43–54
Sovrano VA, Albertazzi L, Salva OR (2015) The Ebbinghaus illusion in a fish (*Xenotoca eiseni*). Anim Cogn 18:533–542
Sovrano VA, Bisazza A (2009) Perception of subjective contours in fish. Perception 38:579–590
Sovrano VA, Da Pos O, Albertazzi L (2016) The Müller-Lyer illusion in the teleost fish *Xenotoca eiseni*. Anim Cogn 19:123–132
Spinozzi G, De Lillo C, Salvi V (2006) Local advantage in the visual processing of hierarchical stimuli following manipulations of stimulus size and element numerosity in monkeys (*Cebus apella*). Behav Brain Res 166:45–54
Suganuma E, Pessoa VF, Monge-Fuentes V, Castro BM, Tavares MCH (2007) Perception of the Müller–Lyer illusion in capuchin monkeys (*Cebus apella*). Behav Brain Res 182:67–72
Sugita Y (1999) Grouping of image fragments in primary visual cortex. Nature 401:269–272
Tanaka HK, Fujita I (2000) Global and local processing of visual patterns in macaque monkeys. Neuroreport 11:2881–2884
Timney B, Keil K (1996) Horses are sensitive to pictorial depth cues. Perception 25:1121–1128
Tudusciuc O, Nieder A (2010) Comparison of length judgments and the Müller-Lyer illusion in monkeys and humans. Exp Brain Res 207:221–231
Tvardíková K, Fuchs R (2010) Tits use amodal completion in predator recognition: a field experiment. Anim Cogn 13:609–615
Ushitani T, Fujita K, Yamanaka R (2001) Do pigeons (*Columba livia*) perceive object. unity? Anim Cogn 4:153–161
Vallortigara G (2012) Core knowledge of object, number, and geometry: a comparative and neural approach. Cogn Neuropsychol 29:213–236
von Uexküll J (1957) A stroll through the worlds of animals and men: a picture book of invisible worlds. In: Schiller CH (ed) Instinctive behavior: the development of a modern concept. International Universities Press, New York, pp 5–80. (Original work published 1934)
Warden CJ, Baar J (1929) The Müller-Lyer illusion in the ring dove, *Turtur risorius*. J Comp Psychol 9:275–292
Watanabe S, Furuya I (1997) Video display for study of avian visual cognition: From psychophysics to sign language. Int J Comp Psychol 10:111–127
Watanabe S, Nakamura N, Fujita K (2011) Pigeons perceive a reversed Zöllner illusion. Cognition 119:137–141

Watanabe S, Nakamura N, Fujita K (2013) Bantams (*Gallus gallus domesticus*) also perceive a reversed Zöllner illusion. Anim Cogn 16:109–115

Weintraub DJ (1979) Ebbinghaus illusion: Context, contour, and age influence the judged size of a circle amidst circles. J Exp Psychol Hum Percept Perform 5:353–364

White KG (1972) Implicit contours in the Zöllner illusion. Am J Psychol 85:421–424

Wyzisk K, Neumeyer C (2007) Perception of illusory surfaces and contours in goldfish. Vis Neurosci 24:291–298

Zylinski S, Darmaillacq AS, Shashar N (2012) Visual interpolation for contour completion by the European cuttlefish (*Sepia officinalis*) and its use in dynamic camouflage. Proc Roy Soc 279:2386–2390

Chapter 3
Comparative Studies on Geometric Illusions: A Review of Methods and Results

Sota Watanabe

Abstract As the characteristics of the visual system we have today are outcomes of our evolutionary history and our ancestors' adaptation to their environment, comparing susceptibility to illusions in humans and nonhuman animals can contribute not only toward discovering previously unknown characteristics of other species' visual systems, but also toward understanding of the evolution of those systems. There is now a sizeable literature describing comparative studies of various species respond to illusion figures. This chapter outlines research on the susceptibility of a range of nonhuman species to five popular illusion figures, namely, Ponzo, Müller-Lyer, Zöllner, Ebbinghaus, and Delboeuf illusions. Careful attention to the methods is critical in studies on visual illusions, given the requirements that nonhuman subjects attend, judge, and execute responses without any linguistic instruction. For this reason, this chapter places particular emphasis on some of the methodological issues in conducting comparative illusion studies, and how many of the difficulties may be overcome.

Keywords Geometric illusions · Visual illusions · Visual perception · Animal vision · Comparative perception · Size perception

Introduction

We humans process much information through vision. Through vision, we capture an object's presence, color, inclination, angle, two-dimensional shape, texture, changes of position or area, among other features. And this visual capture (sometimes integrating other sensory information) can rapidly lead to decisions about the object's size, three-dimensional shape, spatial arrangements including depth and

S. Watanabe (✉)
Division of General Education, Osaka Kyoiku University, Osaka, Japan
e-mail: sotamw@cc.osaka-kyoiku.ac.jp

J. R. Anderson, H. Kuroshima (eds.), *Comparative Cognition*,
https://doi.org/10.1007/978-981-16-2028-7_3

movement, and whether the object represents danger or is otherwise important for survival.

Although our physical world has a three-dimensional spread, our retina can only sense the physical world as two-dimensional information. Thus, our perceptual world is more or less inaccurate. We are generally unaware of this inherent inaccuracy but awareness often emerges when we experience a visual illusion. An illusion figure is not specifically a figure that causes "inaccurate perception; "instead, it simply tells us that our perceptual world is always inaccurate. What matters is not *whether or not* it is inaccurate but *how* it is inaccurate. The inaccuracy reflects the characteristics of our visual system, and visual illusions are useful tools for examining these characteristics.

The characteristics of the visual system we have today are outcomes of our evolutionary history and our ancestors' adaptation to the environment. In other words, we perceive the world with inaccuracies that have likely been advantageous for survival. This hypothesis can be tested by comparing susceptibility to illusions between humans and nonhuman animals (Fujita et al. 2012); not only can this reveal characteristics of other species' visual systems, it also helps toward a better understanding of the likely evolution of these systems.

There have been many comparative studies on perception of illusion figures, among which the best-studied are the classic geometric illusions including the Ponzo illusion (Fig. 3.1a, b), the Muller-Lyer illusion (Fig. 3.1c, d as a variant), the Zöllner illusion (Fig. 3.1e), the Ebbinghaus illusion (Fig. 3.1f), and the Delboeuf illusion (Fig. 3.1g). Particularly in the 1990s, the development and widespread use of PCs facilitated the creation of various versions of illusion figures and implementation of operant experiments; the resulting literature continues to flourish (Table 3.1; see also Appendix; Fujita et al. 2017; Agrillo et al. 2020). There are now many reports of animals perceiving (or not perceiving) the same visual illusion as humans, even animals perceiving the same illusion as humans but in the opposite direction to humans!

Careful attention to the methods is critical in studies on visual illusions, especially as studies on nonhuman species require that the subjects attend, judge, and execute responses without any linguistic instruction. In this chapter I outline research on the susceptibility of various species to the five illusion figures mentioned above, with special reference to the procedures used. For ease of description, I refer to the five illusions as: the *solid* Ponzo illusion, the *one-sided* Müller-Lyer illusion, the *complex* Zöllner illusion, the *hot* Ebbinghaus illusion, and the *advanced* Delboeuf illusion. The precise reasons for these terms will be explained in the following sections.

The *Solid* Ponzo Illusion

Although a precise definition of the Ponzo illusion is elusive, in this chapter, a typical Ponzo illusion figure is one in which multiple identical objects (e.g., horizontal lines or circles) are enclosed within two converging lines. Humans

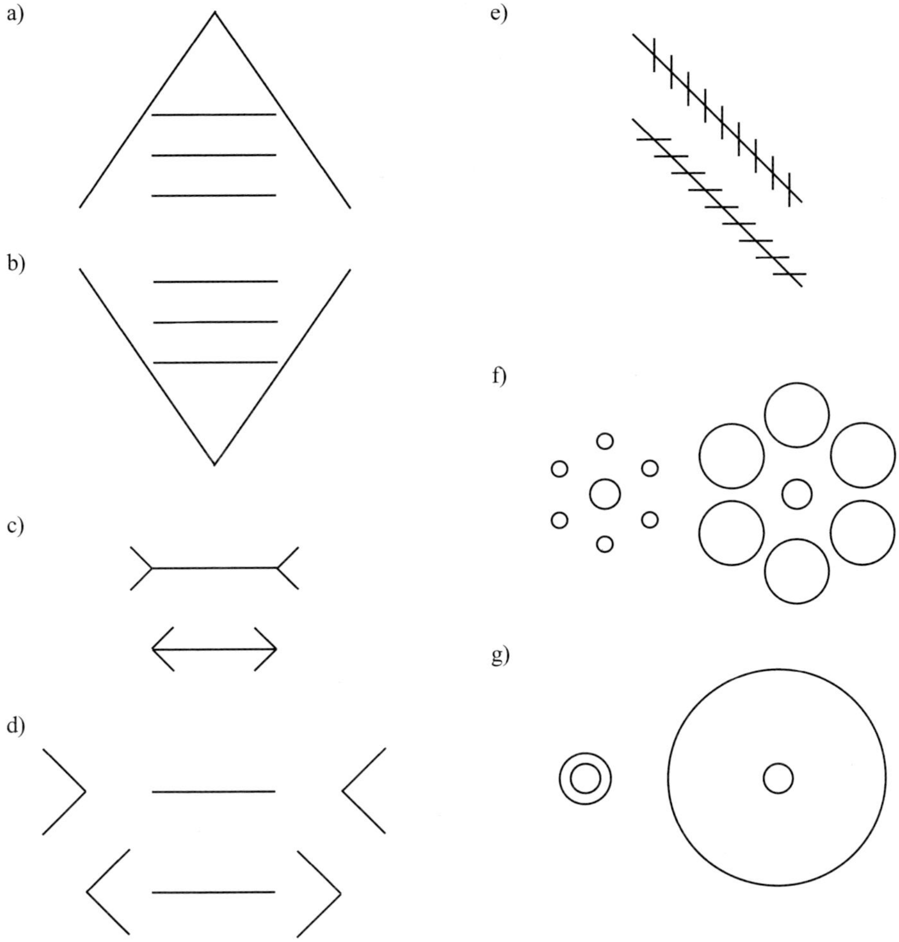

Fig. 3.1 Typical patterns of geometric illusions. (**a**) Ponzo illusion (**b**) the inverted pattern of the Ponzo illusion (**c**) Müller-Lyer illusion (**d**) reversed Müller-Lyer illusion (**e**) Zöllner illusion (**f**) Ebbinghaus illusion (**g**) Delboeuf illusion. All the horizontal lines in **a**, **b**, **c**, **d** are identical in length. All the four inner dots in **f** and **g** are identical in size

overestimate the size of the object that is closer to the outer lines (Fig. 3.1a) (e.g., Robinson 1998). The corridor illusion, which partially overlaps the Ponzo illusion, consists of a photograph or several lines giving the impression of a natural perspective. For simplicity, in this chapter I define Ponzo illusion studies as those that include just the two converging lines.

Many studies have tested Gregory's (1963, 1965, see also 2009) "inappropriate constancy scaling theory" explanation of the Ponzo illusion. This theory proposes that our visual system automatically applies a size constancy scaling to correct the subjective size impression. For example, when we see a Ponzo figure, we interpret the closer object to the outer lines as more distant than the other object(s), and so we

Table 3.1 The number of comparative studies on the Ponzo illusion, the standard and reversed Müller-Lyer illusion, Zöllner illusion, Ebbinghaus illusion, and Delboeuf illusion in each decade. We counted the total number of studied species instead of the number of studies. For example, if one study tested two different species, this was counted as two. If two studies each tested one species, this was counted them as two

	Mammals	Birds	Fish	Other	Total
–1989	3	9	2	1	15
1990–1999	2	2	0	0	4
2000–2009	2	3	1	0	6
2010–	13	5	7	3	28
Sum	20	19	10	4	53

overestimate its size. Several studies have shown that the magnitude of illusion is larger when the whole figure is upright (Fig. 3.1a) and the top object seems "farther away" than when it is inverted (Fig. 3.1b) (e.g., Fisher 1968; Kincade and Wilson 2000). Although these reports can be taken as support for Gregory's account, other studies have reported conflicting results (e.g., Brown and Houssiadas 1965; Robinson 1998).

Fujita et al. (1993) used an original and impressive method for analyzing the Ponzo illusion in nonhuman animals. Their method consisted of two training phases and a test phase. First, they displayed a single target line without any bordering lines and trained pigeons to respond by pecking at it. This response caused two choice-keys to appear. One of them was assigned to "long," and the other was to "short" target lines, with different assignments for different subjects. The subjects had to choose one of the two keys, depending on the lengths of the target lines. Next, the authors added the illusion inducer (the outer lines) and conducted the same training. In this phase, the gap between the target and the inducer was fixed. Each training phase continued until the subjects reached a learning criterion, which in turn led to the test phase, consisting of baseline trials identical to those in the previous training phase and probe test trials. In the test trials, target positions were manipulated to be nearer or farther to the inducer than in baseline trials, and responses to the various target positions were compared. Remembering this training procedure will facilitate reading the rest of this chapter: initial training with a target only, followed by training with the target and inducer(s), and finally testing by manipulating parameters (e.g., length, size, tilt, brightness) of inducer(s). Variants of this procedure have now been used in comparative studies of not only the Ponzo illusion but also the Müller-Lyer, Zöllner, Ebbinghaus, and Delboeuf illusions.

I call the Ponzo illusion in this chapter *solid* for two reasons. First, several comparative studies have been conducted on this illusion in other species: rhesus macaques (Fujita 1996, 1997), pigeons (Fujita et al. 1991, 1993), chimpanzees (Fujita 1997), and dogs (Byosiere et al. 2017a, b; Byosiere et al. 2018). These studies have established that macaques, pigeons, and chimpanzees perceive this illusion in the same direction as humans. By contrast, the studies on dogs have produced only weak evidence that dogs perceive illusions in the same direction as humans. Overall, these findings suggest the possibility that the properties of the visual system that underlie the Ponzo illusion may be widely shared, at least between mammals and birds.

Second, Fujita and his co-researchers carefully investigated various versions of this illusion stimulus. Fujita et al. (1991) presented the entire stimulus upright, inverted, or sideways (Fig. 3.2a), a version in which the outer lines do not intersect each other (Fig. 3.2b), and a version in which the outer lines do not create a depth perspective (Fig. 3.2c). Results suggested that pigeons consistently perceive the Ponzo illusion in the same direction as humans, regardless of how the stimulus was presented. Fujita et al. (1993) systematically manipulated the angle between the outer lines so the entire stimulus varied from being upright to inverted (Fig. 3.2d) and reported that pigeons perceived the illusion in both upright and inverted versions. Fujita (1996) presented typical Ponzo figures superimposed on a photograph with linear perspective to rhesus macaques and humans (Fig. 3.2e). Both species perceived the illusion in both typical Ponzo figures and also photographic linear perspective, but only the rhesus macaques did even when the photograph was inverted. When a deformed Ponzo illusion in which dots sandwiched the horizontal line (Fig. 3.2f) was presented to rhesus macaques (Fujita 1996), chimpanzees, and humans (Fujita 1997), chimpanzees and humans perceived the illusion from this dot pattern, whereas for rhesus monkeys it induced little or no illusion.

Taken together, the results from the studies described above consistently indicate that pigeons, rhesus macaques, and chimpanzees perceive the Ponzo illusion in the same direction as humans, but with chimpanzees in particular giving results very similar to humans. Pigeons and rhesus monkeys, unlike humans, were consistently unaffected by photographs or lines, suggesting species differences in susceptibility to a depth-perception from such pictorial stimuli. Further comparative studies will surely reveal further interesting species-similarities and -differences in Ponzo illusion and depth-perception.

Recently, increasing recognition of the importance of using various versions of a single illusion in comparative studies has been expressed, following the example of Fujita and colleagues (e.g., Qadri and Cook 2019). One reason is that, as seen in the Ponzo illusion research, some presentations of a stimulus may induce different illusionary tendencies in different species. Further comparative analyses can highlight species-typical characteristics of the visual system. Another reason is that sometimes contradictory results about illusion perception have been obtained even within species, as exemplified by comparative studies on the Ebbinghaus illusion, described in a later section. Presenting different patterns of a single illusion can be useful for identifying which parts of the figure subjects are discriminating.

The *One-Sided* Müller-Lyer Illusion

A typical Müller-Lyer figure is a shape in which both ends of one line (shaft) are sandwiched by inward (“> <”) or outward (“< >”) brackets (Fig. 3.1c). Humans overestimate the length of the shaft sandwiched by inward brackets and

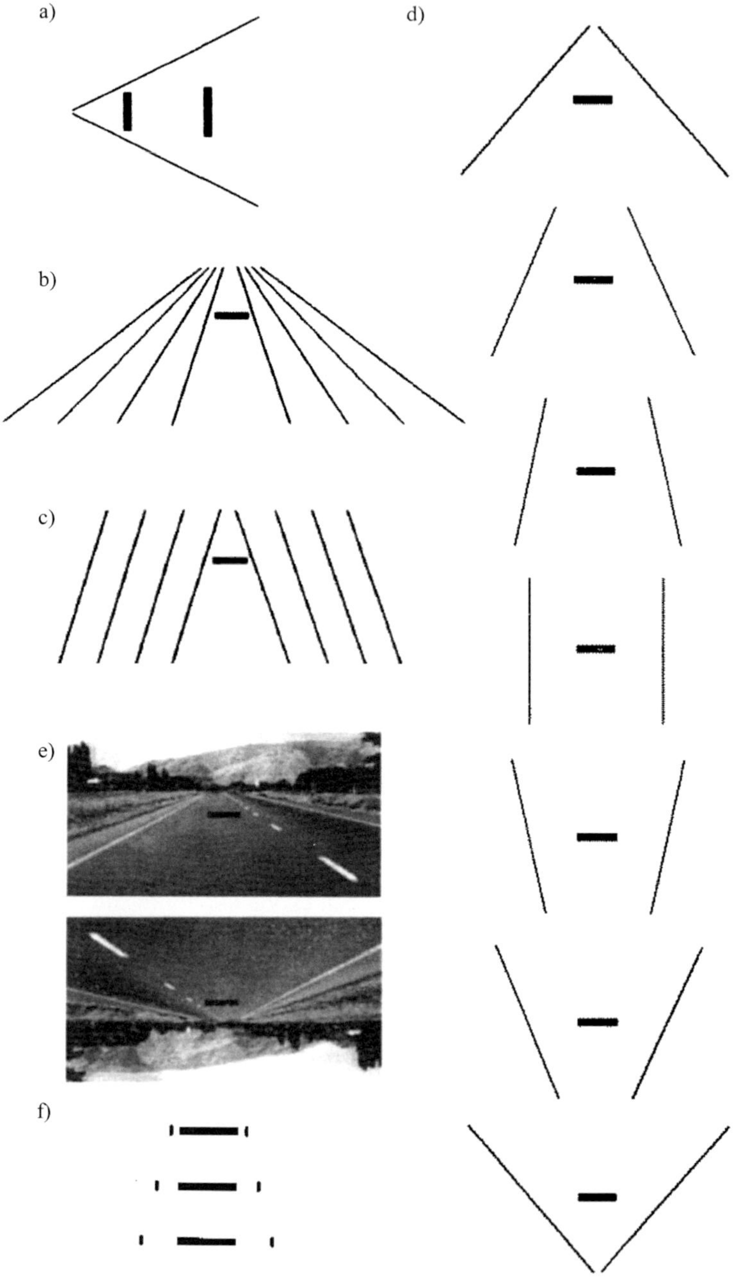

Fig. 3.2 Various versions of the Ponzo illusion. (**a**) one of the patterns used in Fujita et al. (1991) for pigeons; (**b**) one of the patterns used in Fujita et al. (1991) for pigeons and Fujita (1997) for rhesus monkeys and chimpanzees, showing perspective context; (**c**) one of the patterns used in Fujita et al. (1991) for pigeons and Fujita (1997) for rhesus monkeys and a chimpanzee, with no perspective context; (**d**) examples of patterns used in Fujita et al. (1993) for pigeons; (**e**) part of the upright and inverted patterns used in Fujita (1996) for rhesus monkeys and humans, showing only a target line on a picture of a road with perspective; (**f**) patterns used in Fujita (1997) for rhesus monkeys and chimpanzees

underestimate the one sandwiched by outward brackets. Due to the simplicity of the stimulus and the largeness of the illusionary magnitude, the Müller-Lyer illusion is probably the most famous geometric illusion. It is interesting, however, that the mechanism of this illusion is in dispute even now.

The Müller-Lyer illusion has notable similarities to the Ponzo illusion: both can arise simply from a horizontal target line and diagonally converged inducer lines, both are illusions of size, and both are argued to be due to perspective size constancy as proposed by the inappropriate constancy scaling theory (e.g., Gregory 1963, 1965). Moreover, no study has reported animals showing the opposite direction of either the Ponzo illusion or Müller-Lyer illusion experienced by humans.

Despite many comparative studies on the Müller-Lyer illusion and the difficulty of identifying a standard methodology, an appropriate and reliable method was used for pigeons by Nakamura et al. (2006), based on Fujita's (1996) procedure for examining the Ponzo illusion. First, Nakamura et al. (2006) presented only one line (shaft) in each trial and trained the subjects to indicate whether it was longer or shorter than a criterion. After the subjects mastered this task, they received further training in which a baseline stimulus with leftward or rightward brackets was attached to both ends of the shaft. Importantly, unlike in Fujita (1996), across sessions the brackets were gradually faded in from the background color to the target color. The following test phase consisted of baseline trials as in the previous training phase and probe test trials. In a test trial, the Müller-Lyer figure consisted of the shaft and inward ("> <") or outward ("< >") brackets They compared the subjects' responses between the bracket-types. The subjects judged the Müller–Lyer figures with inward-pointing brackets as "long" more frequently than the figures with outward-pointing brackets, suggesting perception of the same Müller-Lyer illusion as humans.

I refer to the procedure just described as a *one-sided* Muller-Lyer illusion, for two reasons. First, to emphasize the *one-sided* direction of this illusion. Most of the many studies on this topic have reported the same direction of this illusion as in humans: in rhesus macaques (Tudusciuc and Nieder 2010); capuchin monkeys (Suganuma et al. 2007); pigeons (Malott et al. 1967; Nakamura et al. 2006, 2008); ring doves (Warden and Baar 1929); chickens (Winslow 1933); a grey parrot (Pepperberg et al. 2008); redtail splitfin (Sovrano et al. 2016); guppies (Santacà and Agrillo 2019). Only two species have shown no evidence of perceiving the illusion: Fuss et al. (2014) reported negative results in sharks, and Wyzisk (2005) did likewise for goldfish.

Second, and more importantly, researchers are likely to arrive at the conclusion that their target species perceives the Müller-Lyer illusion in the same direction as humans. Studies on animals, with one rare exception (Pepperberg et al. 2008), have not used linguistic instruction nor otherwise communicated with their subjects about which parts and features (e.g., length, size, angle, inclination, color) of the illusion figure they should respond to. Therefore, each study of an illusionary figure needs to be carefully examined to determine how the subject responded. Unfortunately, the shaft and brackets of the Müller-Lyer illusion are both linear, and they are in contact with each other. Conceivably, with no explicit instructions, this high similarity might cause animals to confuse the target and the inducers and identify the entire stimulus, including the inducers, as the target. This means that researchers might get data that suggest perception of the illusion in the same direction as humans, but in at least some cases the subjects might simply have confused which part of the stimulus they should judge.

As yet, only a few studies have adequately addressed this problem. Nakamura et al. (2006) first demonstrated that pigeons perceived the Müller-Lyer illusion in the same direction as humans and then proceeded to ask whether the birds might have simply judged the overall lengths of the figures instead of actually perceiving the illusion. They rejected this possibility after plotting the same data against the overall lengths of the figures. Likewise, they also tested and rejected the possibility that the pigeons had based their responses solely on the size of the gap between the ends of the two brackets. Nakamura et al. (2009a) manipulated the length of the brackets and again showed that pigeons perceive the Müller-Lyer illusion similarly to humans, rather than just confusing the target and the inducers. Future comparative studies on Müller-Lyer illusion will require verifications of the type conducted by Nakamura et al. (2006, 2009a). Analysis of the effects of various presentation conditions would be useful, as in the series of Ponzo illusion studies by Fujita and colleagues.

A reversed Müller-Lyer illusion occurs in humans when there is a gap between the shaft and each bracket (Fig. 3.1d). When this occurs, the shaft enclosed by inward ("> <") brackets is perceived as shorter than the same length shaft enclosed by outward ("< >") brackets (Yanagisawa 1939; Fellows 1968). The results of three studies on nonhuman species suggest that neither rhesus macaques (Tudusciuc and Nieder 2010) nor pigeons (Nakamura et al. 2006, 2009b) perceive the reversed Müller-Lyer illusion. There are relatively few studies on the reversed Müller-Lyer illusion in humans, and its mechanism and laws have not been fully elucidated (e.g., Pressey and Bross 1973). Also, since the magnitude of this illusion is small even in humans, I cannot yet be confident about whether pigeons or monkeys do not perceive it, or they do perceive it but at small magnitude. However, it is clear that no study has reported that any other species clearly perceives the opposite direction of either the standard or the reversed Müller-Lyer illusions reported in humans. This contrasts with the situation concerning the Ebbinghaus and the Zöllner illusions, as described in the following section.

The *Complex* Zöllner Illusion

I call the Zöllner illusion *complex* for three reasons. First, the Zöllner illusion has a relatively complicated shape. A typical Zöllner figure consists of several parallel lines (shafts) and several groups of short lines (crosshatches) attached to the shafts at an acute angle (Fig. 3.1e) (e.g., Robinson 1998), and we perceive the shafts as not being parallel. Four of the five illusion figures considered in this chapter typically consist of about 2–7 objects (e.g., lines, circles), whereas the Zöllner illusion typically requires about 16 lines.

Second, along with its shape, the mechanism of the Zöllner illusion is complicated. In a detailed study of the illusion's mechanism, Kitaoka and Ishihara (2000) decomposed its parts and analyzed the magnitude of the effect each time. They hypothesized that the Zöllner illusion consists of three low-level illusions. The first is an overestimation of the inclination difference between a shaft and its crosshatches. The illusion appears larger when the angle between the shaft and each crosshatch is acute (especially about 30°), and it hardly occurs when the angle is obtuse. The second is an underestimation of the same aspects. This underestimation is always slight, regardless of the difference in slope. While these two illusions arise even with just one shaft, the third is an overestimation of inclination difference between each shaft and its crosshatches that arises only when there are multiple shafts to which crosshatches are attached. Based on these findings, we can define the Zöllner illusion as an illusion in which crosshatches make humans overestimate or underestimate the inclination difference between the shafts.

Third, past comparative studies have given rise to conflicting findings and interpretations. Dücker (1966) reported that some crucian carp, birds including domestic chicks, avadavats, splendid starlings, and a mistle thrush, and Guinea pigs were susceptible to this illusion. Agrillo et al. (2014) showed that rhesus macaques also perceive the same Zöllner illusion as humans. Benhar and Samuel (1982) showed that baboons also experienced the illusion, but the authors did not specify the direction of the illusion. By contrast, we have reported that two avian species, namely pigeons (Watanabe et al. 2011) and bantams (Watanabe et al. 2013), perceive the opposite direction of Zöllner illusion to that perceived by humans.

In our studies, we presented birds with two nonparallel shafts whose inclination difference was within 3–15° and trained them to peck at the narrower (or wider, counterbalanced across subjects) of the two gaps between the edges of the lines. First, to clarify where to peck, we also presented red cue dots at both gaps and gradually faded these to the white, background color. After each bird reliably pecked at the correct gap without the cue, we attached ten crosshatches to each shaft, equally spaced, parallel to each other, and at random crossing angles against each shaft across trials. This randomization, which would invoke little apparent illusion to humans, was done to ensure that subjects' attended to the inclination difference between the shafts. These crosshatches were gradually faded in from white to black. After subjects reached a learning criterion with black crosshatches, we moved to a test phase consisting of baseline trials identical to those in the previous training phase, and probe test trials, including trials in which the Zöllner-type crosshatches crossed the shafts at 30° (boost-upside-wider and boost-downside-wider) and trials in which shafts were parallel and each crosshatch-set crossed each shaft at 30°.

The results of these two studies were clear and consistent. Both pigeons and bantams were less likely to judge "the upside gap as wider" when presented with boost-upside-wider stimuli, and more likely to do so when presented with boost-downside-wider stimuli. We also presented the same figures, controlling for the visual angle, to human participants. The latter were more likely to judge "the upside gap as wider" for boost-upside-wider stimuli, and less likely to for boost-downside-wider stimuli. These results suggest that pigeons and chickens perceive a Zöllner illusion, but that the direction is opposite to humans, rhesus macaques (Agrillo et al. 2014), and several other species (Dücker 1966). Whether the strikingly different outcomes of our studies and that by Dücker (1966) are due to the different subject species or to procedural differences is not yet clear. Unfortunately, there has been no further comparative research on the Zöllner illusion since our studies, possibly due to the complex shape of this illusion figure and the difficulty of training nonlinguistic animals to judge the shaft and ignore the crosshatches. Such research is clearly needed, especially on avian species. Although we trained birds to discriminate the invisible gap sizes by using a search task, a conditional position discrimination task as used with rhesus macaques (Agrillo et al. 2014) and an oddity task, used with baboons (Benhar and Samuel 1982) also appear to be potentially useful alternatives.

The *Hot* Ebbinghaus Illusion

The Ebbinghaus illusion is typically composed of a single target circle and inducer circles arranged at equal intervals around the circle (Fig. 3.1e). We humans underestimate the target circle size when the size of an inducer circle is larger than the target circle and vice versa. I describe this illusion as *hot* for three reasons. First, there are many studies of this illusion in both humans and nonhumans, with new studies on humans under various conditions frequently appearing in the literature (e.g., Lavrenteva and Murakami 2018; Mruczek et al. 2015; Takao et al. 2019). There have been studies of the Ebbinghaus illusion on 12 nonhuman species, comparable to research on the Müller-Lyer illusion (see Appendix). Second, in contrast to the *one-sided* Müller-Lyer illusion, the comparative studies have produced conflicting results. Before I refer to the third reason, let me elaborate on the second.

Among mammals, there are reportedly species differences in susceptibility to the Ebbinghaus illusion: A bottlenose dolphin (Murayama et al. 2012) and Guinea pigs (Dücker 1966) perceived the illusion in the same direction as humans, but Guinea baboons (Parron and Fagot 2007) did not perceive this illusion. Dogs perceived the illusion, but the direction was opposite to that in humans (Byosiere et al. 2017a, b). There are differences among aquatic species as well. European carp (Dücker 1966), redtail splitfins (Sovrano et al. 2015), and teleost damselfish were reported to perceive the same Ebbinghaus illusion as humans, whereas in gray bamboo sharks the illusion was reported to be in the opposite direction to that seen in humans. One study reported both the standard and reversed Ebbinghaus illusion by damselfish and shark, using the same method (Fuss and Schluessel 2017). Avian species are the *hottest*. Two studies reported the standard Ebbinghaus illusion in starlings (Dücker 1966) and domestic chicks (Rosa Salva et al. 2013), whereas two others reported the reversed Ebbinghaus illusion in pigeons (Nakamura et al. 2008) and bantams (Nakamura et al. 2014). It is noteworthy that adults of one species, chickens (bantams) and their young (chicks) show opposite tendencies.

Comparative studies of the Ebbinghaus illusion involve the same difficulties as those of Müller-Lyer illusion. In particular, caution is required when results suggest the reversed Ebbinghaus illusion, because when the surrounding circles are smaller, the subjects were more likely to answer that the center circle was "small" and vice versa. Researchers need to demonstrate that the subjects in fact perceive the illusion in the opposite direction to humans, by eliminating the possibility that the subjects simply confuse the surrounding circles and the central circle. In this context, it is useful to look more closely at the experimental design used by Nakamura et al. (2008, 2014), who reported the reversed illusion in pigeons and bantams. Their method was very similar to that in Nakamura et al.'s (2006) study of the Müller-Lyer illusion. They first presented subjects with one target circle of one of the six possible sizes fixed at equal intervals and trained them to respond "larger" or "smaller" than the standard size of 10.0° in diameter. After the subjects reached criterion, the authors added Ebbinghaus-shaped surrounding circles of 10° in diameter. The color of the inducers was initially almost the same as the background, gradually increasing when each bird's performance in one session met criterion. This fade-in process required 14–26 steps before the color of inducers matched that of the target. Each subject proceeded to a test phase after exceeding 307 correct trials in one session of 384 trials (80%) with inducer color the same as the target. In the test phase, test stimuli consisting of a target circle and novel-sized-inducers were presented in probe trials.

Nakamura et al. (2008, 2014) reported that in the test phase, subjects judged the target circle surrounded by larger circles to be larger than it really was and vice versa. These results suggest that pigeons and bantams perceive the Ebbinghaus illusion, but the direction of the illusion is opposite to that in humans. The authors went on to examine alternative explanations of the results. For example, might subjects sometimes have responded to the size of inducers instead of the target circle? Might they have responded based on the total area of the seven circles or based on the average area of the target and one inducer circle? The authors rejected these hypotheses by creating hypothetical graphs simulating the response rates based on each hypothesis and confirming that none of the graphs differed from their experimental results.

The third reason I refer to the *hot* Ebbinghaus illusion is that most comparative studies on the phenomenon are relatively recent, appearing in the last decade. Before Parron and Fagot (2007), only Dücker (1966) existed. The latest (hot) study, by Qadri and Cook (2019), could be particularly impactful, as it directly challenges the interpretation of the reversed Ebbinghaus illusion by Nakamura et al. (2008, 2014). Qadri and Cook (2019) first trained and tested starlings using a method similar to that of Nakamura et al. Starlings judged the target circle surrounded by larger circles to be larger than it was and vice versa, just like pigeons and bantams. The authors then presented various surrounding objects while manipulating dimensions including shape, brightness, number, and spacing and found that the starlings' responses involved integrating the surroundings into their judgment, instead of responding to the target size only. Qadri and Cook (2019) therefore doubted the validity of the training method both they and Nakamura et al. (2008, 2014) had used to examine the perception of the Ebbinghaus illusion in nonhuman animals.

Although the method used by Qadri and Cook (2019) was both careful and robust, I disagree that their results mean that our interpretation of the reversed Ebbinghaus illusion in birds should be rejected. First, the two studies used different avian species. Even if starlings do not perceive the reversed Ebbinghaus illusion, it does not necessarily follow that pigeons and bantams do not perceive it. Nakamura et al. (2008, 2014) not only demonstrated the perception of this illusion by the latter species, they also ruled out alternative explanations by simulating the response rates based on each hypothesis and confirming that none of the graphs differed from their experimental results. Second, in their second training phase and the test phases, Qadri and Cook (2019) switched to using highly novel stimuli. For example, in the second training phase in which the color of the surrounding circles was gradually faded in, Nakamura et al. (2008, 2014) did this in 14–26 steps, whereas Qadri and Cook (2019) did it in only seven steps. As the latter authors pointed out, such a quick change of inducers could divert subjects' attention from the target to the entire stimulus, making it more likely that the subjects would switch to a different strategy, such as selection bias. We have observed such effects in some of our unpublished experiments. Furthermore, the template matching account proposed by Qadri and Cook (2019), in which the subject might shift attention to the entire stimulus group during the course of training, does not seem to fully explain all of their results with starlings. The subjects showed large individual differences when the number of the inducers was manipulated and more often judged "small" for stimuli with a wider interval between the inducers and the target. However, I will return to a crucial point raised by these authors in the General Discussion, about multifaceted investigations with various versions of illusion stimuli.

The *Advanced* Delboeuf Illusion

The fifth visual illusion to be discussed is the Delboeuf illusion, which typically consists of two concentric circles (Fig. 3.1g). Humans overestimate the size of the inner circle when the two circles are approximately the same size and underestimate it as the gap between the circles gets larger. I call it the *advanced* Delboeuf illusion (Fig. 3.1g) for two reasons. First, the illusion is temporally "advanced" in that comparative studies have recently multiplied rapidly. Since the earliest study, by Parrish and Beran (2014), studies have been conducted on 11 nonhuman species (see Appendix). Chimpanzees (Parrish and Beran 2014), rhesus macaques and capuchin monkeys (Parrish et al. 2015), and bearded lizards (Santacà et al. 2019) were reported to perceive this illusion in the same direction as humans. By contrast, guppies (Lucon-Xiccato et al. 2019) were reported to perceive the illusion in the opposite direction to humans. Dogs, if they are susceptible to the illusion, may perceive the reversed illusion (Byosiere et al. 2017a, b). The illusion in honeybees was strongly influenced by visual angle (Howard et al. 2017). Second, the Delboeuf illusion is probably one of the most challenging figures for comparative researchers to work with. Because the target and the inducer are both circles, it is possible that subjects simply confuse the two, especially in cases of overestimation of the inner circle.

The most popular method for exploring the Delboeuf illusion in animals has involved allowing subjects to choose between two food dishes. In the first such study using this method, Parrish and Beran (2014) presented chimpanzees with two round food portions on two dishes and allowed them to choose one. They arranged each food portion and dish concentrically, in a way that causes the Delboeuf illusion for humans. When one portion was actually larger than the other, the subjects more often chose the larger one. When two portions were the same size and one dish was larger than the other, the subjects preferred the portion on the smaller dish. Parrish and Beran (2014) concluded that chimpanzees perceive the Delboeuf illusion in the same direction as humans.

This method presupposes that nonhuman animals, like human children offered cookies, would naturally choose the larger one. However, when an animal is presented with two food portions that are quite similar in size, it may not be sufficiently motivated or cognitively able to discriminate between them, and it might simply show some kind of reaching bias. Furthermore, even if the subject showed a preference for the larger (or seemingly larger) portion, this might not be based on wanting to choose the larger one; for example, the subject might simply choose the one whose edge is closer and easier to reach. Conversely, failures to demonstrate the Delboeuf illusion in animals do not exclude the possibility that other procedures might provide positive evidence of perceiving the illusion. Despite failures to find this illusion in lemurs (Santacà et al. 2017), dogs (Miletto Petrazzini et al. 2017), cats (Szenczi et al. 2019), and turtles (Santacà et al. 2019), the basic question about whether these species can perceive this illusion remains open.

Failures to demonstrate the Delboeuf illusion are much more frequent compared to the other four illusions that I have discussed. It is our view that forced two-choice tasks are unsuitable for testing whether animals experience this illusion. In Watanabe et al. (2016), budgerigars and humans were first requested to select the larger (or smaller, counterbalanced across subjects) one of the two target graphics, presented left and right. Next, we added one concentric frame over one target and confirmed that the subjects' discrimination was not affected by the frame. In the next test, we presented stimulus combinations varying in size and found that humans perceived normal Delboeuf illusion, suggesting the validity of stimulus sets. When the size difference between the two targets was large enough for easy discrimination, the budgerigars maintained a high level of correct performance, unaffected by the frame. However, when the difference was small, they tended to select targets embedded in the frame, regardless of frame size. Conceivably, the budgerigars switched to a strategy of choosing the closer-to-the-center stimulus set (i.e., a target embedded in a frame) only when the task was more challenging. Troublingly, this strategy is a generic one that could be used in all two-choice tasks, not just this study. To avoid this strategy, a single presentation such as a conditional discrimination task would be desirable.

General Discussion and Conclusions

In this chapter I have reviewed the perceptions of animals to five illusions. Researchers have not yet fully identified the cause of the perception of each illusion by humans, but they have proposed multiple hypotheses and theories for each one (see Ninio 2014; Feng et al. 2017). One of the leading explanations is the effect of assimilation and contrast. We can describe the assimilation effect and the contrast effect as "decrease of difference" and "increase of difference," respectively (Goto et al. 2007). Several researchers claim that the perception of the Ponzo illusion (e.g., Girgus and Coren 1982; Pressey and Epp 1992) and the Müller-Lyer illusion (e.g., Pressey and Bross 1973; Jordan and Uhlarik 1986) are due to assimilation effects. By contrast, others claim that perception of the Ebbinghaus (e.g., Massaro and Anderson 1971; Coren and Miller 1974) and Zöllner illusions (Kitaoka and Ishihara 2000) depends on contrast effects. The Delboeuf illusion is particularly suitable for testing both effects, because the overestimation of its inner circle nicely illustrates the assimilation effect, and the underestimation, the contrast effect (e.g., Goto et al. 2002, 2007). Thus, we can analyze both of these effects simultaneously, simply by manipulating the parameters of this one illusion figure.

The findings in the comparative literature concerning the five geometric illusions present a mixed picture. Why are species differences sometimes seen but at other times reported to be absent? The "assimilation and contrast hypothesis" (e.g., Nakamura et al. 2008, 2009b; Watanabe et al. 2011) proposes that unlike humans, some species including birds are susceptible to the assimilation effect but not the contrast effect. Due to the assimilation effect birds perceive the Ponzo illusion and the Müller-Lyer illusion in the same direction as humans. Due to the lack of the contrast effect, birds do not perceive the reversed Müller-Lyer illusion (Nakamura et al. 2009b). Moreover, susceptibility to the assimilation effect means that birds perceive the opposite direction of the Ebbinghaus and Zöllner illusions (Watanabe et al. 2011, 2013) than humans. Thus, this hypothesis provides a unified explanation for the results our team has obtained in birds (but see Rosa Salva et al. 2013; Qadri and Cook 2019); however, clearly even this account requires further scrutiny involving more species and illusion figures.

In fact, results of several studies challenge the existence of a mechanism for a unified explanation of optical illusions in humans (Axelrod et al. 2017; Grzeczkowski et al. 2017, 2018; Cretenoud et al. 2019), notably, the assimilation and contrast hypothesis. However, as the Zöllner illusion is reported to consist of three factors that have opposite effects (Kitaoka and Ishihara 2000), it is possible that more than one factor is even in a single illusion. If so, it is unlikely that a single factor can be found by comparing visual illusions. Thorough investigation of the existence of common mechanisms requires careful analysis of, including studies of nonhuman animals. The assimilation and contrast hypothesis provides a good starting point.

However, it is difficult to devise and validate methods for comparative illusion studies, particularly because, studies with other species require purely behavioral tasks, with no verbal reports. Each behavioral task requires proof that (1) subjects focus on the relevant area of the figure and (2) subjects' performance depends on the mechanism of interest to the experimenter. A typical procedure in comparative studies in this field (the Delboeuf illusion excepted) consists of the following three steps: (1) discrimination training: presenting only the target portion of an illusion figure; (2) further training introducing inducer(s) that barely give rise to the illusion; (3) presentation of various exemplars of the illusion figure. I consider this is the most robust method for comparative illusion studies, and in fact it is widely used. However, the second step—training with inducer(s) added—is particularly tricky, as subjects may well change their strategy at this point (see Qadri and Cook 2019 for criticism of this method).

Further steps will be needed to ensure that subjects focus on the area that interests researchers, and that they perform the task using the strategy that is of interest. I propose three techniques. The first is to add verification processes. For example, when studying the Müller-Lyer and Ebbinghaus illusions, Nakamura et al. (2006, 2008) plotted graphs and asked whether their subjects integrated the inducer with the target, instead of expressing illusionary perception. In fact, their estimated and actual graphs differed. It could also be useful to analyze data such as response coordinates and latency for each trial.

The second technique, as proposed by Qadri and Cook (2019), is to analyze response tendencies to various presentations of an illusion figure by manipulating the parts. Fujita et al. (1991, 1993) did this in their series of Ponzo illusion studies and discovered that both pigeons and then rhesus macaques (Fujita 1996, 1997) perceive the Ponzo illusion in the same direction as humans, despite differences in their visual systems. As even a single illusion figure has multiple components, this technique could help provide important information.

The third technique is to devise a task for judging a target feature that is different from the dimension of the illusion. For example, the Ebbinghaus illusion is a *size* illusion caused by the relationship between the size of the center circle and surrounding circles. Therefore, the task of reporting the size of the target makes it difficult to prove that subjects consistently judge only the target size, rather than integrating or confusing it with the inducer size. Watanabe et al. (2011, 2013) required their subjects to judge the size of gaps between edges of two shafts, to investigate the Zöllner illusion as an illusion of *tilt* or *angle*. It is unlikely that the tilt (or angle between each shaft) of the inducer affects size judgment in this task. Of course, before proceeding this way in comparative studies, it is necessary to confirm that humans perceive the illusion as expected.

Comparative illusion studies help to reveal how the visual systems of humans and other species function. But as described above, it is challenging to come up with appropriate methods for studying nonverbal species, and interpretation of the results obtained by those methods requires caution. It is often said that: "Seeing is believing." How we humans perceive an illusion figure can easily be communicated or expressed. By contrast, we cannot directly access the illusionary perception of another species, so we must be skeptical of any report of nonhuman animals experiencing illusions. However, species comparisons often throw up unexpected results; indeed this fact is what makes comparative illusion studies so fascinating. I hope this field of comparative psychology will continue to flourish.

Appendix

Summary of comparative studies on the geometric illusions discussed in this chapter. Italics indicate studies that did not feature in Table 8.1 of Fujita et al. (2017). Numbers after symbols indicate whether the illusion was seen (+), not seen (−), or a reversed illusion was seen. (R) indicates subject numbers for statistical tests. Numbers are included only for studies which conducted statistical tests. Fractions indicate the numbers of positive (+ or R) subjects of total subjects tested, for studies using statistical tests for each subject.

	Ponzo	Müller-Lyer	Zollner	Ebbinghaus-Titchener	Delboeuf
Mammals					
Chimpanzee	Fujita (1997) [+, 1]				Parrish and Beran (2014) [+, 3/3]
Rhesus macaque	Fujita (1997) [+, 3]	Tudusciuc and Nieder (2010) [+] (N2/2)	Agrillo et al. (2014) [+, 3/3]		Parrish et al. (2015) [+, 7]
Guinea baboon				Parron and Fagot (2007) [−, 8]	
Anubis baboon			Benhar and Samuel (1982) [+]		
Capuchin monkey		Suganuma et al. (2007) [+−]			Parrish et al. (2015) [+, 13]
Lemur					Santacà et al. (2017)
Bottlenosed dolphin				Murayama et al. (2012) [+, 1/1]	
Dog	Byosiere et al. (2017a,b) [+−, 2/8]			Byosiere et al. (2017a, b) [R, 8/8]	Byosiere et al. (2017a, b) [−(R), 8 (2/8)]
	Byosiere et al. (2018) [+−, 1/6]				Miletto Petrazzini et al. (2017)
Cat					Szenczi et al. (2019)

(continued)

	Ponzo	Müller-Lyer	Zollner	Ebbinghaus-Titchener	Delboeuf
Guinea pig			Dücker (1966) [+]	Dücker (1966) [+]	
Birds					
African grey parrot		Pepperberg et al. (2008) [+, 1/1]			
Avadavat			Dücker (1966) [+]	Dücker (1966) [+]	
Budgerigar					Watanabe et al. (2016)
Chicken		Winslow (1933) [+]	Dücker (1966) [+]	Nakamura et al. (2014) [R, 3/3]	
			Watanabe et al. (2013) [R, 3/3]	Rosa Salva et al. (2013) [+, 16]	
Mistle thrush			Dücker (1966) [+]		
Pigeon	Fujita et al. (1991) [+, 3/3]	Malott et al. (1967) [+−]	Watanabe et al. (2011) [R, 6]	Nakamura et al. (2008) [R, 5/5]	
	Fujita et al. (1993) [+ 6]	Nakamura et al. (2006) [+, 3/3]			
Ring dove		Warden and Baar (1929) [+]			
Starling	Qadri and Cook (2019)		Dücker (1966) [+]	Dücker (1966) [+]	
Fish				Qadri and Cook (2019)	
Crucian carp			Dücker (1966) [+]	Dücker (1966) [+]	
Damselfish				Fuss and Schluessel (2017) [+, 5]	
Guppy		Santacà and Agrillo (2019) [+, 12]			Lucon-Xiccato et al. (2019) [R, 12]
Goldfish		Wyzisk (2005) [−]			
Redtail splitfin		Sovrano et al. (2016) [+, 6]		Sovrano et al. (2015) [+, 8]	

(continued)

	Ponzo	Müller-Lyer	Zollner	Ebbinghaus-Titchener	Delboeuf
Shark		Fuss et al. (2014) [–, 0/6]		Fuss and Schluessel (2017) [R, 4]	
Others					
Bearded dragon					Santacà et al. (2019) [+, 12]
Turtle					Santacà et al. (2019)
Bee					Howard et al. (2017) [+ – R, 10]
Fly		Geiger and Poggio (1975) [+]			

References

Agrillo C, Parrish AE, Beran MJ (2014) Do rhesus monkeys (*Macaca mulatta*) perceive the Zöllner illusion? Psychon Bull Rev 21:986–994. https://doi.org/10.3758/s13423-013-0573-2

Agrillo C, Santacà M, Pecunioso A, Petrazzini MEM (2020) Everything is subjective under water surface, too: visual illusions in fish. Anim Cogn 23:1–14. https://doi.org/10.1007/s10071-019-01341-7

Axelrod V, Schwarzkopf DS, Gilaie-Dotan S, Rees G (2017) Perceptual similarity and the neural correlates of geometrical illusions in human brain structure. Sci Rep 7(1):1–16. https://doi.org/10.1038/srep39968

Benhar E, Samuel D (1982) Visual illusions in the baboon (*Papio anubis*). Anim Learn Behav 10:115–118. https://doi.org/10.3758/BF03212056

Brown LB, Houssiadas L (1965) The perception of illusions as a constancy phenomenon. Br J Psychol 56:135–141. https://doi.org/10.1111/j.2044-8295.1965.tb00952.x

Byosiere SE, Feng LC, Woodhead JK, Rutter NJ, Chouinard PA, Howell TJ, Bennett PC (2017a) Visual perception in domestic dogs: susceptibility to the Ebbinghaus–Titchener and Delboeuf illusions. Anim Cogn 20(3):435–448. https://doi.org/10.1007/s10071-016-1067-1

Byosiere SE, Feng LC, Rutter NJ, Woodhead JK, Chouinard PA, Howell TJ, Bennett PC (2017b) Do dogs see the Ponzo illusion. Anim Behav Cogn 4(4):396–412. https://doi.org/10.26451/abc/04.04.01.2017

Byosiere SE, Feng LC, Wuister J, Chouinard PA, Howell TJ, Bennett PC (2018) Do dogs demonstrate susceptibility to a vertically presented Ponzo illusion. Anim Behav Cogn 5 (3):254–267. https://doi.org/10.26451/abc.05.03.01.2018

Coren S, Miller J (1974) Size contrast as a function of figural similarity. Percept Psychophys 16 (2):355–357. https://doi.org/10.3758/BF03203955

Cretenoud AF, Karimpur H, Grzeczkowski L, Francis G, Hamburger K, Herzog MH (2019) Factors underlying visual illusions are illusion-specific but not feature-specific. J Vis 19(14):12–12. https://doi.org/10.1167/19.14.12

Dücker G (1966) Untersuchungen über geometrisheoptische Tauschungen bei Wirbeltieren [Studies on geometric-optical illusion s in vertebrates]. Z Tierpsychol 23:452–496. https://doi.org/10.1111/j.1439-0310.1966.tb01607.x

Fellows BJ (1968) The reverse Müller-Lyer illusion and 'enclosure'. Br J Psychol 59(4):369–372. https://doi.org/10.1111/j.2044-8295.1968.tb01151.x

Feng LC, Chouinard PA, Howell TJ, Bennett PC (2017) Why do animals differ in their susceptibility to geometrical illusions? Psychon Bull Rev 24(2):262–276. https://doi.org/10.3758/s13423-016-1133-3

Fisher GH (1968) Gradients of distortion seen in the context of the Ponzo illusion and other contours. Q J Exp Psychol 20:212–217. https://doi.org/10.1080/14640746808400153

Fujita K (1996) Linear perspective and the Ponzo illusion: a comparison between rhesus monkeys and humans. Jpn Psychol Res 38:136–145. https://doi.org/10.1111/j.1468-5884.1996.tb00018.x

Fujita K (1997) Perception of the Ponzo illusion by rhesus monkeys, chimpanzees, and humans: similarity and difference in the three primate species. Percept Psychophys 59:284–292. https://doi.org/10.3758/BF03211896

Fujita K, Blough DS, Blough PM (1991) Pigeons see the Ponzo illusion. Anim Learn Behav 19:283–293. https://doi.org/10.3758/BF03197888

Fujita K, Blough DS, Blough PM (1993) Effects of the inclination of context lines on perception of the Ponzo illusion by pigeons. Anim Learn Behav 21:29–34. https://doi.org/10.3758/BF03197972

Fujita K, Nakamura N, Sakai A, Watanabe S, Ushitani T (2012) Amodal completion and illusory perception in birds and primates. In: Lazareva O, Shimizu T, Wasserman E (eds) How animals see the world: behavior, biology, and evolution of vision. Oxford University Press, New York, pp 101–116

Fujita, K., Nakamura, N., Watanabe, S., Ushitani, T. (2017). Comparative visual illusions: evolutionary, cross-cultural, and developmental perspectives. In Call, J., Burghardt, G., Pepperberg, I., Snowdon, C., Zentall, T. Shapiro, A., Todorovic, D. APA handbook of comparative psychology. 2. Perception, learning, and cognition. Washington: American Psychological Association. 163-181

Fuss T, Bleckmann H, Schluessel V (2014) The brain creates illusions not just for us: sharks (*Chiloscyllium griseum*) can "see the magic" as well. Front Neural Circ 8:24. https://doi.org/10.3389/fncir.2014.00024

Fuss T, Schluessel V (2017) The Ebbinghaus illusion in the gray bamboo shark (*Chiloscyllium griseum*) in comparison to the teleost damselfish (*Chromis chromis*). Fortschr Zool 123:16–29. https://doi.org/10.1016/j.zool.2017.05.006

Geiger G, Poggio T (1975) The Muller-Lyer figure and the fly. Science 190(4213):479–480. https://doi.org/10.1126/science.1166319

Girgus JS, Coren S (1982) Assimilation and contrast illusions: differences in plasticity. Percept Psychophys 32(6):555–561

Goto T, Uchiyama I, Imai A, Takahashi SY, Hanari T, Nakamura S, Kobari H (2007) Assimilation and contrast in optical illusions. Jpn Psychol Res 49(1):33–44. https://doi.org/10.1111/j.1468-5884.2007.00330.x

Goto T, Uchiyama I, Kobari H (2002) Effects of width-variation in slanted stripe-patterns on simultaneous assimilation and contrast. Psychol Forsch 45(1):24–33. https://doi.org/10.2117/psysoc.2002.24

Gregory RL (1963) Distortion of visual space as inappropriate constancy scaling. Nature 199:678–680. https://doi.org/10.1038/199678a0

Gregory RL (1965) Seeing in depth. Nature 207:16–19. https://doi.org/10.1038/207016a0

Gregory RL (2009) Seeing through illusions. Oxford University Press, Oxford

Grzeczkowski L, Clarke AM, Francis G, Mast FW, Herzog MH (2017) About individual differences in vision. Vision Res 141:282–292. https://doi.org/10.1016/j.visres.2016.10.006

Grzeczkowski L, Roinishvili M, Chkonia E, Brand A, Mast FW, Herzog MH, Shaqiri A (2018) Is the perception of illusions abnormal in schizophrenia? Psychiatry Res 270:929–939. https://doi.org/10.1016/j.psychres.2018.10.063

Howard SR, Avarguès-Weber A, Garcia JE, Stuart-Fox D, Dyer AG (2017) Perception of contextual size illusions by honeybees in restricted and unrestricted viewing conditions. Proc R Soc B Biol Sci 284(1867):20172278. https://doi.org/10.1098/rspb.2017.2278

Jordan K, Uhlarik J (1986) Length contrast in the Müller-Lyer figure: functional equivalence of temporal and spatial separation. Percept Psychophys 39(4):267–274. https://doi.org/10.3758/BF03204934

Kincade S, Wilson AE (2000) Effects of method, orientation, and size of angle on the Ponzo illusion. Percept Mot Skills 91(3):837–847. https://doi.org/10.2466/pms.2000.91.3.837

Kitaoka A, Ishihara M (2000) Three elemental illusions determine the Zöllner illusion. Percept Psychophys 62:569–575. https://doi.org/10.3758/BF03212108

Lavrenteva S, Murakami I (2018) The Ebbinghaus illusion in contrast-defined and orientation-defined stimuli. Vision Res 148:26–36. https://doi.org/10.1016/j.visres.2018.04.006

Lucon-Xiccato T, Santacà M, Petrazzini MEM, Agrillo C, Dadda M (2019) Guppies, *Poecilia reticulata*, perceive a reversed Delboeuf illusion. Anim Cogn 22(3):291–303. https://doi.org/10.1007/s10071-019-01237-6

Malott RW, Malott MK, Pokrzywinski J (1967) The effects of outward-pointing arrowheads on the Mueller-Lyer illusion in pigeons. Psychon Sci 9(1):55–56. https://doi.org/10.3758/BF03330756

Massaro DW, Anderson NH (1971) Judgmental model of the Ebbinghaus illusion. J Exp Psychol 5:147–151. https://doi.org/10.1037/h0031158

Miletto Petrazzini ME, Bisazza A, Agrillo C (2017) Do domestic dogs (*Canis lupus familiaris*) perceive the Delboeuf illusion? Anim Cogn 20:427–434. https://doi.org/10.1007/s10071-016-1066-2

Mruczek RE, Blair CD, Strother L, Caplovitz GP (2015) The dynamic Ebbinghaus: motion dynamics greatly enhance the classic contextual size illusion. Front Hum Neurosci 9:77. https://doi.org/10.3389/fnhum.2015.00077

Murayama T, Usui A, Takeda E, Kato K, Maejima K (2012) Relative size discrimination and perception of the Ebbinghaus illusion in a bottlenose dolphin (*Tursiops truncatus*). Aquat Mamm 38(4):333–342. https://doi.org/10.1578/AM.38.4.2012.333

Nakamura N, Fujita K, Ushitani T, Miyata H (2006) Perception of the standard and the reversed Müller-Lyer figures in pigeons (*Columba livia*) and humans (*Homo sapiens*). J Comp Psychol 120(3):252–261. https://doi.org/10.1037/0735-7036.120.3.252

Nakamura N, Watanabe S, Fujita K (2008) Pigeons perceive the Ebbinghaus-Titchener circles as an assimilation illusion. J Exp Psychol Anim Behav Process 34(3):375–387. https://doi.org/10.1037/0097-7403.34.3.375

Nakamura N, Watanabe S, Fujita K (2009a) Further analysis of perception of the standard Müller-Lyer figures in pigeons (*Columba livia*) and humans (*Homo sapiens*): effects of length of brackets. J Comp Psychol 123(3):287–294. https://doi.org/10.1037/a0016215

Nakamura N, Watanabe S, Fujita K (2009b) Further analysis of perception of reversed Müller-Lyer figures for pigeons (*Columba livia*). Percept Mot Skills 108(1):239–250. https://doi.org/10.2466/pms.108.1.239-250

Nakamura N, Watanabe S, Fujita K (2014) A reversed Ebbinghaus–Titchener illusion in bantams (*Gallus gallus domesticus*). Anim Cogn 17(2):471–481. https://doi.org/10.1007/s10071-013-0679-y

Ninio J (2014) Geometrical illusions are not always where you think they are: a review of some classical and less classical illusions, and ways to describe them. Front Hum Neurosci 8:856. https://doi.org/10.3389/fnhum.2014.00856

Parrish AE, Beran MJ (2014) When less is more: like humans, chimpanzees (*Pan troglodytes*) misperceive food amounts based on plate size. Anim Cogn 17(2):427–434. https://doi.org/10.1007/s10071-013-0674-3

Parrish AE, Brosnan SF, Beran MJ (2015) Do you see what I see? A comparative investigation of the Delboeuf illusion in humans (*Homo sapiens*), rhesus monkeys (*Macaca mulatta*), and capuchin monkeys (*Cebus apella*). J Experiment Psychol 41(4):395–405. https://doi.org/10.1037/xan0000078

Parron C, Fagot J (2007) Comparison of grouping abilities in humans (*Homo sapiens*) and baboons (*Papio papio*) with the Ebbinghaus illusion. J Comp Psychol 121(4):405–411. https://doi.org/10.1037/0735-7036.121.4.405

Pepperberg IM, Vicinay J, Cavanagh P (2008) Processing of the Müller-Lyer illusion by a Grey parrot *(Psittacus erithacus)*. Perception 37(5):765–781. https://doi.org/10.1068/p5898

Pressey AW, Bross M (1973) Assimilation theory and the reversed Müller—Lyer illusion. Perception 2(2):211–218. https://doi.org/10.1068/p020211

Pressey AW, Epp D (1992) Spatial attention in Ponzo-like patterns. Percept Psychophys 52:211–221. https://doi.org/10.3758/BF03206774

Qadri MAJ, Cook RG (2019) Perception of Ebbinghaus–Titchener stimuli in starlings (*Sturnus vulgaris*). Anim Cogn 22:973–989. https://doi.org/10.1007/s10071-019-01289-8

Robinson JO (1998) The psychology of visual illusion. Dover Publications, Mineola

Rosa Salva O, Rugani R, Cavazzana A, Regolin L, Vallortigara G (2013) Perception of the Ebbinghaus illusion in four-day-old domestic chicks (*Gallus gallus*). Anim Cogn 16 (6):895–906. https://doi.org/10.1007/s10071-013-0622-2

Santacà M, Agrillo C (2019) Perception of the Müller–Lyer illusion in guppies. Curr Zool 66 (2):205–213. https://doi.org/10.1093/cz/zoz041

Santacà M, Miletto Petrazzini ME, Agrillo C, Wilkinson A (2019) Can reptiles perceive visual illusions? Delboeuf illusion in red-footed tortoise (*Chelonoidis carbonaria*) and bearded dragon (*Pogona vitticeps*). J Comp Psychol 133(4):419–427. https://doi.org/10.1037/com0000176

Santacà M, Regaiolli B, Miletto Petrazzini ME, Spiezio C, Agrillo C (2017) Preliminary study to investigate the Delboeuf illusion in ring-tailed lemurs *(Lemur catta)*: methodological challenges. Anim Behav Cogn 4(3):365–377. https://doi.org/10.26451/abc.04.03.13.2017

Sovrano VA, Albertazzi L, Salva OR (2015) The Ebbinghaus illusion in a fish (*Xenotoca eiseni*). Anim Cogn 18(2):533–542. https://doi.org/10.1007/s10071-014-0821-5

Sovrano VA, Da Pos O, Albertazzi L (2016) The Müller-Lyer illusion in the teleost fish *Xenotoca eiseni*. Anim Cogn 19(1):123–132. https://doi.org/10.1007/s10071-015-0917-6

Suganuma E, Pessoa VF, Monge-Fuentes V, Castro BM, Tavares MCH (2007) Perception of the Müller–Lyer illusion in capuchin monkeys (*Cebus apella*). Behav Brain Res 182(1):67–72. https://doi.org/10.1016/j.bbr.2007.05.014

Szenczi P, Velázquez-López ZI, Urrutia A, Hudson R, Bánszegi O (2019) Perception of the Delboeuf illusion by the adult domestic cat (*Felis silvestris catus*) in comparison with other mammals. J Comp Psychol 133(2):223–232. https://doi.org/10.1037/com0000152

Takao S, Clifford CW, Watanabe K (2019) Ebbinghaus illusion depends more on the retinal than perceived size of surrounding stimuli. Vision Res 154:80–84. https://doi.org/10.1016/j.visres.2018.10.010

Tudusciuc O, Nieder A (2010) Comparison of length judgments and the Müller-Lyer illusion in monkeys and humans. Exp Brain Res 207(3–4):221–231. https://doi.org/10.1007/s00221-010-2452-7

Warden CJ, Baar J (1929) The Müller-Lyer illusion in the ring dove, *Turtur risorius*. J Comp Psychol 9(4):275–292. https://doi.org/10.1037/h0071052

Watanabe S, Hase Y, Nakamura N (2016) Do budgerigars *(Melopsittacus undulatus)* perceive the Delboeuf illusion?: A preliminary study with a simultaneous discrimination task. Psychol Forsch 59(2–3):121–135. https://doi.org/10.2117/psysoc.2016.121

Watanabe S, Nakamura N, Fujita K (2011) Pigeons perceive a reversed Zöllner illusion. Cognition 119:137–141. https://doi.org/10.1016/j.cognition.2010.10.020

Watanabe S, Nakamura N, Fujita K (2013) Bantams (*Gallus gallus domesticus*) also perceive reversed Zöllner illusion. Anim Cogn 16:109–115. https://doi.org/10.1007/s10071-012-0556-0

Winslow CN (1933) Visual illusions in the chick. Arch Psychol 153:80

Wyzisk K (2005) Experimente zur Form- und Größenwahrnehmung beim Goldfisch (*Carassius auratus*) unter Verwendung von Scheinkonturen und Größentäuschungen. Ph.D. thesis, Johannes-Gutenberg-Universität Mainz

Yanagisawa N (1939) Reversed illusion in the Müller–Lyer illusion. Jpn J Psychol 14:321–326. https://doi.org/10.1068/p120049

Part II
Face and Body

Chapter 4
It Takes One to Know One: Do Human and Nonhuman Primates Share Similar Face Processing?

Olivier Pascalis, Fabrice Damon, Kun Guo, and David Méary

Abstract The abilities to identify individuals within the group, and to interpret their expressions and intentions are essential for many social animals. Face recognition in human and nonhuman primates stems from a conjunction of evolutionary inheritance and experience via exposure to faces present in the environment. Individuation is clearly a vital mechanism for any social species. By uncovering similarities across primate face systems, comparative studies allow us to better understand the evolution of face processing capabilities in humans. Some researchers have argued that primates, including humans, may possess an innate face processing system that is predisposed to respond to conspecifics. The argument is supported by a study showing that monkeys raised without experience of own-species faces still prefer to look at faces of conspecifics (Fujita, Int J Primatol 11:553–573, 1990). However, this proposal does not fit well with findings from the human infant literature (Pascalis et al., Science 296:1321–1323, 2002) or with data on monkeys raised without seeing faces (Sugita, Proc Natl Acad Sci, 105, 394–398, 2008) which suggest that face processing is highly shaped by experience at an early age. We argue that human and nonhuman primates possess an evolved system for processing faces that becomes specialized as a consequence of predominant exposure to faces from a single species. According to this interpretation, a limitation of the face processing expertise to own species should be observed.

Keywords Face · Nonhuman primates · Humans · Development

O. Pascalis (✉) · D. Méary
Laboratoire de Psychologie et Neurocognition, Université Grenoble Alpes, Grenoble, France
e-mail: Olivier.Pascalis@univ-grenoble-alpes.fr; David.Meary@univ-grenoble-alpes.fr

F. Damon
Centre des Sciences du Gout et de l'Alimentation, Université Bourgogne Franche-Comté, Dijon, France

K. Guo
School of Psychology, University of Lincoln, Lincoln, UK
e-mail: kguo@lincoln.ac.uk

J. R. Anderson, H. Kuroshima (eds.), *Comparative Cognition*,
https://doi.org/10.1007/978-981-16-2028-7_4

Introduction

The ability to identify individuals within the group, and to interpret their expressions and intentions is essential for many social animals. In humans, it is well established that face processing abilities underpin such functions, and comparative studies suggest a common evolutionary route for this important skill (Leopold and Rhodes 2010). Nonhuman primates have often been used as a model of human face processing. Fujita (1987) has been interested in nonhuman primate models for many years, and has continued to publish influential work on this topic. In this chapter, we will document some similarities in face recognition but also differences that make this cognitive system well adapted for each species.

Faces are omnipresent in our environment, and are crucial for our everyday social life. It has been suggested that face recognition might somehow be «special» (Farah 1996) compared to recognition of other objects. Indeed, faces provide an early channel of communication, operating prior to the onset of language between infant and caretaker. It has been shown that infants preferentially orient to faces during the first week of life (Goren et al. 1975; Johnson et al. 1991; Valenza et al. 1996). By the time they reach adulthood they can be considered as face experts. Diamond and Carey (1986) suggested that we learn to process faces differently than other visual stimuli. Information on a face can be construed as hierarchically organized with facedness information at the top (e.g., faces vs. objects), followed by species information (e.g., human vs. monkey), then by subordinate information such as gender and race, and at the lowest level, individual identity information. Individuating information can in turn be further divided into featural information (e.g., the shape or size of the eyes), configural information (e.g., the distance between the eyes), and holistic information (the face gestalt that binds featural and configural information into an unbreakable whole) (Leder and Bruce 2000). Despite the role of early visual experience in the development of configural and holistic processing, some researchers speculate that both are adult-like at a later age relative to featural face processing. This speculation was first proposed in the late 1970s with the introduction of the controversial encoding switch hypothesis (Carey and Diamond 1977; Diamond and Carey 1977). The encoding switch hypothesis proposes that young children encode faces in a piecemeal fashion (i.e., individual facial features), whereas older children—beginning at around the age of 10—encode the spatial relations between individual facial features. This hypothesis was based on findings regarding the effects of paraphernalia and inverting faces on children's recognition. This configural processing is responsible for the striking *inversion effect* (Yin 1969): inverted faces are less rapidly and accurately processed than upright faces (for reviews, see Rakover 2013; Rossion 2008). Carey and Diamond (1977) used inversion as a measure of children's use of spatial relations among features (i.e., involved in both configural and holistic processing). They found that the recognition of faces in 10-year-olds was disproportionately more impaired by inversion than the recognition of houses, whereas 6- and 8-year-olds showed similar impairment for both categories. Recent data suggest, however, that from a young age infants are sensitive to configuration (Anzures et al. 2019). Diamond and Carey (1986) showed that this processing mode is a consequence of experience, and can be extended to other categories in which subjects are expert.

How Do Other Species Process Faces?

A variety of nonhuman social species possesses remarkable face discrimination abilities. Tibbetts (2002) reported that wasps, primarily reliant on chemical signatures for communication and identification of conspecifics, are also capable of hierarchical categorization (e.g., "worker") based on facial visual patterns only. Dyer et al. (2005) found that honeybees can learn the picture of a human face and recognize it when it is paired with a novel face (for similar findings with wasps, see also Avarguès-Weber et al. 2017). This result is consistent with what we already know about the visual ability of bees. However, Pascalis et al. (2006) pointed out that in the study by Dyer et al. (2005), recognition could have been performed on the basis of very simple pattern processing, and might not be related to face processing at all. Thus, for now there is no compelling evidence of human-like face processing in insects. Recently, sheep (*Ovis aries*) have been found to present advanced face recognition skills, including the ability to recognize and discriminate individual human faces (Knolle et al. 2017). However, the claim that sheep also possess human-like face processing seems too strong (for a discussion, see Kendrick 2019; Peirce 2019; Towler et al. 2019). Sheep and human performances were, for instance, not compared on equivalent tasks, as face recognition tasks typically require that humans learn new face identities in few seconds and in one trial, whereas sheep are trained over several consecutive days and throughout many trials (Towler et al. 2019). In addition, the strong advantage in processing familiar over unfamiliar faces, thought of as a key feature of human face processing (Young and Burton 2018), has not been found in sheep, suggesting qualitatively different mechanisms between two species (Towler et al. 2019). It is worth noting, however, that sheep are clearly better at processing sheep faces than human faces—as humans are at processing own- vs. other-species faces—hence, for a true comparison, both human and sheep faces should be used in future comparative studies (Kendrick 2019) to determine to what extent the face recognition mechanisms overlap between the species.

The situation is quite different for nonhuman primates, and comparative studies on face recognition in various primate species have burgeoned during the last 20 years. The adult face processing system of nonhuman primates shares several similarities with that of humans: eye scanning, region of interest, individual recognition of face pictures, sensitivity to face inversion (see Pascalis et al. 1999 for a review). Humphrey (1974) used a habituation task to show that rhesus monkeys can distinguish different individuals from their own species—but this performance did not extend to individuals from other, nonprimate species. Using a classic habituation paradigm, Dahl et al. (2007) found that individual recognition by macaques was better for conspecific faces than for other-species faces, and that, like humans, macaques show holistic processing. Gothard et al. (2009) found the same result, and their detailed analysis of the eye scanning of the faces showed that whereas monkeys use both configural and feature-based processing to recognize the faces of conspecifics, they use primarily feature-based strategies to recognize human faces. These studies suggest the use of different strategies while processing other-species faces. Another study by Dahl et al. (2009) compared scanning strategies by humans and macaques when looking at conspecific or heterospecific faces. Both species displayed the same pattern of scanning for their own- versus other-species faces: more attention was directed to the eye region for their own species than for other species. Findings of better performance in face recognition tasks with own- versus other-species faces extend to other primates species (e.g., brown capuchin monkeys and Tonkean macaques, Dufour et al. 2006; chimpanzees, Martin-Malivel and Okada 2007; cotton-top tamarins, Neiworth et al. 2007; Japanese monkeys, Sugita 2008).

Although both humans and macaques can learn to individuate conspecifics by faces, the former outperform the latter in individual recognition of own-species faces (Rossion and Taubert 2019), and macaques need extensive operant conditioning to achieve even modest recognition performance. As noted by Parr (2011), studies using macaques have reported mixed evidence regarding the extent of human-like face processing (e.g., inversion effect, configural processing, etc.), with results varying depending on the methods used, perhaps suggesting that humans and monkeys have evolved different face processing strategies. By contrast, compared to macaques, chimpanzees have been found to present face processing mechanisms more homologous with those of humans (Parr 2011; Taubert et al. 2017; Wilson and Tomonaga 2018, but see Griffin 2020). Although the macaque–chimpanzee difference may be related to the latter's greater phylogenetic proximity to humans, it may also be related to species-specific social structures: chimpanzees and humans share a "fission–fusion" social organization (Parr 2011), whereas macaque monkeys live in more stable social groups with more or less strict, often linear dominance hierarchies. In fission–fusion society, group composition is dynamic and subject to frequent fracture and growth over time; hence, robust face processing mechanisms are highly advantageous, and there is a strong social pressure to recognize the faces of conspecifics. Strikingly, it has been reported that spider monkeys (*Ateles geoffroyi*), New World species with a fission–fusion social structure, also present more human-like face processing abilities than macaque monkeys (i.e., holistic processing for high-experienced face categories, Taubert 2010; Taubert and Parr 2009), further suggesting that socioecological pressures might contribute to the emergence of a specialized face processing system.

Beyond these differences, visual biases for own-species seem to be especially common across primates species (Scott and Fava 2013), including humans (Heron-Delaney et al. 2011). Fujita (1987), using an operant lever-pressing task, explored the extent to which different species showed conspecific preference, by comparing portrait preference in five macaques species (*Macaca fuscata, M. mulatta, M. radiata, M. nemestrina*, and *M. arctoides*). Each monkey was trained to press a lever to show a picture on a screen. The picture remained on-screen for as long as the subject held the lever down. Using this simple method it was possible to assess the time spent looking at a picture of a conspecific compared to one showing a different species. In this task, it is assumed that habituation will occur more quickly (i.e., reduced looking time) with a picture of a heterospecific, which is less attractive for the subject. Fujita found that monkeys showed conspecific preference, with the exception of the stumptailed macaque *M. arctoides* (but see Demaria and Thierry 1988). Most of the subjects discriminated conspecifics in pictures, and preferred to look at their own species than macaques of a different species. Overall, Fujita's studies demonstrated that preference for pictures of own species is common across multiple macaque species (Fujita 1987, 1990, 1993a, Fujita and Watanabe 1995). Preferences were diluted when either the head, or head and tail, were removed from the picture (Fujita 1993b), suggesting that the head is a crucial cue driving the preference. Similarly, the face was found to be important in conspecific discrimination by longtailed macaques (Dittrich 1994), although it should be noted that this last study used line drawings, not halftone images of real faces.

Although own-species preference seems to be widespread in primates, its origin may differ between species. For instance, own-species preference can be spontaneously present at birth, or emerge from social experience with conspecifics during infancy. Cross-fostering studies can be especially informative to disambiguate this issue. Fujita showed that Japanese monkeys raised by rhesus macaque mother presented a preference for rhesus macaque faces (Fujita 1990, 1993a), while Japanese monkeys raised by their mother presented own-species preferences. By contrast, rhesus macaques showed own-species preferences regardless of their early social experience. Fujita concluded that rhesus have a hard-wired preference for their own species, whereas Japanese macaques have a more flexible preference that is influenced by experience. Related findings were reported in great apes, as adult chimpanzees that had been reared by humans preferred pictures of human faces over chimpanzee faces (Tanaka 2003, 2007). These biases may, in turn, influence face discrimination processes, as human-reared chimpanzees were found to be better at processing human faces than chimpanzee faces, compared to a group of chimpanzees who grew up in their own-species group (Martin-Malivel and Okada 2007).

From a theoretical point of view, preferences for own-species faces may have emerged from some ancient general perceptual bias shared among primates (Damon et al. 2017a; Rhodes 2006). This development might be linked to the ease of processing (i.e., perceptual fluency, Whittlesea and Leboe 2003), thus facilitating the development of familiarity preferences. Such frameworks extend beyond mere face processing mechanisms, and have been proposed as relevant for a general theory of object representation and recognition (Wallis 2013; Wallis et al. 2008). Interestingly, Fujita provided indirect evidence for such shared perceptual mechanisms between human and nonhuman primates, in reporting preferences for regular and symmetrical patterns in capuchin and squirrel monkeys (Anderson et al. 2005). These findings indicated that monkeys prefer visual stimuli that humans find aesthetically pleasing, as visual symmetry is perceived as attractive in a variety of domains (Little 2014). Furthermore, when it comes to faces, ratings of rhesus macaque face attractiveness by human judges can predict implicit visual preferences in rhesus macaques (Damon et al. 2019).

However, mate selection mechanisms probably operate in addition to general perceptual biases for symmetry, as the effect of face attractiveness turned out to be species-specific. When two faces from the same species but differing in attractiveness were presented side-by-side, humans showed a visual preference for human faces (but not macaque faces) rated as attractive, and macaques displayed a visual preference for macaque faces (but not human faces) rated as attractive. Findings of similar preferences in human and nonhuman primates have also been reported for curved contours (humans and chimpanzees: Munar et al. 2015), and for prototypicality (human and macaque infants: Damon et al. 2017b), further highlighting shared underlying face/object processing mechanisms. Interestingly, face familiarity further shows differential impact on face preference in humans and rhesus macaques (Méary et al. 2014). When two faces of different species/races were presented side-by-side, Caucasian humans demonstrated clear visual preference for Caucasian faces over Asian faces (own-race bias), and human faces (regardless of races) over nonhuman primate (e.g., chimpanzee, Barbary, and rhesus macaque) faces (own-species bias). Rhesus macaques also showed visual preference for nonhuman primate faces over human faces regardless of race, but by contrast displayed preference for chimpanzee or Barbary macaque faces over own-species faces. This pattern suggests that face preference in macaques, unlike humans, is modulated by factors beyond familiarity, such as species.

How Does Face Processing Change Early in Life?

A developmental approach may help in determining what the common processes in different species are before experience influences the face processing system, if the developmental trend is similar across primates, and how flexible the system is for handling other-species faces. Faces represent a highly attractive stimulus for infant primates, including humans (Goren et al. 1975), pigtailed macaques (*Macaca nemestrina*, Lutz et al. 1998), gibbons (*Hylobates agilis*, Myowa-Yamakoshi and Tomonaga 2001), rhesus monkeys (*Macaca mulatta*, Kuwahata et al. 2004), and chimpanzees (*Pan troglodytes*, Tomonaga et al. 2004). Some researchers have argued that primates may possess an innate face processing system that is predisposed to respond to conspecifics (e.g., Sackett 1966). Two-month-old pigtailed macaques already demonstrate a strong preference for their own species (Kim et al. 1999). The argument is also supported by several studies demonstrating that rhesus macaques raised without any experience of monkeys still prefer to look at faces of conspecifics than other species (Fujita 1987, 1993a). This is not true for Japanese macaques; however, as mentioned above, rhesus monkeys may have a hard-wired preference for their own species, while Japanese macaques have a more malleable preference, influenced by experience.

An influential study by Sugita (2008) illustrated the critical role of experience in the development of face processing in macaques. Infant Japanese macaques separated from their parents at birth and reared by humans in a faceless environment for a minimal period of 6–24 months showed a preference for both monkey and human faces over objects, but no preference for either category of faces when tested in a visual preference task. Furthermore, they were able to recognize both monkey and human faces. Following the face deprivation period, half of the monkeys were introduced to macaques, and the other half to humans. When tested following a minimum of 1 month of exposure to faces, monkeys exposed to human faces displayed a preference for human faces over objects, and also for human over monkey faces. They displayed no preference between monkey faces and objects. Consistent with this pattern of results, monkeys exposed to monkey faces preferred monkey faces over objects and over human faces, but showed no preference when human faces and objects were presented simultaneously. Furthermore, when tested for recognition, monkeys were only able to recognize faces from the category to which they were exposed. These results highlight the crucial role visual experience can play in the specialization of the face system toward own-species faces. Japanese macaques are not displaying hard-wired preference for their own species, but an experience-dependent preference.

As mentioned above, Japanese macaque monkeys looked more at faces than objects regardless of species, even though they were deprived of face exposure. This suggests that a broad face detection mechanism was still functional despite the lack of exposure to faces, as if initial perceptual sensitivities for faces were preserved, and waiting for refinement through individualized experiences (i.e., experience-expectant processes, Nelson 2003). Are rhesus different from Japanese macaques as suggested by Fujita? Recent studies with nursery-reared infant rhesus macaques (with no exposure to adult macaque faces and limited face exposure in general) showed a "coarse to fine" pattern of face detection (Simpson et al. 2017). Three-week-old rhesus infants with very limited face exposure showed a bias for both own- and other-species faces over objects. Three-, and 6-month-olds, after peer exposure, also showed a similar bias for faces over objects, but also looked faster toward conspecific faces compared to heterospecific faces. Taken together, these results slightly qualify Fujita's earlier findings suggesting hard-wired preference for own-species faces in rhesus macaques, and indicate that the face processing system of rhesus monkeys undergoes a transition from experience-expectant to experience-dependent mechanisms. This developmental trajectory is not limited to nonhuman primates, but seems to extend to humans as well.

Nelson (2001) hypothesized that in humans the representation of faces at birth is broad and that it develops according to the type of facial input received, tuning toward the predominant faces in the environment. To test whether experience tunes face processing, Pascalis et al. (2002) investigated the ability of 6- and 9-month-old human infants to recognize faces from their own species and another species (rhesus macaques) using a standard recognition paradigm. Infants at both ages showed individual recognition with human faces, looking longer at a new face compared to a previously seen face. However, when tested with the monkey faces, only the

6-month-old group showed evidence of individual recognition. By contrast, 9-month-olds were unable to recognize which monkey face they had seen before. These findings suggest that the face system becomes "tuned" to human faces between 6- and 9-months of age (Pascalis et al. 2002). In a follow-up study, Pascalis et al. (2005) investigated the impact of experience with other-species faces on the development of face processing in relation to this species. They provided 6-month-olds' parents with a book containing a selection of monkey faces, which they were instructed to show to their infant following a fixed schedule. Tests showed that this exposure was sufficient to preserve recognition capabilities for monkey faces when the infants returned for testing at 9 months of age. Training experienced near the end of the tuning period appears to be effective for maintaining the ability to discriminate between individuals of other species. Thus, it is possible to alter the development of a cognitive system (for face processing) by providing training and learning via pictorial stimuli. However, it is still unclear how much training is necessary to produce other-species face processing skills at 9 months and for how long long-term training might affect the face system (for a review, see Maurer and Werker 2014).

Conclusion

Overall, the studies reviewed here illustrate that the similarity of the face processing systems of human and nonhuman primates is best found in their plasticity. The experience-expectant/dependent mechanisms progressively adjust the infant primate to its environment, refining the face processing system as the infant accrues experience with the facial characteristics of the individuals from its social group (Pascalis et al. 2014). Despite the indisputable importance of faces in many primate species, however, it is probably too far-fetched to conclude that cognitive and neural face processing mechanisms are continuous across all primates (Rossion and Taubert 2019). Some face processing mechanisms appeared to be shared among primate species whereas others have turned out to be species-specific, possibly reflecting solutions to evolutionary challenges imposed by particular ecological niches. However, these mechanisms always share a similar purpose: to subserve social cognition.

Acknowledgments This work was supported by the French "Investissements d'Avenir" program, project ISITE-BFC (contract ANR-15-IDEX-0003) to F.D.

References

Anderson JR, Kuwahata H, Kuroshima H, Leighty KA, Fujita K (2005) Are monkeys aesthetists? Rensch (1957) revisited. J Exp Psychol: Anim Behav Process 31:71–78. https://doi.org/10.1037/0097-7403.31.1.71

Anzures G, Quinn PC, Pascalis O, Slater A, Lee K (2019) Development of face processing ability in childhood. In: Zelazo P (ed) Oxford handbook of developmental psychology. Oxford University Press, Oxford, UK. https://doi.org/10.1093/oxfordhb/9780199958450.013.0012_update_001
Avarguès-Weber A, D'Amaro D, Metzler M, Garcia JE, Dyer AG (2017) Recognition of human face images by the free flying wasp *Vespula vulgaris*. Anim Behav Cogn 4:314–323. https://doi.org/10.26451/abc.04.03.09.2017
Carey S, Diamond R (1977) From piecemeal to configurational representation of faces. Science 195:312–314
Dahl CD, Logothetis NK, Hoffman KL (2007) Individuation and holistic processing of faces in rhesus monkeys. Proc R Soc B: Biol Sci 274:2069–2076. https://doi.org/10.1098/rspb.2007.0477
Dahl CD, Wallraven C, Bülthoff HH, Logothetis NK (2009) Humans and macaques employ similar face-processing strategies. Curr Biol 19:509–513. https://doi.org/10.1016/j.cub.2009.01.061
Damon F, Li Z, Yan Y, Li W, Guo K, Quinn PC et al (2019) Preference for attractive faces is species-specific. J Comp Psychol 133:262–271. https://doi.org/10.1037/com0000148
Damon F, Mottier H, Méary D, Pascalis O (2017a) A review of attractiveness preferences in infancy: from faces to objects. Adapt Hum Behav Physiol 3:321–336. https://doi.org/10.1007/s40750-017-0071-2
Damon F, Méary D, Quinn PC, Lee K, Simpson EA, Paukner A et al (2017b) Preference for facial averageness: evidence for a common mechanism in human and macaque infants. Sci Rep 7:46303. https://doi.org/10.1038/srep46303
Demaria C, Thierry B (1988) Responses to animal stimulus photographs in stumptailed macaques (*Macaca arctoides*). Primates 29:237–244. https://doi.org/10.1007/BF02381125
Diamond R, Carey S (1977) Developmental changes in the representation of faces. J Exp Child Psychol 23:1–22
Diamond R, Carey S (1986) Why faces are and are not special: an effect of expertise. J Exp Psychol Gen 115:107–117
Dittrich W (1994) How monkeys see others: discrimination and recognition of monkeys' shape. Behav Process 33(1–2):139–154. https://doi.org/10.1016/0376-6357(94)90063-9
Dufour V, Pascalis O, Petit O (2006) Face processing limitation to own species in primates: a comparative study in brown capuchins, Tonkean macaques and humans. Behav Process 73:107–113. https://doi.org/10.1016/j.beproc.2006.04.006
Dyer AG, Neumeyer C, Chittka L (2005) Honeybee (*Apis mellifera*) vision can discriminate between and recognise images of human faces. J Exp Biol 208:4709–4714. https://doi.org/10.1242/jeb.01929
Farah MJ (1996) Is face recognition 'special'? Evidence from neuropsychology. Behav Brain Res 76:181–189. https://doi.org/10.1016/0166-4328(95)00198-0
Fujita K (1987) Species recognition by five macaque monkeys. Primates 28:353–366. https://doi.org/10.1007/BF02381018
Fujita K (1990) Species preference by infant macaques with controlled social experience. Int J Primatol 11:553–573. https://doi.org/10.1007/BF02197057
Fujita K (1993a) Development of visual preference for closely related species by infant and juvenile macaques with restricted social experience. Primates 34:141–150. https://doi.org/10.1007/BF02381385
Fujita K (1993b) Role of some physical characteristics in species recognition by pigtail monkeys. Primates 34:133–140. https://doi.org/10.1007/BF02381384
Fujita K, Watanabe K (1995) Visual preference for closely related species by Sulawesi macaques. Am J Primatol 37:253–261. https://doi.org/10.1002/ajp.1350370307
Goren CC, Sarty M, Wu PYK (1975) Visual following and pattern discrimination of face-like stimuli by newborn infants. Pediatrics 56:544–549
Gothard KM, Brooks KN, Peterson MA (2009) Multiple perceptual strategies used by macaque monkeys for face recognition. Anim Cogn 12:155–167. https://doi.org/10.1007/s10071-008-0179-7

Griffin JW (2020) Quantifying the face inversion effect in nonhuman primates: a phylogenetic meta-analysis. Anim Cogn 23:237–249. https://doi.org/10.1007/s10071-019-01340-8

Heron-Delaney M, Wirth S, Pascalis O (2011) Infants' knowledge of their own species. Phil Trans R Soc Lond B: Biol Sci 366:1753–1763. https://doi.org/10.1098/rstb.2010.0371

Humphrey NK (1974) Species and individuals in the perceptual world of monkeys. Perception 3:105–114. https://doi.org/10.1068/p030105

Johnson MH, Dziurawiec S, Ellis H, Morton J (1991) Newborns' preferential tracking of face-like stimuli and its subsequent decline. Cognition 40:1–19

Kendrick KM (2019) Response to 'Sheep recognize familiar and unfamiliar human faces from two-dimensional images'. R Soc Open Sci 6(5):182187. https://doi.org/10.1098/rsos.182187

Kim JH, Gunderson VM, Swartz KS (1999) Humans all look alike: cross-species face recognition in infant pigtailed macaque monkeys. Biennial meeting of the Society for Research in Child Development, April 15–18, Albuquerque, New Mexico

Knolle F, Goncalves RP, Morton AJ (2017) Sheep recognize familiar and unfamiliar human faces from two-dimensional images. R Soc Open Sci 4(11):171228. https://doi.org/10.1098/rsos.171228

Kuwahata H, Adachi I, Fujita K, Tomonaga M, Matsuzawa T (2004) Development of schematic face preference in macaque monkeys. Behav Process 66:17–21. https://doi.org/10.1016/j.beproc.2003.11.002

Leder H, Bruce V (2000) When inverted faces are recognized: the role of configural information in face recognition. Q J Exp Psychol A: Hum Exp Psychol 53:513–536. https://doi.org/10.1080/713755889

Leopold DA, Rhodes G (2010) A comparative view of face perception. J Comp Psychol 124:233–251. https://doi.org/10.1037/a0019460.A

Little AC (2014) Domain specificity in human symmetry preferences: symmetry is most pleasant when looking at human faces. Symmetry 6:222–233. https://doi.org/10.3390/sym6020222

Lutz CK, Lockard JS, Gunderson VM, Grant KS (1998) Infant monkeys' visual responses to drawings of normal and distorted faces. Am J Primatol 44:169–174

Martin-Malivel J, Okada K (2007) Human and chimpanzee face recognition in chimpanzees (*Pan troglodytes*): role of exposure and impact on categorical perception. Behav Neurosci 121:1145–1155. https://doi.org/10.1037/0735-7044

Maurer D, Werker JF (2014) Perceptual narrowing during infancy: a comparison of language and faces. Dev Psychobiol 56:154–178. https://doi.org/10.1002/dev.21177.121.6.1145

Méary D, Li Z, Li W, Guo K, Pascalis O (2014) Seeing two faces together: preference formation in humans and rhesus macaques. Anim Cogn 17:1107–1119. https://doi.org/10.1007/s10071-014-0742-3

Munar E, Gomez-Puerto G, Call J, Nadal M (2015) Common visual preference for curved contours in humans and great apes. PLoS One 10:1–15. https://doi.org/10.1371/journal.pone.0141106

Myowa-Yamakoshi M, Tomonaga M (2001) Development of face recognition in an infant gibbon (*Hylobates agilis*). Infant Behav Dev 24:215–227. https://doi.org/10.1016/S0163-6383(01)00076-5

Neiworth JJ, Hassett JM, Sylvester CJ (2007) Face processing in humans and new world monkeys: the influence of experiential and ecological factors. Anim Cogn 10:125–134. https://doi.org/10.1007/s10071-006-0045-4

Nelson CA (2001) The development and neural bases of face recognition. Infant Child Dev 10:3–18. https://doi.org/10.1002/icd.239

Nelson CA (2003) The development of face recognition reflects an experience-expectant and activity-dependent process. In: Pascalis O, Slater A (eds) The development of face processing in infancy and early childhood: current perspectives. Nova Science, New York, NY, pp 79–97

Parr LA (2011) The evolution of face processing in primates. Phil Trans R Soc Lond B: Biol Sci 366:1764–1777. https://doi.org/10.1098/rstb.2010.0358

Pascalis O, de Haan M, Nelson CA (2002) Is face processing species-specific during the first year of life? Science 296:1321–1323. https://doi.org/10.1126/science.1070223

Pascalis O, Kelly DJ, Caldara R (2006) What can bees really tell us about the face processing system in humans? J Exp Biol 209:3266–3266. https://doi.org/10.1242/jeb.02411

Pascalis O, Loevenbruck H, Quinn PC, Kandel S, Tanaka JW, Lee K (2014) On the links among face processing, language processing, and narrowing during development. Child Dev Perspect 8:65–70. https://doi.org/10.1111/cdep.12064

Pascalis O, Petit O, Kim JH, Campbell R (1999) Picture perception in primates: the case of face perception. Curr Psychol Cogn 18:889–922

Pascalis O, Scott LS, Kelly DJ, Shannon RW, Nicholson E, Coleman M, Nelson CA (2005) Plasticity of face processing in infancy. Proc Natl Acad Sci USA 102:5297–5300. https://doi.org/10.1073/pnas.0406627102

Peirce JW (2019) Do sheep really recognize faces? Evidence from previous studies. Response to "Sheep recognize familiar and unfamiliar human faces from two-dimensional images". R Soc Open Sci 6(7):4–6. https://doi.org/10.1098/rsos.182157

Rakover SS (2013) Explaining the face-inversion effect: the face-scheme incompatibility (FSI) model. Psychon Bull Rev 20:665–692. https://doi.org/10.3758/s13423-013-0388-1

Rhodes G (2006) The evolutionary psychology of facial beauty. Annu Rev Psychol 57:199–226. https://doi.org/10.1146/annurev.psych.57.102904.190208

Rossion B (2008) Picture-plane inversion leads to qualitative changes of face perception. Acta Psychol 128:274–289. https://doi.org/10.1016/j.actpsy.2008.02.003

Rossion B, Taubert J (2019) What can we learn about human individual face recognition from experimental studies in monkeys? Vis Res 157:142–158. https://doi.org/10.1016/j.visres.2018.03.012

Sackett GP (1966) Monkeys reared in isolation with pictures as visual input: evidence for an innate releasing mechanism. Science 154(3755):1468–1473. https://doi.org/10.1126/science.154.3755.1468

Scott LS, Fava E (2013) The own-species face bias: a review of developmental and comparative data. Vis Cogn 21:1364–1391. https://doi.org/10.1080/13506285.2013.821431

Simpson EA, Jakobsen KV, Damon F, Suomi SJ, Ferrari PF, Paukner A (2017) Face detection and the development of own-species bias in infant macaques. Child Dev 88:103–113. https://doi.org/10.1111/cdev.12565

Sugita Y (2008) Face perception in monkeys reared with no exposure to faces. Proc Natl Acad Sci USA 105:394–398. https://doi.org/10.1073/pnas.0706079105

Tanaka M (2003) Visual preference by chimpanzees (*Pan troglodytes*) for photos of primates measured by a free-choice-order task: implication for influence of social experience. Primates 44:157–166. https://doi.org/10.1007/s10329-002-0022-8

Tanaka M (2007) Development of the visual preference of chimpanzees (*Pan troglodytes*) for photographs of primates: effect of social experience. Primates 48:303–309. https://doi.org/10.1007/s10329-007-0044-3

Taubert J (2010) Evidence of human-like, holistic face processing in spider monkeys (*Ateles geoffroyi*). J Comp Psychol 124:57–65. https://doi.org/10.1037/a0017704

Taubert J, Parr LA (2009) Visual expertise does not predict the composite effect across species: a comparison between spider (*Ateles geoffroyi*) and rhesus (*Macaca mulatta*) monkeys. Brain Cogn 71:187–195. https://doi.org/10.1016/j.bandc.2009.09.002

Taubert J, Weldon KB, Parr LA (2017) Robust representations of individual faces in chimpanzees (*Pan troglodytes*) but not monkeys (*Macaca mulatta*). Anim Cogn 20:321–329. https://doi.org/10.1007/s10071-016-1054-6

Tibbetts EA (2002) Visual signals of individual identity in the wasp *Polistes fuscatus*. Proc R Soc Lond Ser B: Biol Sci 269(1499):1423–1428. https://doi.org/10.1098/rspb.2002.2031

Tomonaga M, Tanaka M, Matsuzawa T, Myowa-Yamakoshi M, Kosugi D, Mizuno Y et al (2004) Development of social cognition in infant chimpanzees (*Pan troglodytes*): face recognition, smiling, gaze, and the lack of triadic interactions. Jpn Psychol Res 46:227–235. https://doi.org/10.1111/j.1468-5584.2004.00254.x

Towler A, Kemp RI, Bruce V, Burton AM, Dunn JD, White D (2019) Are face recognition abilities in humans and sheep really "comparable"? R Soc Open Sci 6(1):1–3. https://doi.org/10.1098/rsos.180772

Valenza E, Simion F, Macchi Cassia V, Umiltà C (1996) Face preference at birth. J Exp Psychol: Hum Percept Perform 22:892–903

Wallis G (2013) Toward a unified model of face and object recognition in the human visual system. Front Psychol 4:1–25. https://doi.org/10.3389/fpsyg.2013.00497

Wallis G, Siebeck UE, Swann K, Blanz V, Bülthoff HH (2008) The prototype effect revisited: evidence for an abstract feature model of face recognition. J Vis 8:1–15. https://doi.org/10.1167/8.3.20

Whittlesea BWA, Leboe JP (2003) Two fluency heuristics (and how to tell them apart). J Mem Lang 49:62–79. https://doi.org/10.1016/S0749-596X(03)00009-3

Wilson DA, Tomonaga M (2018) Visual discrimination of primate species based on faces in chimpanzees. Primates 59:243–251. https://doi.org/10.1007/s10329-018-0649-8

Yin RK (1969) Looking at upside-down faces. J Exp Psychol 81:141–145. https://doi.org/10.1037/h0027474

Young AW, Burton AM (2018) Are we face experts? Trends Cogn Sci 22:100–110. https://doi.org/10.1016/j.tics.2017.11.007

Chapter 5
Factors Affecting Facial Recognition in Capuchin Monkeys

Chihiro Hiramatsu

Abstract In many species the face mediates social communication, conveying meaningful information to other individuals. The face is thus a highly important visual stimulus, and many factors are involved in recognizing individuals from facial information. Capuchin monkeys are excellent subjects for studying facial recognition from a comparative perspective because of their intelligence and their highly social nature. Furthermore, they show polymorphic color vision, with groups containing both dichromatic and trichromatic individuals, which provides an opportunity to examine potential effects of color on facial recognition. In this chapter, I describe experimental research on the relationships between color vision and detection of face color modulations in capuchin monkeys. Using face images of familiar individuals, the color of either the left eye, the right eye, or the nose was shifted in a red or blue direction. Face orientation (upright and inverted) and face identity were also manipulated. Trichromatic monkeys showed no greater sensitivity to face color modulation, but face parts, orientation, and identity all affect their accuracy and reaction times in the detection task. These results indicate that facial stimuli appear equally salient to dichromatic and trichromatic individuals, and that social factors such as dominance and kinship affect face processing in capuchin monkeys.

Keywords Capuchin monkeys · Visual perception · Facial recognition · Face identity · Social relationship

Introduction

Recognizing identity and condition of group members is an important cognitive task in many species of social animals. In human and nonhuman primates, the face plays a highly important role in this aspect of social cognition: it conveys visual

---

C. Hiramatsu (✉)
Department of Human Science, Faculty of Design, Kyushu University, Fukuoka, Japan
e-mail: chihiro@design.kyushu-u.ac.jp

J. R. Anderson, H. Kuroshima (eds.), *Comparative Cognition*,
https://doi.org/10.1007/978-981-16-2028-7_5

information about the owner's familiarity, emotion, age, gender, health and reproductive status, gaze direction, etc. A specialized neural network for face processing has been found in the visual cortex of various primate species including marmosets (Hung et al. 2015), macaques (Chang and Tsao 2017), and humans (Kanwisher and Yovel 2006), indicating that at least some underlying face processing mechanisms are shared among species. Recognition of conspecific faces has in fact been reported in species as disparate as paper wasps (Sheehan and Tibbetts 2011), cichlid fish (Kohda et al. 2015), and sheep (Kendrick et al. 2001). Furthermore, animals kept as companions by humans (dogs: Bognar et al. 2018; cats: Takagi et al. 2019) have been shown to attend to human faces and process them depending on their relationship with the owner of the face. Regardless of the extent to which facial recognition abilities are shared or have evolved independently in different taxa, in many visually capable species intra- and inter-specific communication is largely mediated through facial signals.

Several studies on visual social perception and communication in tufted capuchin monkeys (*Sapajus apella*) conducted in the laboratory of Kazuo Fujita have conspecifics and humans as social stimuli (Anderson et al. 2009; Hattori et al. 2007; Kawaguchi et al. 2019; Kuroshima et al. 2002; Matsuno and Fujita 2018; Morimoto and Fujita 2012; Takimoto and Fujita 2011). Capuchin monkeys, highly social platyrrhine monkeys endemic to South America (Fragaszy et al. 2004), are known to extract a range of information from the faces of conspecifics, including familiarity (Talbot et al. 2016), age (Kawaguchi et al. 2019), symmetry (Paukner et al. 2017), and emotional expression (Calcutt et al. 2017).

Capuchin monkeys, like most other platyrrhines, show polymorphic color vision: dichromatic and trichromatic individuals coexist in the same group (Jacobs 2007). This intriguing diversity is caused by the multiallelic trait of the L/M opsin gene on the X-chromosome (Jacobs et al. 1993). The L/M opsin gene codes photopigments sensitive to the middle- to long-wavelength of visible light. Species of the family Cebidae, which includes capuchin monkeys, usually have three variants of the L/M opsin gene in the gene pool of the population, with maximal absorbance at around 530, 545, and 560 nm, respectively (Hiramatsu et al. 2005; Saito et al. 2005). Together with the S opsin gene for short-wavelength sensitivity coded on an autosome, all males (with a single X-chromosome, hence hemizygous) and homozygous females on this allele become dichromats, while heterozygous female become trichromats. By contrast, in catarrhine primates (including humans), tandemly repeated L and M opsin genes on the X-chromosome enable routine trichromatic vision in both males and females (Jacobs 1996). Regardless of the underlying genetic mechanisms, the evolution of trichromatic vision in primates is likely to have had important implications for species' visual cognition.

The shift from nocturnal to diurnal activity patterns might be one primary etho-ecological factor in the prevalence of trichromatic vision in primates, as nocturnal primates have remained dichromatic—the normal form of color vision for mammals (Heesy and Ross 2001). Another likely driving force in the evolution of trichromacy is frugivory (Mollon 1989; Regan et al. 2001; Sumner and Mollon 2000), while the "social signal" hypothesis Changizi et al. (2006) proposes that trichromatic vision is optimized for detecting skin color modulations with variations in blood amount and blood oxygen levels. Changizi et al. (2006) focused on the fact that the bare skin region of faces is visually salient for diurnal primates with polymorphic or routinely trichromatic vision, whereas most of the face of nocturnal mammals is covered with fur. In humans, face coloration can be an honest signal of health and emotional state (Stephen et al. 2009; Thorstenson et al. 2019). In macaque monkeys, it is likely that face color modulation communicates reproductive states (Dubuc et al. 2009, 2014; Higham et al. 2011). Although it is arguable that detecting skin color modulations has driven the evolution of trichromacy, it is worthwhile to evaluate the role of trichromatic vision in this ability in social animals. Capuchin monkeys are clearly an excellent nonhuman primate model for investigating the effect of face color modulations on conspecific behavior. Currently, there are no objective data that suggest face color modulates in capuchin monkeys. However, red facial skin is more prominent in males than females in the bald uakari, a platyrrhine monkey that has a bald head and highly polymorphic color vision (Corso et al. 2016). The relatively large region of bare skin around the eyes and nose in capuchin monkeys supports the possibility that their face color modulates depending on emotional or reproductive states, and that this modulation is detectable by conspecifics with trichromatic vision. In addition, a simulation experiment suggested that face color modulation in macaque monkeys can be detected by trichromatic vision with narrower spectral separation of L and M photopigments, which is frequently observed in platyrrhines (Hiramatsu et al. 2017).

Several years ago, I set out to examine the effects of color in a social context, using capuchin monkeys. The original motivation was to investigate how capuchins with different color vision types would respond to color modulation in the faces of their group members. Furthermore, were they sensitive to color modulations of specific face parts? Were they also sensitive to face orientation, reflecting feature-based or holistic/configural processing of faces? To examine if color-related effects are enhanced in the face context, before our main experiment using intact face images, Fujita and I used randomized images in a titration experiment to adjust for individual differences in color sensitivity. We soon discovered that our experimental design was too complicated—involving too many variables—to provide clear answers to our original question. However, our results did suggest some consistencies and commonalities in capuchins' recognition of faces of familiar individuals. Below, I describe our experiments in some detail, as this information might provide useful pointers for future perceptual and cognitive studies with capuchin monkeys.

A Study of Color Modulation Detection in Capuchin Monkeys

At the start of our study, the group containing our subjects consisted of Heiji (19-year-old male), Pigmon (15-year-old male), Zinnia (12-year-old male), Zilla (19-year-old female), Kiki (17-year-old female), Theta (17-year-old female), Zen (9-year-old female), Zephie (3-year-old female), and Kojilo (3-year-old female). Zen and Zephie were the offspring of Zilla. Kojilo was Kiki's daughter, but as an infant she was raised by human caretakers due to Kiki's poor maternal behavior. Zinnia was not tested in these studies. They lived together as a group in a multi-cage, two-level complex spanning two rooms, with several interconnecting doors. Color vision type was determined for each monkey by genetic analysis as described elsewhere (Hiramatsu et al. 2005). Only Kiki, Zen, and Kojilo had trichromatic vision, with two classes of L/M opsin genes for photopigment with estimated peak sensitivity at 530 and 545 nm. The other five monkeys were dichromats, with L/M photopigments with peak sensitivity at either 530 or 545 nm. All monkeys had extensive experience of experimental visual stimuli being presented on a touch-sensitive monitor, and except for Pigmon, all had participated in a study involving visual categorization of surface materials (Hiramatsu and Fujita 2015). The monkeys received a portion of their daily diet during experimental sessions and the remainder in their home cage after the experiment each day. Water was always freely available in the home environment.

Stimulus Manipulations

The faces of all nine capuchin monkeys in the group were photographed under a full spectrum fluorescent light using a color-calibrated camera (Fig. 5.1a). Each monkey's facial expression was neutral (at least to human observers). A color modulation towards red or blue was made on three bare skin parts of the pictures—the nose, around the left side of the eye (hereafter: left eye) or around the right side of the eye (hereafter: right eye) from the viewpoint of observer—by increasing R or B in the 8-bit RGB values (Fig. 5.1b). Area sizes of the modulated parts were almost the same. For the color titration experiment described below, the color modulation was gradually increased in 100 steps. At the highest level (100), the R or B value was 255 (maximum value); at the lowest level (0), there was no color modulation and each part maintained its original color. The gamma of the monitor (relationship between input RGB values and output monitor luminance) was set to 2.2 and the luminance of RGB values was modulated linearly for a visually typical human observer.

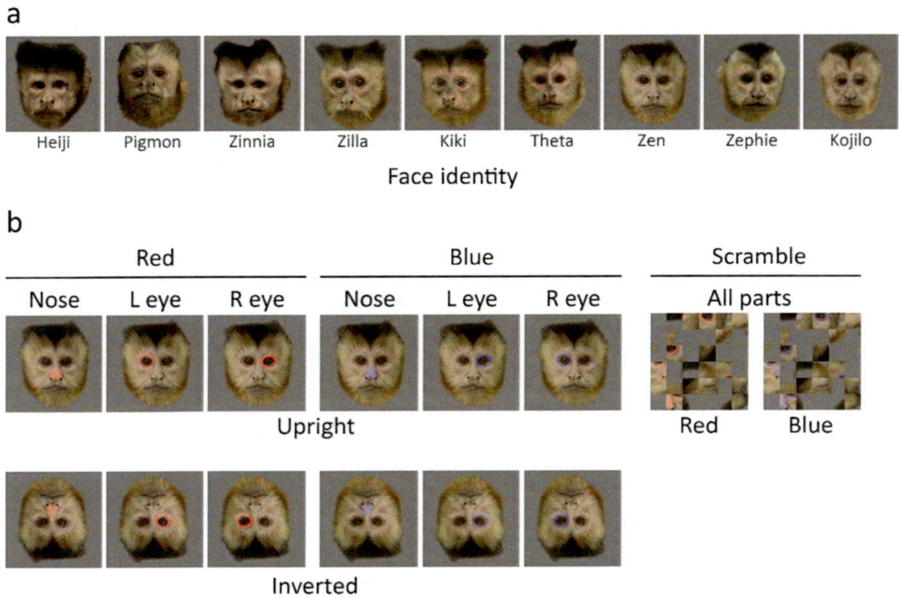

Fig. 5.1 Example of facial stimuli used in the color modulation detection experiment. (**a**) Neutral faces of nine monkeys. (**b**) Color-modulated pictures of one face (monkey Zen). The color of the nose, and around the left eye and right eye was modulated toward red and blue. Inverted stimuli were included in the main experiment. Scrambled images containing all the parts of the modulated faces were used in the titration experiment. Note that stimuli with the mid-level color modulation (50) are shown; these correspond approximately to the middle level red color modulation for dichromats in the main experiment. Middle color modulation level of red for trichromats and blue for all monkeys in the main experiment was lower (see Results) than the color modulation level shown here

Experimental Procedures

A four-alternative forced-choice (4AFC) procedure was used in which monkeys had to choose the stimulus picture that differed from the other three when all four stimuli were simultaneously presented on a calibrated touch-sensitive LCD monitor. The monkeys, all tested individually in a familiar operant chamber, responded by touching one stimulus. To prevent confusion arising from changes in the target stimulus on every trial, we used a non-color-modulated stimulus as the target (oddball procedure). Each stimulus image was 300 × 300 pixels (ca. 16 × 16 degree) and the background was uniformly gray (x = 0.311 and y = 0.330, 30 cd/m^2). The four stimuli were randomly assigned to four areas of the monitor, and to maintain the monkey's attention they were presented in a slightly different position in each area across trials. A trial started after the monkey touched a simple square (start image) that appeared in the center of the monitor (Fig. 5.2).

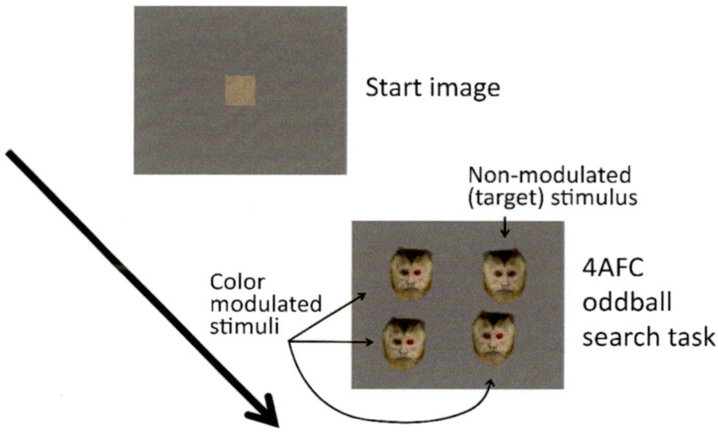

Fig. 5.2 Schematic of the 4AFC oddball search task. The same task design (4AFC oddball search task) was used for both the titration and the main experiment

Color Titration

To adjust for individual differences in color sensitivity, we first conducted a color titration experiment by using scrambled images of all face parts (both eyes, nose) of the modulated stimuli (Fig. 5.1b). Seven evenly separated color modulation levels were selected from the 100 steps of the stimuli and 630 4AFC trials with randomized face images (9 faces $\times$ 2 color direction $\times$ 7 levels $\times$ 5 times) were conducted on 5 consecutive days. The monkeys' task was to detect an oddball stimulus of original color from three identical color modulated stimuli. The monkeys' performances were plotted as a function of color modulation level and fitted to a logistic function. This procedure was repeated until the fit curve came within the range of the mean performance $\pm$ CI (confidence interval) by narrowing the range of color levels. In 4AFC, the accuracy threshold is 0.625 (1/4 + (1–1/4)/2) correct (Kingdom and Prins 2010). In this task, threshold indicates the minimum color modulation that can be detected. We estimated the color modulation level at threshold and four more levels where the estimated proportion of correct responses deviated from the threshold by 0.15 or 0.3, i.e., proportions correct at 0.325, 0.475, 0.775, and 0.925, from the psychometric function for each monkey. These values were then used for the main experiment.

Face Color Modulation Experiment

To examine if sensitivity to color modulation is enhanced in the face context, we used intact face images in the main experiment. Based on the titration experiment, five levels of color-modulated stimuli for each face part were recreated for each individual (Fig. 5.1). In addition to the face parts, color direction (red and blue), color level (five), face identity (nine individuals), face orientation (upright and inverted) were included as stimulus parameters in the analysis. Therefore, there were 540 stimulus conditions in the main experiment. We added 90 scrambled trials used in the final titration experiment to transfer the 4AFC task to face context (Fig. 5.2). Each test session consisted of 630 trials, with the trial order randomized across the five experimental days. Each monkey received 126 trials per day, and 3150 trials in total in the five sessions, each condition being run five times to obtain a more accurate sample of their responses.

Differences in Color Sensitivity between Dichromats and Trichromats

Seven monkeys (the exception was Pigmon) completed the titration experiment with scrambled images, allowing estimation of the threshold color modulation level for red and blue directions. The mean threshold for the red direction appeared higher in dichromats (47.5 ± 11.3) than in trichromats (12.7 ± 1.53), underscoring the higher sensitivity for red in the latter. The mean threshold for blue in dichromats (8 ± 1.63) also tended to be higher than in trichromats (5.3 ± 0.58). Statistical analysis indicated lower threshold in trichromats than dichromats for both red ($t\ (5) = 6.14$, $p < 0.01$) and blue ($t\ (5) = 3.03$, $p < 0.05$), although caution is required due to small sample size (four dichromats and three trichromats).

Effect of Color, Face Part, and Face Orientation

For the seven individuals who completed the main experiment, accuracy at the mid-color modulation level (the threshold in the titration experiment) did not improve in the face context (Fig. 5.3). However, some tendencies appeared via generalized linear mixed model (GLMM) analysis. We evaluated the effects of various factors on accuracy at the threshold color modulation levels for each individual. Color vision type, color direction, face part, face orientation, face identity, the interaction between color vision type and color direction, and the interaction between face part and face orientation were included in the models as fixed effects. Participant identities were included as random effects. For the analysis of accuracy, the response variable was correctness (1: correct or 0: incorrect) in each trial, and a model was fitted to a binomial distribution with the logit link function.

The GLMM analysis and type II test showed effects on accuracy of color direction ($p < 0.001$), face part ($p < 0.001$), face identity ($p < 0.05$), and an interaction between face part and face orientation ($p < 0.001$). There was no effect of color vision type. Post-hoc comparison showed that accuracy under the blue

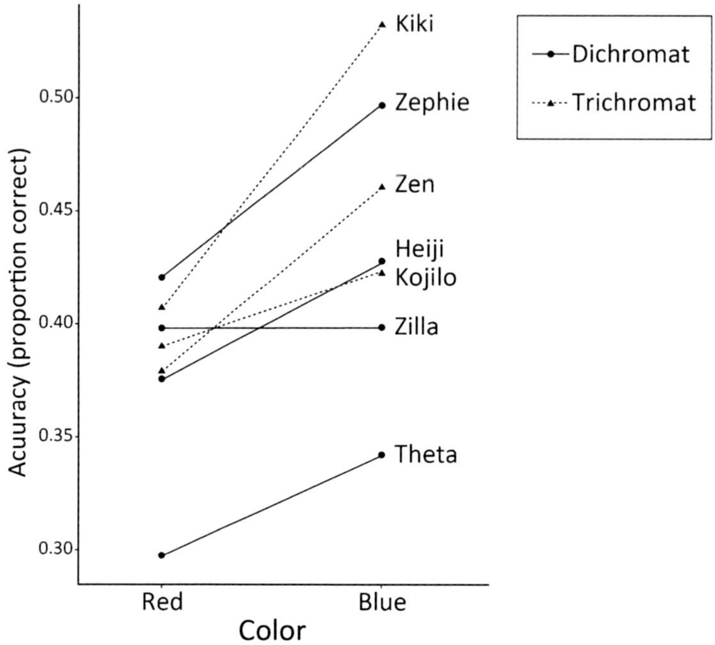

Fig. 5.3 Mean accuracy for each monkey under red and blue conditions at the mid-color modulation in the main experiment. Accuracy was averaged over the levels for face parts, face orientation, and face identity. Dichromat: circle with solid line; trichromat: triangle with dotted line. Note that mean accuracies did not exceed 0.625, the expected accuracy from the titration experiment

condition was higher than under the red condition ($p < 0.001$) (Fig. 5.3). Regarding face parts, accuracy of color modulation on the left eye was higher than on the right eye and nose under the upright condition ($p < 0.001$), whereas accuracy for right eye was higher than nose under the inverted face condition ($p < 0.001$) (Fig. 5.4).

Effect of Face Identity on Reaction Time

To analyze reaction time (RT), a GLMM model was fitted to a Gaussian distribution with the log link function for the same fixed and random effects as used in the accuracy analysis. Trials for all subjects with long RTs (1.5 times longer than the overall interquartile range) were removed as outliers (the same trials were also removed from the accuracy analysis). The GLMM analysis revealed a significant effect of face parts ($p < 0.05$). Post-hoc comparison showed that RT for the nose was shorter than for the left eye ($p < 0.01$).

To reveal individual tendencies, a linear mixed model (LMM) was used to analyze effects of color direction, face part, face orientation, face identity, and

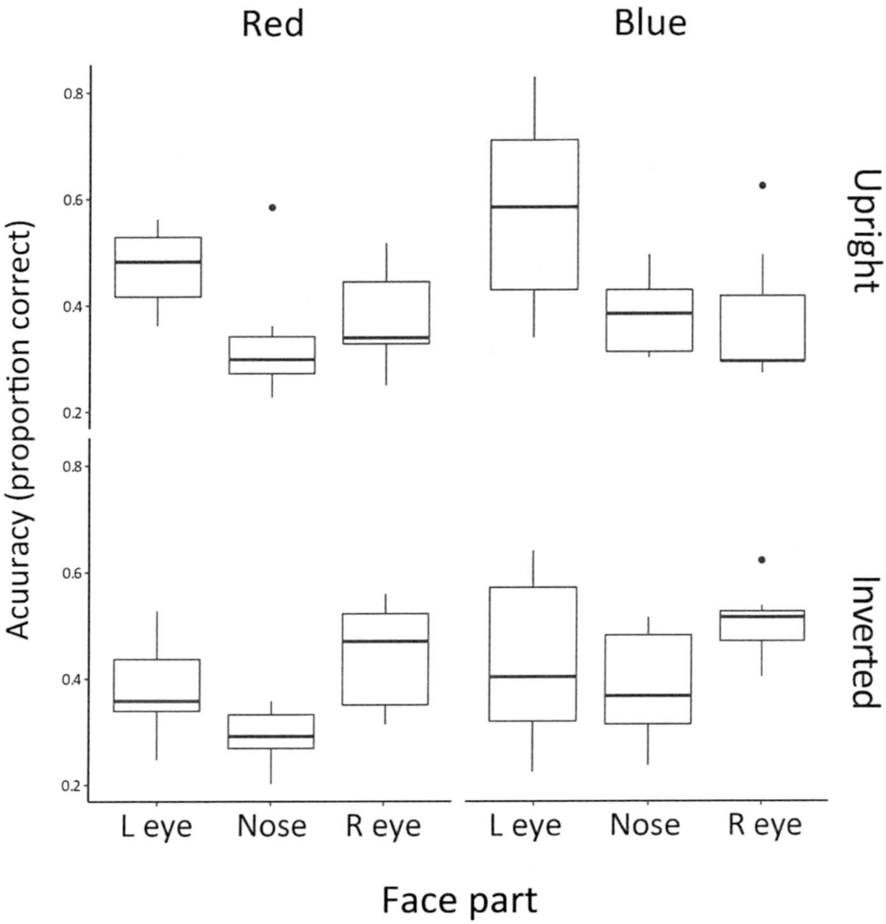

Fig. 5.4 Effect of face part, face orientation, and color direction. The boxplot shows the median, lower and upper quantiles, and the minimum and maximum values for accuracy under each condition across all monkeys (n = 7). Dots indicate outliers

interactions between color direction, face part, and face orientation on RTs of individual monkeys. The LMMs showed that face identity affected RTs significantly in Heiji ($p < 0.01$), Zilla ($p < 0.05$), and Theta ($p < 0.001$). Post-hoc analyses showed that the Heiji's RTs were longer for the faces of Zen and Kojilo than those of Zinnia, Zilla, and himself. Theta's RTs were longer for the faces of Pigmon and Zinna than those of Zephie and Kojilo. Zilla showed the longest RTs for the face of Zephie, who in turn showed the longest RTs for the face of Zilla, although Zephie showed no significant effect of face identity (Fig. 5.5).

The analysis of each individual's RTs also showed significant to marginally significant effects in Kiki (interaction between face orientation and color: $p = 0.034$), Theta (interaction between face part and color: $p = 0.097$), Zephie

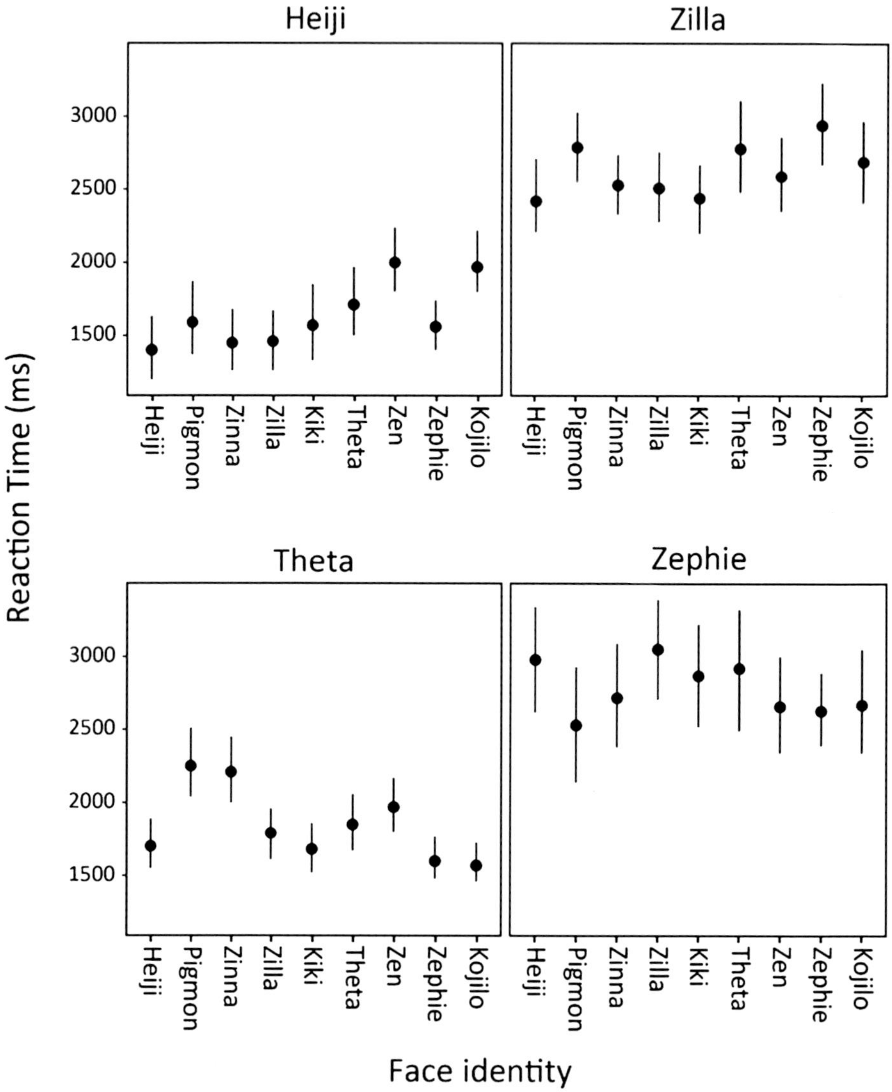

Fig. 5.5 Effect of face identity on RT for each monkey. Points indicate mean RTs, lines indicate lower- and upper- 95% confidence interval limits calculated by the bootstrapping method. Face identities are ordered (from left to right) according to dominance information from Takimoto and Fujita (2011)

(color: $p = 0.084$), and Kojilo (interaction between face orientation and color: $p = 0.062$). However, the relationship between color and face orientation or face part was inconsistent across the individuals.

Discussion

Below, I review the results of our study on capuchin monkeys' responses to manipulated facial stimuli, and discuss both possible explanations and implications for the effects or lack of effects of the variables that we manipulated.

Color Vision and Color Modulation in the Face Context

Accuracy in the face context was higher than chance level (0.25) but lower than 0.625 (the performance for scrambled images at threshold color modulation) in both dichromatic and trichromatic monkeys (Fig. 5.3). There are several possible explanations for this result. First, the color of only one part of the face was modulated in the main experiment, whereas in the titration experiment with scrambled images all parts were modulated. The small size of the area of the color modulation might be a reason for the lower performance. Second, since the RGB values of the stimuli were modulated artificially, our stimuli did not reflect truly natural color modulations in monkeys. Our modulation scheme was based on increasing R or B values in the pixels of face parts, and this treatment increased both the color saturation and the brightness of the pixels. In real macaque monkeys, the reddish color modulation associated with the reproductive status also decreases lightness (Higham et al. 2010; Hiramatsu et al. 2017). Although we did not measure the true color modulation of capuchin monkey faces, our modulation scheme may not have been ecologically valid. Furthermore, since the monitor RGB spectra were not optimized for capuchin monkeys' spectral sensitivity, which differs from our own, the face coloration might have appeared unnatural to the monkeys.

In both dichromats and trichromats, accuracy was higher for blue color modulation than red, although in trichromats we expected a higher or comparable performance for red. The larger chromatic contrast between modulated parts and the surrounding skin under the blue condition compared to red might explain this result. Capuchin monkeys might be more sensitive to unnatural modulation in the blue direction with large chromatic contrast; this possibility remains to be tested.

Nonetheless, the color modulation scheme in our experiments failed to reveal a face enhancement effect in trichromatic capuchin monkeys. It is also possible that our experimental design was too complicated and artificial to obtain such an effect. A study that used face pictures of rhesus macaque monkeys taken under natural conditions and a color substitution paradigm in which humans experienced various color vision types, found enhanced detection of ecologically valid face color modulation in participants who experienced trichromatic vision (Hiramatsu et al. 2017). This needs to be examined with monkey participants. Their combination of color vision polymorphism and cognitive competence points to capuchin monkeys as an excellent model for studies of cognitive aspects of color vision in primates.

Face Parts and Face Orientation

Most of our monkeys' data showed significant effects of face part or an interaction between face parts and face orientation. Accuracy was generally greater for the left eye in the upright condition. The finding of upright superiority is consistent with the widely accepted hypothesis of configural processing of faces in humans (Maurer et al. 2002). Studies with nonhuman primates including capuchins (Calcutt et al. 2017), squirrel monkeys (Nakata and Osada 2012), tamarins (Neiworth et al. 2007), macaques (Adachi et al. 2009), and chimpanzees (Tomonaga 2007) have shown a similar phenomenon. Configural processing of bodies is also suggested by body inversion effects in capuchin monkeys (Matsuno and Fujita 2018). A left eye bias may be consistent with a left gaze bias reported in humans, macaques, and dogs (Guo et al. 2009). In humans, the leftward face bias is at least partly attributable to right hemisphere dominance in processing faces (Megreya and Havard 2011) but this is debated for monkeys (Zangenehpour and Chaudhuri 2005; Tsao et al. 2008). In fact, the results of a recently published meta-analysis led to the conclusion that the inversion effect is not a reliable phenomenon in nonhuman primates (Griffin 2020). Therefore, the left side or upright bias found in our study might be an effect of our captive monkeys' extensive history of interacting with humans or due to some unidentified procedural aspects of the study. Further research is required to bring greater clarity to the issue of configural face processing in capuchin monkeys.

Face Identity

In three monkeys, all dichromats, RTs varied significantly depending on stimulus face identity. Although there might be no clear relationship with color vision type, dichromatic monkeys with less sensitivity to reddish modulation might have paid more attention to identity during the task. Interestingly, RTs of Heiji, the calm, alpha male in the group, were longer for faces of younger monkeys than older monkeys (including his own). The longer RT's for young monkey faces might reflect interest or concern for those individuals, assuming that he did in fact recognize the pictures as members of his group. By contrast, Theta showed longer RTs in response to adult males (except for Heiji) rather than young individuals. Theta's accuracy in the test trials was markedly lower than in her titration trials with randomized stimuli. She in fact appeared afraid to touch the face stimuli during test trials, and her social position—the most subordinate adult in the group—might have led to low accuracy and slower RTs to faces of dominant individuals (Figs. 5.3 and 5.4). In other words, for Theta face context might have had a negative effect on performance. The longest RTs for Zephie and Zilla were in response to seeing each other's face, possibly an effect of their mother-offspring relationship.

Clearly, many factors might affect RTs, and so the possible explanations offered here are speculative. However, individual differences in the effects of face identity even within the same color vision type exclude the possibility that variations in RT were due to physical properties of face stimuli. By using an eye-tracking technique, Lonsdorf et al. (2019) showed that capuchin monkeys looked longer at unfamiliar conspecific faces of the same sex than the opposite sex. Although our experiment was not set up to examine looking durations, social factors could affect visual cognition of faces especially when those faces are familiar, and those factors might be reflected in RTs. What I hope to have conveyed here is how monkeys living in a social group but tested individually can provide valuable opportunities to investigate how social relationships influence perceptual and cognitive processes.

Conclusion

This chapter describes a study in which we aimed to examine how color modulations on face parts of familiar group members are responded to by capuchin monkeys with different color vision types, namely dichromatic and trichromatic. Monkeys were more sensitive to blue than red modulation in the face context, irrespective of color vision type. In other words, we found that the monkeys detected a color modulation despite it having (to our knowledge) no biological relevance. However, we found no evidence of sensitivity to what we expected might be the more biologically relevant change, that is, in redness. We also observed upright orientation and left side biases, and likely effects of social relationships in face context. These results illustrate the multiplicity of factors associated with face recognition and processing in capuchin monkeys, and this no doubt also applies to a wider range of animals.

Acknowledgements My sincere and deep gratitude goes to Professor Kazuo Fujita and his enthusiasm for exploring the animal mind. I was always encouraged by his expert advice, and his kindness continues to help people and animals. I am grateful to Dr. Takaaki Kaneko, who gave valuable suggestions for our study. I am also grateful to Professor James Anderson for his kind editing of the manuscript and to Dr. Annika Paukner for helpful suggestions for improving this chapter. This research was supported by a Grant-in-Aid for JSPS Fellows (No. 10 J04395) to C. Hiramatsu and JSPS Grants-in-Aide for Scientific Research Nos. 20220004 and 25240020 to K. Fujita.

References

Adachi I, Chou DP, Hampton RR (2009) Thatcher effect in monkeys demonstrates conservation of face perception across Primates. Curr Biol 19(15):1270–1273. https://doi.org/10.1016/j.cub.2009.05.067

Anderson JR, Kuroshima H, Paukner A, Fujita K (2009) Capuchin monkeys (*Cebus apella*) respond to video images of themselves. Anim Cogn 12(1):55–62. https://doi.org/10.1007/s10071-008-0170-3

Bognar Z, Iotchev IB, Kubinyi E (2018) Sex, skull length, breed, and age predict how dogs look at faces of humans and conspecifics. Anim Cogn 21(4):447–456. https://doi.org/10.1007/s10071-018-1180-4

Calcutt SE, Rubin TL, Pokorny JJ, de Waal FBM (2017) Discrimination of emotional facial expressions by tufted capuchin monkeys (*Sapajus apella*). J Comp Psychol 131(1):40–49. https://doi.org/10.1037/com0000055

Chang L, Tsao DY (2017) The code for facial identity in the primate brain. Cell 169(6):1013–1028. https://doi.org/10.1016/j.cell.2017.05.011

Changizi MA, Zhang Q, Shimojo S (2006) Bare skin, blood and the evolution of primate colour vision. Biol Lett 2(2):217–221

Corso J, Bowler M, Heymann EW, Roos C, Mundy NI (2016) Highly polymorphic colour vision in a New World monkey with red facial skin, the bald uakari (*Cacajao calvus*). Proc R Soc B Biol Sci 283(1856):20160067. https://doi.org/10.1098/rspb.2016.0067

Dubuc C, Allen WL, Maestripieri D, Higham JP (2014) Is male rhesus macaque red color ornamentation attractive to females? Behav Ecol Sociobiol 68(7):1215–1224

Dubuc C, Brent LJ, Accamando AK, Gerald MS, Mac Larnon A, Semple S, Heistermann M, Engelhardt A (2009) Sexual skin color contains information about the timing of the fertile phase in free-ranging Macaca mulatta. Int J Primatol 30(6):777–789

Fragaszy DM, Visalberghi E, Fedigan LM (2004) The complete capuchin: the biology of the genus Cebus. Cambridge University Press, Cambridge, UK

Griffin JW (2020) Quantifying the face inversion effect in nonhuman primates: a phylogenetic meta-analysis. Anim Cogn 23(2):237–249. https://doi.org/10.1007/s10071-019-01340-8

Guo K, Meints K, Hall C, Hall S, Mills D (2009) Left gaze bias in humans, rhesus monkeys and domestic dogs. Anim Cogn 12(3):409–418. https://doi.org/10.1007/s10071-008-0199-3

Hattori Y, Kuroshima H, Fujita K (2007) I know you are not looking at me: capuchin monkeys' (*Cebus apella*) sensitivity to human attentional states. Anim Cogn 10(2):141–148. https://doi.org/10.1007/s10071-006-0049-0

Heesy CP, Ross CF (2001) Evolution of activity patterns and chromatic vision in primates: Morphometrics, genetics and cladistics. J Hum Evol 40(2):111–149

Higham JP, Brent LJN, Dubuc C, Accamando AK, Engelhardt A, Gerald MS, Heistermann M, Stevens M (2010) Color signal information content and the eye of the beholder: a case study in the rhesus macaque. Behav Ecol 21(4):739–746. https://doi.org/10.1093/beheco/arq047

Higham JP, Hughes KD, Brent LJ, Dubuc C, Engelhardt A, Heistermann M, Maestriperi D, Santos LR, Stevens M (2011) Familiarity affects the assessment of female facial signals of fertility by free-ranging male rhesus macaques. Proc R Soc B 278(1723):3452–3458

Hiramatsu C, Tsutsui T, Matsumoto Y, Aureli F, Fedigan LM, Kawamura S (2005) Color vision polymorphism in wild capuchins (*Cebus capucinus*) and spider monkeys (*Ateles geoffroyi*) in Costa Rica. Am J Primatol 67(4):447–461. https://doi.org/10.1002/ajp.20199

Hiramatsu C, Fujita K (2015) Visual categorization of surface qualities of materials by capuchin monkeys and humans. Vis Res 115:71–82. https://doi.org/10.1016/j.visres.2015.07.006

Hiramatsu C, Melin AD, Allen WL, Dubuc C, Higham JP (2017) Experimental evidence that primate trichromacy is well suited for detecting primate social colour signals. Proc R Soc B Biol Sci 284(1856):20162458. https://doi.org/10.1098/rspb.2016.2458

Hung C-C, Yen CC, Ciuchta JL, Papoti D, Bock NA, Leopold DA, Silva AC (2015) Functional mapping of face-selective regions in the extrastriate visual cortex of the marmoset. J Neurosci 35 (3):1160–1172. https://doi.org/10.1523/JNEUROSCI.2659-14.2015

Jacobs GH (1996) Primate photopigments and primate color vision. Proc Natl Acad Sci 93 (2):577–581

Jacobs GH (2007) New World monkeys and color. Int J Primatol 28(4):729–759

Jacobs GH, Neitz J, Neitz M (1993) Genetic basis of polymorphism in the color vision of platyrrhine monkeys. Vis Res 33(3):269–274. https://doi.org/10.1016/0042-6989(93)90083-9

Kanwisher N, Yovel G (2006) The fusiform face area: a cortical region specialized for the perception of faces. Phil Trans R Soc B 361(1476):2109–2128. https://doi.org/10.1098/rstb.2006.1934

Kawaguchi Y, Kuroshima H, Fujita K (2019) Age categorization of conspecific and Heterospecific faces in capuchin monkeys (*Sapajus apella*). J Comp Psychol 133(4):502–511. https://doi.org/10.1037/com0000185

Kendrick KM, da Costa AP, Leigh AE, Hinton MR, Peirce JW (2001) Sheep don't forget a face. Nature 414(6860):165–166. https://doi.org/10.1038/35102669

Kingdom FAA, Prins N (2010) *Psychophysics: a practical introduction*. Elsevier Academic Press, Cambridge, MA, p 279

Kohda M, Jordan LA, Hotta T, Kosaka N, Karino K, Tanaka H, Taniyama M, Takeyama T (2015) Facial recognition in a group-living cichlid fish. PLoS One 10(11):e0142552. https://doi.org/10.1371/journal.pone.0142552

Kuroshima H, Fujita K, Fuyuki A, Masuda T (2002) Understanding of the relationship between seeing and knowing by tufted capuchin monkeys (*Cebus apella*). Anim Cogn 5(1):41–48. https://doi.org/10.1007/s10071-001-0123-6

Lonsdorf EV, Engelbert LM, Howard LH (2019) A competitive drive? Same-sex attentional preferences in capuchins. Am J Primatol 81(6):e22998. https://doi.org/10.1002/ajp.22998

Matsuno T, Fujita K (2018) Body inversion effect in monkeys. PLoS One 13(10):e0204353. https://doi.org/10.1371/journal.pone.0204353

Maurer D, Grand RL, Mondloch CJ (2002) The many faces of configural processing. Trends Cogn Sci 6(6):255–260. https://doi.org/10.1016/S1364-6613(02)01903-4

Megreya AM, Havard C (2011) Left face matching bias: right hemisphere dominance or scanning habits? Laterality 16(1):75–92. https://doi.org/10.1080/13576500903213755

Mollon JD (1989) "Tho'she kneel'd in that place where they grew..." the uses and origins of primate colour vision. J Exp Biol 146(1):21–38

Morimoto Y, Fujita K (2012) Capuchin monkeys (*Cebus apella*) use conspecifics' emotional expressions to evaluate emotional valence of objects. Anim Cogn 15(3):341–347. https://doi.org/10.1007/s10071-011-0458-6

Nakata R, Osada Y (2012) The Thatcher illusion in squirrel monkeys (*Saimiri sciureus*). Anim Cogn 15(4):517–523. https://doi.org/10.1007/s10071-012-0479-9

Neiworth JJ, Hassett JM, Sylvester CJ (2007) Face processing in humans and new world monkeys: the influence of experiential and ecological factors. Anim Cogn 10(2):125–134. https://doi.org/10.1007/s10071-006-0045-4

Paukner A, Wooddell LJ, Lefevre CE, Lonsdorf E, Lonsdorf E (2017) Do capuchin monkeys (*Sapajus apella*) prefer symmetrical face shapes? J Comp Psychol 131(1):73–77. https://doi.org/10.1037/com0000052

Regan BC, Julliot C, Simmen B, Viénot F, Charles–Dominique, P., and Mollon, J. D. (2001) Fruits, foliage and the evolution of primate colour vision. Philos Trans R Soc London Ser B Biol Sci 356(1407):229–283

Saito A, Kawamura S, Mikami A, Ueno Y, Hiramatsu C, Koida K, Fujita K, Kuroshima H, Hasegawa T (2005) Demonstration of a genotype-phenotype correlation in the polymorphic color vision of a non-callitrichine new world monkey, capuchin (*Cebus apella*). Am J Primatol 67(4):471–485. https://doi.org/10.1002/ajp.20201

Sheehan MJ, Tibbetts EA (2011) Specialized face learning is associated with individual recognition in paper wasps. Science 334(6060):1272–1275. https://doi.org/10.1126/science.1211334

Stephen ID, Smith MJL, Stirrat MR, Perrett DI (2009) Facial skin coloration affects perceived health of human faces. Int J Primatol 30(6):845–857

Sumner P, Mollon JD (2000) Catarrhine photopigments are optimized for detecting targets against a foliage background. J Exp Biol 203(13):1963–1986

Takagi S, Arahori M, Chijiiwa H, Saito A, Kuroshima H, Fujita K (2019) Cats match voice and face: cross-modal representation of humans in cats (*Felis catus*). Anim Cogn 22(5):901–906. https://doi.org/10.1007/s10071-019-01265-2

Takimoto A, Fujita K (2011) I acknowledge your help: capuchin monkeys' sensitivity to others' labor. Anim Cogn 14(5):715–725. https://doi.org/10.1007/s10071-011-0406-5

Talbot CF, Leverett KL, Brosnan SF (2016) Capuchins recognize familiar faces. Anim Behav 122:37–45. https://doi.org/10.1016/j.anbehav.2016.09.017

Thorstenson CA, Pazda AD, Young SG, Elliot AJ (2019) Face color facilitates the disambiguation of confusing emotion expressions: toward a social functional account of face color in emotion communication. Emotion 19(5):799–807. https://doi.org/10.1037/emo0000485

Tomonaga M (2007) Visual search for orientation of faces by a chimpanzee (*Pan troglodytes*): face-specific upright superiority and the role of facial configural properties. Primates 48(1):1–12. https://doi.org/10.1007/s10329-006-0011-4

Tsao DY, Moeller S, Freiwald WA (2008) Comparing face patch systems in macaques and humans. Proc Natl Acad Sci 105(49):19514–19519. https://doi.org/10.1073/pnas.0809662105

Zangenehpour S, Chaudhuri A (2005) Patchy organization and asymmetric distribution of the neural correlates of face processing in monkey inferotemporal cortex. Curr Biol 15 (11):993–1005. https://doi.org/10.1016/j.cub.2005.04.031

Chapter 6
Visual Body Perception in Primates: From Individual to Social Dyad

Toyomi Matsuno

Abstract The body is an important and unique visual object among other objects in our visual environment. We observe others' bodies every day, anticipate their upcoming behavior, and estimate their explicit and implicit intentions based on body postures and movements. Consistent with the ecological importance of body perception, recent studies have revealed the special attunement of the human visual system to body shapes and movements. The body is also a significant research subject in comparative perception studies, given the unique locomotor patterns of the human body, and species variations in morphological features. In addition, comparative research on visual tuning to social stimuli provides insight into the perceptual basis of social cognitive abilities, widely held to be more advanced in humans than in other animals. In this chapter, I highlight the significance of comparative cognitive studies on bodies and review the experimental evidence regarding basic properties of visual body processing in nonhuman primates.

Keywords Social vision · Visual body perception · Primate · Configural body processing · Identity recognition · Perception of social interaction

Introduction

Superior visual functions, such as fine visual acuity, color vision, and depth perception, are prominent characteristics of anthropoid primate species relative to other mammalian species (Matsuno and Fujita 2009). This could be a result of the earlier adaptation of those primates to a diurnal and arboreal lifestyle, where such visual functions would have been advantageous for locating food and moving from branch to branch in three-dimensional space (Nityananda and Read 2017; Dominy and Lucas 2001). In addition to the physical environment, adapting to the social

---

T. Matsuno (✉)
Faculty of Economics, Hosei University, Tokyo, Japan
e-mail: toyomi.matsuno@hosei.ac.jp

J. R. Anderson, H. Kuroshima (eds.), *Comparative Cognition*,
https://doi.org/10.1007/978-981-16-2028-7_6

environment could also have been a significant challenge influencing primate vision. Many primate species, especially humans, are social animals, living in complex social environments surrounded by competing and collaborating conspecifics, and some theories have argued that the social rather than the physical environment likely contributed to the primary pressures that drove the evolution of our cognitive abilities (Byrne and Whiten 1988; Dunbar 1998). Considering that perceptual processing of social information underlies these cognitive abilities for managing social relations with others, our perceptual system might have become specially attuned to social information. Social vision studies on humans have accumulated evidence that are consistent with this idea (Adams et al. 2011).

Faces and bodies are the principal media through which we express and read communicative signals. In particular, studies on humans have paid much attention to the visual processing of faces, and have revealed that faces capture and guide our attention, are processed in a different, more holistic manner from other objects, relying on specialized neural mechanisms, and that they convey various types of social information including identity, attractiveness, sex, and emotional and attentional states (e.g., Tsao and Livingstone 2008; Oruc et al. 2019). The body has long been left behind the face in social perception studies, but it has been attracting more attention in the past two decades (Knoblich et al. 2006; De Gelder 2009). Those studies have revealed that bodies are also processed in a specialized manner (e.g., Peelen and Downing 2007; Reed et al. 2003, 2006) and that they complement or sometimes supersede faces in conveying social information (Aviezer et al. 2012; Aviezer et al. 2008).

Comparative studies on nonhuman vision have followed the early trend of the studies in humans, with most focusing on faces as stimuli. Various species including nonhuman primates (Parr 2011), other vertebrates (Coulon et al. 2009; Brown and Dooling 1992; Kendrick et al. 2001), and even invertebrates (Tibbetts 2002; Van der Velden et al. 2008), have been tested for their abilities to discriminate conspecific faces. Apart from individual identity discrimination, face processing to gain information about current states or social attributes of conspecifics (e.g., emotion, age, sex, direction of attention) has also featured in this body of work (e.g., Parr 2001; Kawaguchi et al. 2019; Tomonaga 2007; Koba et al. 2009). These studies have shown that the face is a primary social communication medium in many animals, although the extent to which mechanisms underlying visual face processing are shared across species is still debated.

In contrast to the growing body of comparative research on visual face processing, studies on visual body perception in nonhuman animals remain sporadic. However, it can be argued that more studies should especially focus on the body, perhaps even more than the face. I will do this in this chapter. A brief overview of perspectives on human body perception will be followed by addressing comparative research on body perception. Thereafter, I review relevant experiments in primates, focusing on the four research topics: configural perception of body shape, perception of biological motion, recognition of individual identity, and perception of social dyads.

Why Bodies in Comparative Cognition?

Two main issues make the body significant and unique in recent human visual perception studies. First, bodies are regarded as a special visual object. Bodies in a scene receive higher processing priority by our visual system than inanimate objects: they capture our attention and have privileged access to our awareness (Downing et al. 2004; Stein et al. 2012). Like faces, bodies are processed differently from inanimate visual objects. They are visually discriminated depending on the relative spatial relations of their parts rather than the specific visual feature of each body part. This way of processing information, known as configural processing, can be advantageous for detecting even very slight structural differences among objects composed of multiple parts. Neurophysiological and neuroimaging studies also have revealed that there is a specific cortical network dedicated to processing other people's bodies and actions (e.g., Peelen and Downing 2007; Kiani et al. 2007). Although the origin of this specialized processing of social stimuli is still under debate, based on inherent biological constraints or expertise due to extensive daily social exposure (e.g., Gauthier et al. 2014; Hirai and Senju 2020; Powell et al. 2018), body perception is important for our understanding of the organization of our visual system and the perception of ecologically important visual objects. Second, visual recognition of another's body may not be purely visual. Recent theories of perception-action coupling suppose that perception of others' bodies and actions is achieved not solely by visual analysis but also by processing related to the observer's own bodily sensations and actions (Gallese and Sinigaglia 2011). Several studies have demonstrated that the perception of others' bodies interferes with the observers' own body representations and that, conversely, one's own bodily state affects perceptual processing, supporting the idea of bodily simulation (e.g., Kilner et al. 2003; Grosjean et al. 2007). Some theories go further and propose that such a simulation system works not just for perceptual analyses but also for understanding others' mental states including intentions and emotions (Blakemore and Decety 2001; Iacoboni 2009). Investigations on visual body perception in relation to bodily sensations would help us to understand whether and how we integrate internal and external information to represent our social world.

A comparative approach to body perception gives a biological and evolutionary perspective on these issues. It also has a distinctive significance compared to other topics in the field of comparative cognition. Below, I present four reasons why visual body perception merits further study in a range of species.

First, the body might surpass the voice and the face as an important communication medium in nonhuman primates. Many communicative signals of other species are redundant and stereotypic, partly depending on the constraints of the mediating physical environments and the species' cognitive limitations. In particular, vocal repertoires are narrow and their usage has limited flexibility (Zuberbühler 2003). In contrast, gestural communication in nonhuman primates (Fig. 6.1) is more variable, flexible, and generative (Pollick and De Waal 2007; Tomasello and Zuberbühler 2002). The efficacy of bodily communication signals is such that humans can

Fig. 6.1 A 5-year-old chimpanzee stands bipedally and continues to twirl her extended left wrist in front of her mother until the latter allows her to breastfeed. The original video is available at https://osf.io/nm4at/

communicate much better with others through gestures than non-linguistic vocalizations when they are not allowed to use language (Fay et al. 2013). Based on such advantages, some theories of language evolution have argued that bodily gestures rather than vocal communication could be at the origin of human language (e.g., Fadiga and Craighero 2006; Corballis 2012). Variations in visual intensity of bodily actions can also be advantageous in animal communication. Changes in posture are more salient and easier to recognize than subtle changes of face morphology in distant communications or limited visual environments (e.g., leafy forest). Bodily expression remains reliable even when expressed signal intensities are high and concurrent facial expressions are hard to discriminate from each other due to morphological constraints (Aviezer et al. 2012). Furthermore, some studies suggest that, compared to humans, nonhuman primates are less reliant on facial information for communication. For example, the creation of the muscular movement-based facial action coding system (FACS) for apes and monkeys identified a smaller number of action units than in humans, suggesting that their facial movements are probably less salient (Parr et al. 2010; Vick et al. 2007). An eye-tracking study also reported that rhesus macaques preferred to look at the bodies rather than the heads of conspecifics showing different affective states (Bliss-Moreau et al. 2017).

Second, humans are unique among primates with respect to bipedal locomotion. The human bipedal posture maintains the vertical order of body part positions, with the head at the top, followed by the arms, torso, then legs. In addition, we humans are highly terrestrial animals and usually look at conspecifics from the same vertical plane. The fixed spatial configuration of the body and uniformly constrained observation perspective bring the patterned stimulation to our visual system, and that might develop the sensitivity to the fixed body structure and specific perceptual strategy for social stimuli. For example, the body is processed configurally in humans, as mentioned above. The stable spatial relationship among body parts might be a prerequisite for the mechanism that selectively processes such highly probable configurations. The stable vertical order of the body parts also privileges the head over other body parts. Because the head is almost always at the top of the

Fig. 6.2 (**a**) Body silhouettes of a human and monkeys. Monkeys' heads are not always at the top and are visually inseparate from the torso. (**b**) Bodies of vertebrate and invertebrate animals. Hierarchical body structures (a torso and limbs) and the headward orientation of various animals are clear at a glance

body space in bipedal locomotion and level with the observers' eyes, it can be easily localized. After human observers detect a body-like shape in the scene, their gaze is guided to the head by simply sliding it to the upper part (Bindemann et al. 2010). A round head is positioned on a vertically stretched and thinner neck, and is therefore easily separable as a spherical component in our visual analyses (Fig. 6.2a; Biederman 1987; Marr and Nishihara 1978). Efficient head detection supports the active use of head and facial cues in human visual communication. In contrast, it is perceptually more costly to detect the head of a quadrupedal, nonhuman animal body, because the spatial relationship among body parts varies depending on posture, and head components are not easily isolated based on the contour concavities (Fig. 6.2a). In agreement with this, behavioral experiments on face or eye perception in nonhuman primates sometimes fail to replicate the robust perceptual effects known in humans and report relative insensitivity to facial cues (Tomonaga 2007; Rossion and Taubert 2019; Parr 2011). In addition to postural variations, arboreal species do not necessarily perceive others' bodies from "canonical" angles as assumed in many psychological experiments in humans. These factors increase the variation in spatial relationships among observed body parts. Species differences in body size and observation distances also change the typical retinal images of bodies, and therefore require image processing on different spatial scales in visual analyses (Campbell and Robson 1968). Considering both human uniqueness and species differences, the examination of shared and species-specific aspects of body perception across species can help us to understand the prerequisites of adaptation of our visual system to social stimuli.

Third, the scope of comparative studies on visual body perception could be phylogenetically extended to a broader range of species. This would yield insights on how relatively simple to complex visual systems resolve similar adaptive problems that require analysis of patterned visual objects with hierarchical structures. The basic body structure is shared among many animals including vertebrate and invertebrate species (Davidson and Erwin 2006), despite the variation in shape and number of body parts, size, and locomotory specializations. Animal bodies share visual characteristics such as symmetrical structure, the existence of body axes, visually dividable segments, and hierarchical organization with global configuration and local parts that move with and without obvious coherence (Fig. 6.2b). Because efficient visual detection and accurate recognition of conspecific and other bodies are always prime tasks for animals both in social and nonsocial contexts, their visual systems have conceivably evolutionarily shared or converged sensitivity to such a basic animal body plan (New et al. 2007). Considering that no animal exists without a body, whereas some species lack an identifiable facial configuration, and that flat faces of humans and other primates (advantageous for face-to-face communication) are not necessarily shared even with other mammals, the adaptive value of body processing might be phylogenetically traceable back further than face processing. Although researchers discovered the cortical face area before the body area (Kanwisher et al. 1997; Downing et al. 2001), neural networks tuned to process specific important visual categories might have originally emerged as an adaptation for body processing, with mechanisms for face processing developing later. The visual attunement to body structure could also show functional convergence. Comparative studies of visual body processing in relation to neural architectures of multiple species with different bodily constraints and behavioral repertoires would be helpful for better understanding social vision in terms of algorithm and implementation (Marr 1982).

Fourth, body perception studies in nonhuman animals are important for understanding the perceptual basis of the evolution of social cognition. The latter is a central topic in current comparative cognition. Humans are credited with uniquely sophisticated cognitive skills in this domain (Byrne and Whiten 1988). Our attributions of mental states to others and attempts to infer their thoughts, intentions, goals, and emotions play crucial roles in our social interactions. These skills are only slowly acquired during children's cognitive development (Wimmer and Perner 1983; Heyes and Frith 2014; Premack 1988). However, the requisite cognitive processes start with lower-level analyses of sensory input from others' bodies. Perceptual processing constrains and frames subsequent analyses for making social cognitive inferences. Furthermore, recent studies have revealed that some aspects of such analyses might be intrinsically more perceptual than previously thought (Samson et al. 2010; Ward et al. 2019; Neri et al. 2006; Papeo et al. 2017). Our visual system is attuned to efficiently analyze the structure of others' social interactions, and we infer internal visual contents from another's perspectives in an implicit and automatic manner even if we see task-irrelevant body features of the other person. Studies of similarities and differences between perceptual properties of humans and other species including nonhuman primates can clarify perceptual constraints of each species and thereby the continuum of the social information processing, from the lower sensory level to more deliberate cognitive processing.

Experimental Studies on Visual Body Perception in Primates

In this section, I review studies on how humans and nonhuman primates perceive others' bodies from four perspectives. The first and second subsections focus on basic visual processing of body shape and movement. A few studies have tested visual discrimination of conspecific and non-conspecific body stimuli to ask whether visual attunement to bodies is shared between humans and nonhuman primates despite their different postural and locomotory constraints. Results show both the commonality and variation between humans and other primates. The third subsection addresses studies on visual recognition of individuals in primate species. Individual recognition is fundamental to primate social interactions, and in humans not only faces but also bodies contribute to individual identification. Studies have shown that nonhuman primates also rely on body features to identify companions. The final subsection explores perception of social interactions. Recent discrimination experiments have demonstrated that both humans and monkeys are sensitive not only to visual scenes of bodies but also to their spatial relationships, in social dyads. These four aspects of body perception constitute the essential functions that support the comprehension of individual group mates and their social relationships, and they are cornerstones for our social lives.

Perception of Body Shape

As mentioned above, one prominent characteristic of visual perception of bodies compared to nonsocial objects is that bodies are processed configurally. We recognize bodies based on spatial relationships among body parts rather than local features of the body parts. The body inversion effect (BIE) is regarded as a hallmark of such configural body processing (Reed et al. 2003, 2006); Body discrimination is impaired by inversion, which disrupts the spatial configuration of body parts by putting the legs above the torso and the head below. The basic inversion effect has been replicated in humans and the detailed aspects such as the role of specific body parts (e.g., head) and the contribution of the observer's own body representation have been examined and debated (Arizpe et al. 2017; Axelsson et al. 2019; Barra et al. 2017; Tao et al. 2014).

As quadrupeds, nonhuman primates do not share the first-order spatial configuration of body parts with bipedal humans. Therefore, their studies on this issue will contribute to further understanding of the constraints of the configural body processing. Unfortunately, however, behavioral investigations in nonhuman primates have so far been limited to only two species, namely tufted capuchin monkeys and chimpanzees.

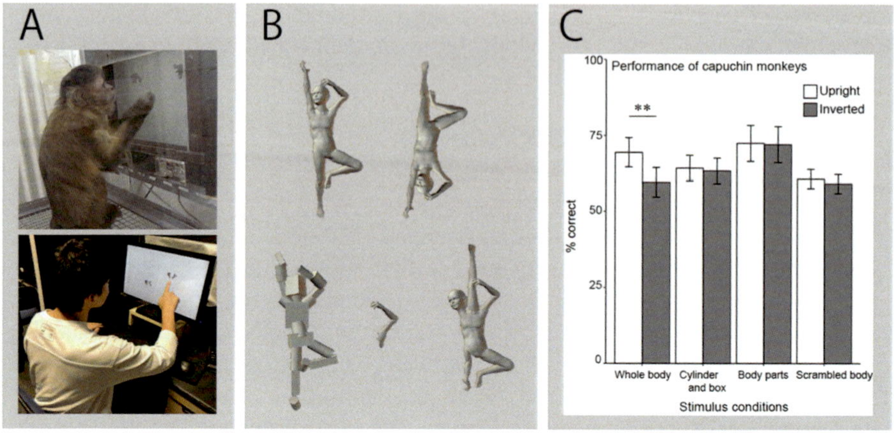

Fig. 6.3 (**a**) Experimental setup for a comparative perception study in capuchin monkeys and humans using a touch sensitive monitor. (**b**) Examples of body stimuli used in the BIE experiments. (**c**) Performance of capuchin monkeys in matching-to-sample tasks using intact and transfigured body stimuli (Redrawn based on Matsuno and Fujita 2018)

Matsuno and Fujita (2018) compared humans and capuchin monkeys, a small arboreal platyrrhine primate, in a visual matching-to-sample task using intact and transfigured human body figures as stimuli (Fig. 6.3a, b). They found that monkeys, like humans, showed a BIE, i.e., body posture discrimination was impaired by inversion of intact human body figures (Fig. 6.3b, c). The effect of stimulus orientation was abolished when body parts that should be discriminated (an arm or a leg) were isolated from the other contextual body parts and when the positions of the contextual body parts were scrambled. These results suggest that configural body processing is a common mechanism in bipedal and quadrupedal primates, and that visual attunement to social signals mediated by body postures is conserved through the evolution of primate vision. In addition, Matsuno and Fujita (2018) found a slight species difference: a response time analysis reveals that humans showed an inversion effect even when intact human body parts were replaced with cubic and cylindrical shapes (Fig. 6.3b), whereas monkeys did not. This difference may indicate the importance of visual experience in configural body processing for analyzing body-like stimuli. Humans have a much more experience with abstract and geometric representations of human bodies and can easily perceive them as "human"; this may be hard for monkeys to do if they lack relevant visual experience.

Other studies in humans and capuchin monkeys reported another species difference in the processing of human body postures that relates to the biomechanical constraints of the body. It is known that humans are sensitive to violations of the biomechanical constraints of the body, and our visual perception of others' body shapes is constrained by the anatomical limitations (Reed et al. 2003; Shiffrar and Freyd 1990). Matsuno and Fujita (2014) compared humans and capuchins in visual discrimination tasks and found a species difference: whereas humans exhibited BIE only for biomechanically possible body structures (Fig. 6.4), as reported in previous

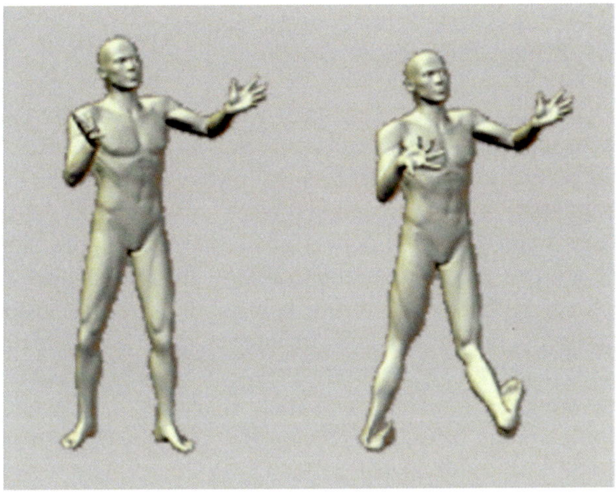

Fig. 6.4 Body stimuli with biomechanically possible and impossible postures

studies (e.g., Reed et al. 2003), monkeys expressed BIE for both possible and impossible body postures, suggesting that they configurally perceived bodies that humans did not. These results indicate that configural body processing and visual sensitivity to anatomical constraints might be independent processes in our visual system. Humans and captive capuchin monkeys share considerable visual exposure to human body shapes, but their own body experiences clearly differ. Therefore, configural body processing might be a species-general process shaped by daily visual exposure and relatively tolerant of relative spatial positions of body parts, not restricted by a rigid body template that includes joint kinematics. In contrast, sensitivity to violations of the biomechanical constraints might be species-specific, possibly involving some embodied processes and perception of conspecifics.

Chimpanzees, our closest evolutionary relatives, also show BIE. Gao and Tomonaga (2018, 2020) tested chimpanzees in a visual matching-to-sample task using photographic images of conspecific bodies and their silhouettes. They found that like capuchin monkeys, chimpanzees more accurately discriminated upright than inverted body stimuli. They also found that chimpanzees showed the BIE even when the relative proportions of the body parts were abnormal, but not when the positions of the parts were scrambled. They concluded that the configural body processing of chimpanzees relies on first-order spatial relations of body parts, as in humans (Reed et al. 2006). In another study, Kret and Tomonaga (2016) pointed out that chimpanzees might also configurally process conspecific body parts in a species-specific manner. They examined the effect of stimulus orientation in a visual matching-to-sample task using photos of faces, hands, and behinds in humans and chimpanzees. Chimpanzees showed inversion effects for visual patterns of females' swollen behinds but neither faces nor hands, whereas humans only showed face inversion effects. The swollen skin located at the rear end of the female chimpanzee's quadrupedal body is a kind of counterpart of the face when in front of the observer's face, and it shares a communicative function with the bare-skinned face.

Based on their results, Kret and Tomonaga (2016) proposed that visual attunement to the behind may be an evolutionary precursor of our modern configural face processing.

In sum, nonhuman primates with quadrupedal locomotion show humanlike BIE with some of the range of stimuli that produce the effect in humans. This means that the spatially vertical arrangement of the body parts is probably not a prerequisite of configural processing and that visual sensitivity to the spatial configuration of body parts may be shared across many species. Future studies that specify necessary conditions for configural processing in species with different bodily constraints and social behaviors should help us to better understand the nature of configural body processing. It is also of interest to know whether others show special visual attunement to certain body parts (as in chimpanzees and sexual swellings). For example, we humans are visually sensitive to the hands of self and others and the space around them, probably due to their functional significance as effectors (Niimi 2019; Ross and Flack 2019; Tseng et al. 2012). Investigations of the topic in nonhuman animals, extending to the other body parts such as toes and tails that have different functions across species, would help clarify relations between the visual body processing and body functions.

Perception of Body Movement

Autonomic and self-propelled movement is another significant feature that distinguishes animate from inanimate objects (Scholl and Tremoulet 2000). Even when visual information about body contour shape is unavailable, we can detect conspecifics or other animals and also adequately identify them or what they are doing, based on body motion cues. Psychophysical studies in humans have typically used the so-called point-light displays to elucidate the kinematic aspects of body perception (Johansson 1973). The point-light displays are degraded body stimuli that consist of a few dots located and moving typically at the joints of the body. Hundreds of studies have investigated the perceptual characteristics of point-light stimuli, from which we can easily perceive a human body as well as other information including walking directions, action categories, emotional states, intentions, person identities, age, and sex of the depicted individual (Troje 2013). Recent models propose that biological motion perception involves hierarchically structured multi-stage processing, including an early stage for detecting local motion signals derived from limb movements and later stages that integrate the local motion signals into global shape structure and movement and thus lead to recognition of the agent and action (Hirai and Senju 2020; Troje 2008). These models supposed that the earlier stage, at least, could be a biologically inherent mechanism.

A few behavioral studies have explored perception of point-light displays in nonhuman primates. However, there is no clear evidence that nonhuman primates interpret point-light or similarly impoverished biological motion stimuli (i.e., cylinders or stick figures) as equivalents of intact original body movements (Tomonaga 2001;

Parron et al. 2007; Vangeneugden et al. 2010). For example, Parron et al. (2007) trained baboons to discriminate point-light biological motion displays from scrambled stimuli and conducted transfer tests with new stimuli. Baboons failed to transfer their training performance to novel biological motion displays, but performed well with disarticulated stimuli that retained the local kinematic information of the trained stimuli, suggesting that they were insensitive to the body configuration. Vangeneugden et al. (2010) also reported that rhesus monkeys took more than 10,000 trials to learn a direction discrimination task with cylindrical "walker" stimuli and showed very limited transfer performance to new types of biological motion such as point-light and stick walkers. These results are inconsistent with some neurophysiological studies reporting that in macaque monkeys the passive observation of point-light displays evoked neural cortical responses in areas known to be involved in social vision (e.g., Oram and Perrett 1994).

Nonhuman primates' inability to interpret biological motion from motion signals could be due to the degraded and abstract nature of the displays. Interpretation of point-light displays involves integrating spatially and temporally local visual elements into articulated body configuration, but nonhuman primates are known for prioritizing local visual information and relatively poor spatiotemporal perceptual integration (Parron et al. 2007; Matsuno and Fujita 2009). In addition, point-light displays are abstract representations of real body figures, which may be harder for nonhuman animals to recognize (Tanaka 2007), as shown in the BIE study introduced above. Consequently, alternative methods are required for degrading body shape information, ones that place less cognitive load on nonhuman animals. One promising alternative could involve blurring of original video footage of bodies. The blurring technique allows us to examine visual processing of body motion while controlling the degree of the gradation of static shape information and avoiding unnatural fragmentation of the body into abstract local elements (Thornton 2005). In a similar vein, pixel-based partial occlusion of video images (Fujita 2006), which degrades the contour information in each frame and requires temporal integration of frames to recognize the original images, could be another valuable approach.

Individual Identity Recognition

Individual identification of conspecifics others is critical for social interactions in animals who construct stable social relationships. The face is the prominent media for that purpose in humans, but bodies also include identifying information. For instance, we can recognize acquaintances from behind, without seeing their faces. Several studies have shown that humans can recognize others based on headless body cues, although face cues tend to be more diagnostic (O'Toole et al. 2011; Simhi and Yovel 2016; Rice et al. 2013b; Robbins and Coltheart 2012). For example, Rice et al. (2013a) examined person identification accuracy using image sets in which the face was not informative enough for person identity discrimination. They found that observers used body features for person identification without awareness about relying on bodies rather than faces. Other studies have revealed that body-based person identification might at least partly depend on the same mechanisms as face-based identification. Ghuman et al. (2010) reported that visual adaptation to faceless body stimuli influenced subsequent face identity perception in the same manner as the face aftereffect, suggesting that the body stimulated cells that code facial identity. A developmental study also showed that the capacity for person identification based on bodies is slowly acquired during childhood, at the same pace as that for faces (Bank et al. 2015).

Comparative studies have provided experimental evidence that visual body features are also an effective information source for conspecific identification in nonhuman primates. In an early study, Dasser (1987) trained one long-tailed macaque on a simultaneous matching to sample task, where the monkey was required to discriminate face pictures of its group members, and then examined matching performance using pictures of the face and faceless body parts in a generalization test. The monkey's generalization scores in 25 test trials was above chance, suggesting that she identified the group members by their body features.

Matsuno and Fujita (2015) investigated whether capuchin monkeys identify familiar individuals' bodies, and if so, what are the diagnostic visual features. They used a task that has the same logical structure as the implicit association test (Nosek et al. 2007). The implicit association test assesses conceptual and categorical knowledge by measuring the amount of interference between stimulus-response associations. In experiments using face and body images of groupmates, capuchins were trained to learn a correct response position (left or right) for each picture. The correct positions for face and body pictures of the same monkey were set to be identical in the "congruent" condition and to be at the opposite side from each other in the "incongruent" condition. The monkeys showed better performance when required to learn congruent than incongruent stimulus-response mappings between the face and body of the same individuals. This tendency was observed even when body pictures without a head were used, but not when they were inverted or when only the upper or lower half of the body was presented. These results indicate that capuchin monkeys can recognize familiar individuals by means of configural body information.

There is also evidence that nonhuman primates use species-specific body features to recognize groupmates. De Waal and Pokorny (2008) reported that chimpanzees matched pictures of the behind of familiar, but not unfamiliar, conspecifics to face pictures of the same individual. These findings indicate that not only is the bare-skin area of the behind configurally processed (Kret and Tomonaga, 2016), but it also supports individual identity recognition in chimpanzees.

Together, the studies presented above provide evidence that nonhuman primates, like humans, can visually identify their companions even when they cannot see their faces. Further studies are awaited that will examine the potential roles of other diagnostic body features, typical postures, or habitual body movements in identity recognition, and to determine how such cues are integrated with other information such as faces, scenes, and social contexts. This line of research can improve our understanding of shared and unique visual strategies for individual identity recognition in species living in diverse physical and social environments and with different body constraints.

Perception of the Social Dyad

Regardless of whether the context in which other individuals are interacting is competitive or cooperative, it is a potentially rich information source for bystanders. From the scene observers can estimate both temporary and stable states of social relations, such as the strength of social bonds or competition, with sometimes critical implications for the observers' subsequent behavior selections. Unsurprisingly therefore, social interactions often capture our attention easily and are visually processed efficiently. Studies have shown that the human visual system is highly sensitive to socially interacting bodies and perceive them in a specific, highly integrative, manner that cannot be reduced to a simple, additive outcome of multiple processing of a body in isolation. Neri et al. (2006) compared the detection thresholds of point-light biological motion stimuli between conditions where two agents were interacting naturally and where movements of the two agents were unsynchronized. Although the biological motion components of each agent were identical between conditions, detection was worse when the natural temporal dynamics of the social interaction was disrupted. Subsequent studies have replicated and extended these findings, revealing the privileged access of such social interaction to our consciousness (Su et al. 2016) and the importance of communicative signals for predictive extraction of biological movement from the interactive scenes (Manera et al. 2011a, 2011b).

Recent studies have demonstrated privileged visual processing of social dyads even when we perceive others who are just standing face-to-face without any movement. Papeo and her colleagues (Papeo et al. 2017; Papeo and Abassi 2019; Papeo et al. 2019) found that facing body dyads are more efficiently detected and more integrally processed than nonfacing bodies in visual classification and visual search tasks. Vestner et al. (2019) further revealed that two bodies facing each other led to their faster visual detection, more accurate retrieval of visual features of the bodies in an accidental memory task, and spatially proximate or bound representation of the two bodies. In addition to the rapid detection of the spatial structure of the dyad, gist information about the nature of the interaction can also be recognized with a single fixation (Hafri et al. 2013; Hafri et al. 2018). These studies suggest that our visual system can efficiently parse the structure of a social relationship by recruiting more attention to the interacting dyad, perceptually grouping the agents as a unit, and starting to analyze the respective social roles. Such visual framing of social scenes may result in advantages on subsequent social cognitive tasks.

Perceiving socially interacting others and judging the observed social relations upon the observation is also widespread in the daily lives of group-living nonhuman primates (Fig. 6.5a). Recent studies have clarified some aspects of the visual processing of social dyads in nonhuman primates. For example, an eye-tracking

Fig. 6.5 (**a**) Social dyad of capuchin monkeys. (**b**) Depictions of stimulus displays used in the visual discrimination experiment

study compared the gaze patterns of rhesus macaques and humans as they observed photographs of natural scenes that contained affiliative or aggressive social interactions between members of their own species and between members of four other species (McFarland et al. 2013). The researchers analyzed observers' gaze allocations in relation to the social roles and body parts of the interacting agents. Although preferentially looked-at agent roles and body parts varied among image conditions, the pattern of variations showed both similarities and differences between macaque and human observers. When looking at pictures of affiliative acts both species focused on the head and face much more than other body parts, whereas negative social interactions also attracted gazing at the bodies. The latter tendency was more remarkable in monkeys than humans; monkeys observed bodies as long as the head with negative social interaction, while humans consistently viewed the head much longer than the other body parts.

Another comparative study showed similar visual sensitivity to social dyads in monkeys and humans (Matsuno and Kuroshima 2019). Capuchin monkeys received visual detection tasks in which they were required to visually discriminate the facing orientation of two upright or inverted body forms of monkeys or humans (Fig. 6.5b). The target stimuli consisted of two bodies that were facing each other or facing away from each other. Distractor stimuli were created by mirroring one of the bodies, so that the two bodies were oriented in the same direction. Like humans, the monkeys detected upright bodies that were facing each other more efficiently than the other types of stimuli. This privileged facing-dyad detection was also evident for human body stimuli with quadrupedal but not bipedal postures, suggesting species-specific influences on the process.

An fMRI study in monkeys has led to the proposal that the neural substrates for the perception and interpretation of social dyads could be a direct precursor of the theory of mind system in humans. Sliwa and Freiwald (2017) showed video images of conspecifics to rhesus macaques and mapped cortical areas activated during visual analyses of social interactions. They found that perceived social interactions selectively activated much broader areas across temporal, parietal, and prefrontal cortices than non-interacting bodies and nonsocial stimuli. In addition to shared activation in face and body areas and premotor areas known as the mirror neuron system with other stimuli, videos of social interactions exclusively activated specific regions including a cluster in the medial prefrontal cortex and anterior cingulate cortex. These areas overlap the counterparts of human cortical regions known as the network for the theory of mind (Gallagher and Frith 2003).

Summary and Conclusion

In this chapter, I introduced the comparative approach to visual body perception. Studies have revealed that visual attunement to individual body shapes and their interaction characterizes humans and nonhuman primates in general despite variations in bodily features including locomotory style, size, and anatomical constraints. Like humans, nonhuman primates configurally process bodies but not inanimate objects, recognize individuals based on body shapes without face, and are especially sensitive to social dyads. In parallel, however, some studies also suggest that visual body perceptual processes may be diverse, reflecting species-specific postures and body features. More empirical work is required before we can propose any definitive conclusions. A range of important and interesting questions are yet to be resolved. First, the basic perceptual properties of bodies as discussed above should be examined more in detail. Does the visual system of nonhuman primates become attuned to any specific body postures, shapes, or parts (other than the behind), or to the space around some body parts, as occurs in humans? Second, how do other species perceive the relations between others' bodies and nearby visual objects, other agents, and contextual environments? Studies addressing such topics will help clarify various species' perceptual basis of social cognitive abilities including perspective-taking, predicting others' goals and future actions, and inferring others' internal states. Third, the issue of how nonhuman primates relate representations of their own and others' bodies is directly relevant to some popular topics in comparative cognitive research, such as recognition of self and others. Comparative, quantitative assessments of various species' social vision capabilities and constraints are necessary for a comprehensive understanding of our social mind; such an approach appears preferable to any simple dichotomous view that focuses on the presence or absence of certain "higher-order" cognitive abilities in nonhuman animals.

Acknowledgments I would like to express my sincere gratitude to Dr. Kazuo Fujita for his kind supervision of my research projects including those I introduced in this chapter. His early studies on perception in nonhuman animals, always finely designed and rigorous, and his first text book on the comparative cognition have been the best guide for my studies in nonhuman animals from the start of my research career. I would also like to thank Drs. Hika Kuroshima, James R. Anderson, Makoto

Takahashi, Yuko Hattori, Yo Morimoto, Mrs. Ayumi Hirose, Drs. Ayaka Takimoto, Reiki Kishimoto, Akane Nagano, Mr. Maxime Bourgeois, Ms. Hyangsun Chin, and the laboratory members in Kyoto University for their support for my studies and their management of the health of the animals. The author was financially supported by the MEXT KAKENHI grant (No. 19 K12739).

References

Adams RB Jr, Ambady N, Nakayama K, Shimojo S (2011) The science of social vision. Oxford University Press, New York

Arizpe JM, McKean DL, Tsao JW, Chan AW (2017) Where you look matters for body perception: preferred gaze location contributes to the body inversion effect. PLoS One 12(1):e0169148. https://doi.org/10.1371/journal.pone.0169148

Aviezer H, Hassin RR, Ryan J, Grady C, Susskind J, Anderson A, Moscovitch M, Bentin S (2008) Angry, disgusted, or afraid? Studies on the malleability of emotion perception. Psychol Sci 19 (7):724–732. https://doi.org/10.1111/j.1467-9280.2008.02148.x

Aviezer H, Trope Y, Todorov A (2012) Body cues, not facial expressions, discriminate between intense positive and negative emotions. Science 338(6111):1225–1229. https://doi.org/10.1126/science.1224313

Axelsson EL, Robbins RA, Copeland HF, Covell HW (2019) Body inversion effects with photographic images of body oostures: is it about faces? Front Psychol 10:2686. https://doi.org/10.3389/fpsyg.2019.02686

Bank S, Rhodes G, Read A, Jeffery L (2015) Face and body recognition show similar improvement during childhood. J Exp Child Psychol 137:1–11. https://doi.org/10.1016/j.jecp.2015.02.011

Barra J, Senot P, Auclair L (2017) Internal model of gravity influences configural body processing. Cognition 158:208–214. https://doi.org/10.1016/j.cognition.2016.10.018

Biederman I (1987) Recognition-by-components: a theory of human image understanding. Psychol Rev 94:115–147. https://doi.org/10.1037/0033-295X.94.2.115

Bindemann M, Scheepers C, Ferguson HJ, Burton AM (2010) Face, body, and center of gravity mediate person detection in natural scenes. J Exp Psychol Hum Percept Perform 36 (6):1477–1485. https://doi.org/10.1037/a0019057

Blakemore S-J, Decety J (2001) From the perception of action to the understanding of intention. Nat Rev Neurosci 2(8):561–567. https://doi.org/10.1038/35086023

Bliss-Moreau E, Moadab G, Machado CJ (2017) Monkeys preferentially process body information while viewing affective displays. Emotion 17(5):765–771. https://doi.org/10.1037/emo0000292

Brown SD, Dooling RJ (1992) Perception of conspecific faces by budgerigars (*Melopsittacus undulatus*): I. natural faces. J Comp Psychol 106(3):203. https://doi.org/10.1037/0735-7036.106.3.203

Byrne RW, Whiten A (eds) (1988) Machiavellian intelligence: social expertise and the evolution of intellect in monkeys, apes, and humans. Oxford University Press, New York

Campbell FW, Robson JG (1968) Application of Fourier analysis to the visibility of gratings. J Physiol 197(3):551–566. https://doi.org/10.1113/jphysiol.1968.sp008574

Corballis MC (2012) How language evolved from manual gestures. Gesture 12(2):200–226. https://doi.org/10.1075/gest.12.2.04cor

Coulon M, Deputte BL, Heyman Y, Baudoin C (2009) Individual recognition in domestic cattle (*Bos taurus*): evidence from 2D-images of heads from different breeds. PLoS One 4(2):e4441. https://doi.org/10.1371/journal.pone.0004441

Dasser V (1987) Slides of group members as representations of the real animals (*Macaca fascicularis*). Ethology 76(1):65–73. https://doi.org/10.1111/j.1439-0310.1987.tb00672.x

Davidson EH, Erwin DH (2006) Gene regulatory networks and the evolution of animal body plans. Science 311(5762):796–800. https://doi.org/10.1126/science.1113832
De Gelder B (2009) Why bodies? Twelve reasons for including bodily expressions in affective neuroscience. Phil Trans R Soc B: Biol Sci 364(1535):3475–3484. https://doi.org/10.1098/rstb.2009.0190
de Waal F, Pokorny JJ (2008) Faces and behinds: Chimpanzee sex perception. Adv Sci Lett 1(1): 99–103. https://doi.org/10.1166/asl.2008.006
Dominy NJ, Lucas PW (2001) Ecological importance of trichromatic vision to primates. Nature 410:363–366. https://doi.org/10.1038/35066567
Downing PE, Bray D, Rogers J, Childs C (2004) Bodies capture attention when nothing is expected. Cognition 93(1):B27–B38. https://doi.org/10.1016/j.cognition.2003.10.010
Downing PE, Jiang YH, Shuman M, Kanwisher N (2001) A cortical area selective for visual processing of the human body. Science 293(5539):2470–2473. https://doi.org/10.1126/science.1063414
Dunbar RI (1998) The social brain hypothesis. Evol Anthropol 6(5):178–190
Fadiga L, Craighero L (2006) Hand actions and speech representation in Broca's area. Cortex 42 (4):486–490. https://doi.org/10.1016/s0010-9452(08)70383-6
Fay N, Arbib M, Garrod S (2013) How to bootstrap a human communication system. Cogn Sci 37 (7):1356–1367. https://doi.org/10.1111/cogs.12048
Fujita K (2006) Seeing what is not there: illusion, completion, and spatio-temporal boundary formation in comparative perspective. In: Wasserman EA, Zentall TR (eds) Comparative cognition: experimental explorations of animal intelligence. Oxford University Press, New York, pp 29–52
Gallagher HL, Frith CD (2003) Functional imaging of 'theory of mind'. Trends Cogn Sci 7 (2):77–83. https://doi.org/10.1016/s1364-6613(02)00025-6
Gallese V, Sinigaglia C (2011) What is so special about embodied simulation? Trends Cogn Sci 15 (11):512–519. https://doi.org/10.1016/j.tics.2011.09.003
Gao J, Tomonaga M (2018) The body inversion effect in chimpanzees *(Pan troglodytes*). PLoS One 13(10):e0204131. https://doi.org/10.1371/journal.pone.0204131
Gao J, Tomonaga M (2020) Body perception in chimpanzees (*Pan troglodytes*): the effect of body structure changes. J Comp Psychol 134(2):222–231. https://doi.org/10.1037/com0000214
Gauthier I, McGugin RW, Richler JJ, Herzmann G, Speegle M, Van Gulick AE (2014) Experience moderates overlap between object and face recognition, suggesting a common ability. J Vis 14 (8):7. https://doi.org/10.1167/14.8.7
Ghuman AS, McDaniel JR, Martin A (2010) Face adaptation without a face. Curr Biol 20(1):32–36. https://doi.org/10.1016/j.cub.2009.10.077
Grosjean M, Shiffrar M, Knoblich G (2007) Fitts's law holds for action perception. Psychol Sci 18 (2):95–99. https://doi.org/10.1111/j.1467-9280.2007.01854.x
Hafri A, Papafragou A, Trueswell JC (2013) Getting the gist of events: recognition of two-participant actions from brief displays. J Exp Psychol Gen 142(3):880–905. https://doi.org/10.1037/a0030045
Hafri A, Trueswell JC, Strickland B (2018) Encoding of event roles from visual scenes is rapid, spontaneous, and interacts with higher-level visual processing. Cognition 175:36–52. https://doi.org/10.1016/j.cognition.2018.02.011
Heyes CM, Frith CD (2014) The cultural evolution of mind reading. Science 344(6190):1243091. https://doi.org/10.1126/science.1243091
Hirai M, Senju A (2020) The two-process theory of biological motion processing. Neurosci Biobehav Rev 111:114–124. https://doi.org/10.1016/j.neubiorev.2020.01.010
Iacoboni M (2009) Imitation, empathy, and mirror neurons. Annu Rev Psychol 60:653–670. https://doi.org/10.1146/annurev.psych.60.110707.163604
Johansson G (1973) Visual perception of biological motion and a model for its analysis. Percept Psychophys 14(2):201–211. https://doi.org/10.3758/BF03212378
Kanwisher N, McDermott J, Chun MM (1997) The fusiform face area: a module in human extrastriate cortex specialized for face perception. J Neurosci 17(11):4302–4311. https://doi.org/10.1523/JNEUROSCI.17-11-04302.1997

Kawaguchi Y, Kuroshima H, Fujita K (2019) Age categorization of conspecific and heterospecific faces in capuchin monkeys (*Sapajus apella*). J Comp Psychol 133(4):1939–2087. https://doi.org/10.1037/com0000185

Kendrick KM, da Costa AP, Leigh AE, Hinton MR, Peirce JW (2001) Sheep don't forget a face. Nature 414(6860):165–166. https://doi.org/10.1038/35102669

Kiani R, Esteky H, Mirpour K, Tanaka K (2007) Object category structure in response patterns of neuronal population in monkey inferior temporal cortex. J Neurophysiol 97(6):4296–4309. https://doi.org/10.1152/jn.00024.2007

Kilner JM, Paulignan Y, Blakemore SJ (2003) An interference effect of observed biological movement on action. Curr Biol 13(6):522–525. https://doi.org/10.1016/S0960-9822(03)00165-9

Knoblich G, Thornton IM, Grosjean M, Shiffrar M (eds) (2006) Human body perception from the inside out. Oxford University Press, New York

Koba R, Izumi A, Nakamura K (2009) Sexual dimorphism in facial shapes and their discrimination in Japanese monkeys (*Macaca fuscata*). J Comp Psychol 123(3):326–333. https://doi.org/10.1037/a0016219

Kret ME, Tomonaga M (2016) Getting to the bottom of face processing. Species-specific inversion effects for faces and behinds in humans and chimpanzees (*Pan troglodytes*). PLoS One 11(11): e0165357. https://doi.org/10.1371/journal.pone.0165357

Manera V, Becchio C, Schouten B, Bara BG, Verfaillie K (2011a) Communicative interactions improve visual detection of biological motion. PLoS One 6(1):e14594. https://doi.org/10.1371/journal.pone.0014594

Manera V, Del Giudice M, Bara BG, Verfaillie K, Becchio C (2011b) The second-agent effect: communicative gestures increase the likelihood of perceiving a second agent. PLoS One 6(7): e22650. https://doi.org/10.1371/journal.pone.0022650

Marr D (1982) Vision: a computational investigation into the human representation and processing of visual information. Henry Holt, New York

Marr D, Nishihara HK (1978) Representation and recognition of the spatial organization of three-dimensional shapes. Proc R Soc Lond B Biol Sci 200(1140):269–294. https://doi.org/10.1098/rspb.1978.0020

Matsuno T, Fujita K (2009) A comparative psychophysical approach to visual perception in primates. Primates 50(2):121–130. https://doi.org/10.1007/s10329-008-0128-8

Matsuno T, Fujita K (2014) Visual discrimination of biomechanically possible and impossible postures in tufted capuchin monkeys and humans. Paper presented at the 33th annual meeting of the Japanese Psychonomic society, Tokyo Metropolitan University, 6–7 December, 2014

Matsuno T, Fujita K (2015) Body identity recognition in capuchin monkeys (*Cebus apella*): a study using an association test. Paper presented at the 75th annual meeting of the Japanese Society of Animal Psychology, Japan Women's university, 10–12 September, 2015

Matsuno T, Fujita K (2018) Body inversion effect in monkeys. PLoS One 13(10):e0204353. https://doi.org/10.1371/journal.pone.0204353

Matsuno T, Kuroshima H (2019) Dyad inversion effect in capuchin monkeys (*Sapajus apella*) and humans (*Homo sapiens*). Paper presented at the 79th annual meeting of the Japanese Society of Animal Psychology, Senshu university, 11–13 October, 2019

McFarland R, Roebuck H, Yan Y, Majolo B, Li W, Guo K (2013) Social interactions through the eyes of macaques and humans. PLoS One 8(2):e56437. https://doi.org/10.1371/journal.pone.0056437

Neri P, Luu JY, Levi DM (2006) Meaningful interactions can enhance visual discrimination of human agents. Nat Neurosci 9(9):1186–1192. https://doi.org/10.1038/nn1759

New J, Cosmides L, Tooby J (2007) Category-specific attention for animals reflects ancestral priorities, not expertise. Proc Natl Acad Sci U S A 104(42):16598–16603. https://doi.org/10.1073/pnas.0703913104

Niimi R (2019) Interacting hands draw attention during scene observation. Atten Percept Psychophys 82:1088–1098. https://doi.org/10.3758/s13414-019-01881-7

Nityananda V, Read JC (2017) Stereopsis in animals: evolution, function and mechanisms. J Exp Biol 220(14):2502–2512. https://doi.org/10.1242/jeb.143883
Nosek BA, Greenwald AG, Banaji MR (2007) The implicit association test at age 7: a methodological and conceptual review. In: Bargh JA (ed) Automatic processes in social thinking and behavior. Psychology Press, New York, pp 265–292
Oram M, Perrett D (1994) Responses of anterior superior temporal polysensory (STPa) neurons to "biological motion" stimuli. J Cogn Neurosci 6(2):99–116. https://doi.org/10.1162/jocn.1994.6.2.99
Oruc I, Balas B, Landy MS (2019) Face perception: a brief journey through recent discoveries and current directions. Vis Res 157:1–9. https://doi.org/10.1016/j.visres.2019.06.005
O'Toole AJ, Phillips PJ, Weimer S, Roark DA, Ayyad J, Barwick R, Dunlop J (2011) Recognizing people from dynamic and static faces and bodies: dissecting identity with a fusion approach. Vis Res 51(1):74–83. https://doi.org/10.1016/j.visres.2010.09.035
Papeo L, Abassi E (2019) Seeing social events: the visual specialization for dyadic human–human interactions. J Exp Psychol Hum Percept Perform 45(7):877–888
Papeo L, Goupil N, Soto-Faraco S (2019) Visual search for people among people. Psychol Sci 30 (10):1483–1496. https://doi.org/10.1037/xhp0000646
Papeo L, Stein T, Soto-Faraco S (2017) The two-body inversion effect. Psychol Sci 28(3):369–379. https://doi.org/10.1177/0956797616685769
Parr LA (2001) Cognitive and physiological markers of emotional awareness in chimpanzees (*Pan troglodytes*). Anim Cogn 4:223–229. https://doi.org/10.1007/s100710100085
Parr LA (2011) The evolution of face processing in primates. Phil Trans R Soc Lond Ser B Biol Sci 366(1571):1764–1777. https://doi.org/10.1098/rstb.2010.0358
Parr LA, Waller BM, Burrows AM, Gothard KM, Vick S-J (2010) MaqFACS: a muscle-based facial movement coding system for the rhesus macaque. Am J Phys Anthropol 143(4):625–630. https://doi.org/10.1002/ajpa.21401
Parron C, Deruelle C, Fagot J (2007) Processing of biological motion point-light displays by baboons (*Papio papio*). J Exp Psychol Anim Behav Process 33(4):381–391. https://doi.org/10.1037/0097-7403.33.4.381
Peelen MV, Downing PE (2007) The neural basis of visual body perception. Nat Rev Neurosci 8 (8):636–648. https://doi.org/10.1038/nrn2195
Pollick AS, De Waal FB (2007) Ape gestures and language evolution. Proc Natl Acad Sci USA 104(19):8184–8189. https://doi.org/10.1073/pnas.0702624104
Powell LJ, Kosakowski HL, Saxe R (2018) Social origins of cortical face areas. Trends Cogn Sci 22 (9):752–763. https://doi.org/10.1016/j.tics.2018.06.009
Premack D (1988) 'Does the chimpanzee have a theory of mind' revisited. In: Byrne RW, Whiten A (eds) Machiavellian intelligence: social expertise and the evolution of intellect in monkeys, apes, and humans. Clarendon Press/Oxford University Press, New York, pp 160–179
Reed CL, Stone VE, Bozova S, Tanaka J (2003) The body-inversion effect. Psychol Sci 14 (4):302–308. https://doi.org/10.1111/1467-9280.14431
Reed CL, Stone VE, Grubb JD, McGoldrick JE (2006) Turning configural processing upside down: part and whole body postures. J Exp Psychol Hum Percept Perform 32(1):73–87. https://doi.org/10.1037/0096-1523.32.1.73
Rice A, Phillips PJ, Natu V, An X, O'Toole AJ (2013a) Unaware person recognition from the body when face identification fails. Psychol Sci 24(11):2235–2243. https://doi.org/10.1177/0956797613492986
Rice A, Phillips PJ, O'Toole A (2013b) The role of the face and body in unfamiliar person identification. Appl Cogn Psychol 27(6):761–768. https://doi.org/10.1002/acp.2969
Robbins RA, Coltheart M (2012) The effects of inversion and familiarity on face versus body cues to person recognition. J Exp Psychol Hum Percept Perform 38(5):1098–1104. https://doi.org/10.1037/a0028584

Ross P, Flack T (2019) Removing hand form information specifically impairs emotion recognition for fearful and angry body stimuli. Perception 49(1):98–112. https://doi.org/10.1177/0301006619893229
Rossion B, Taubert J (2019) What can we learn about human individual face recognition from experimental studies in monkeys? Vis Res 157:142–158. https://doi.org/10.1016/j.visres.2018.03.012
Samson D, Apperly IA, Braithwaite JJ, Andrews BJ, Bodley Scott SE (2010) Seeing it their way: evidence for rapid and involuntary computation of what other people see. J Exp Psychol Hum Percept Perform 36(5):1255–1266. https://doi.org/10.1037/a0018729
Scholl BJ, Tremoulet PD (2000) Perceptual causality and animacy. Trends Cogn Sci 4:299–309. https://doi.org/10.1016/S1364-6613(00)01506-0
Shiffrar M, Freyd JJ (1990) Apparent motion of the human body. Psychol Sci 1(4):257–264. https://doi.org/10.1111/j.1467-9280.1990.tb00210.x
Simhi N, Yovel G (2016) The contribution of the body and motion to whole person recognition. Vis Res 122:12–20. https://doi.org/10.1016/j.visres.2016.02.003
Sliwa J, Freiwald WA (2017) A dedicated network for social interaction processing in the primate brain. Science 356(6339):745–749. https://doi.org/10.1126/science.aam6383
Stein T, Sterzer P, Peelen MV (2012) Privileged detection of conspecifics: evidence from inversion effects during continuous flash suppression. Cognition 125(1):64–79. https://doi.org/10.1016/j.cognition.2012.06.005
Su J, Van Boxtel JJ, Lu H (2016) Social interactions receive priority to conscious perception. PLoS One 11(8):e0160468. https://doi.org/10.1371/journal.pone.0160468
Tanaka M (2007) Recognition of pictorial representations by chimpanzees (*Pan troglodytes*). Anim Cogn 10(2):169–179. https://doi.org/10.1007/s10071-006-0056-1
Tao W, Zeng W, Sun H (2014) Behavioral and electrophysiological measures of the body inversion effect: the contribution of the limb configurations. Neuroreport 25(14):1099–1108. https://doi.org/10.1097/WNR.0000000000000234
Thornton I (2005) Biological motion: point-light walkers and beyond. In: Knoblich G, Thornton I, Grosjean M, Shiffrar M (eds) Human body perception from the inside out. Oxford University Press, New York, pp 271–303
Tibbetts EA (2002) Visual signals of individual identity in the wasp *Polistes fuscatus*. Proc R Soc Lond B Biol Sci 269(1499):1423–1428. https://doi.org/10.1098/rspb.2002.2031
Tomasello M, Zuberbühler K (2002) Primate vocal and gestural communication. In: Bekoff M, Allen C, Burghardt GM (eds) The cognitive animal: empirical and theoretical perspectives on animal cognition. MIT Press, Cambridge, UK, pp 293–229
Tomonaga M (2001) Visual search for biological motion patterns in chimpanzees (*Pan troglodytes*). Psychologia 44:46–59
Tomonaga M (2007) Is chimpanzee (*Pan troglodytes*) spatial attention reflexively triggered by gaze cue? J Comp Psychol 121(2):156–170. https://doi.org/10.1037/0735-7036.121.2.156
Troje NF (2008) Biological motion perception. In: Basbaum AI, Kaneko A, Shepherd GG, Westheimer G (eds) The senses: a comprehensive reference, vol 2. Elsevier, Boston, MA, pp 231–238
Troje NF (2013) What is biological motion? Definition, stimuli and paradigms. In: Rutherford MD, Kuhlmeier VA (eds) Social perception: detection and interpretation of animacy, agency, and intention. The MIT Press, Cambridge, MA, pp 13–36
Tsao DY, Livingstone MS (2008) Mechanisms of face perception. Annu Rev Neurosci 31:411–437. https://doi.org/10.1146/annurev.neuro.30.051606.094238
Tseng P, Bridgeman B, Juan C-H (2012) Take the matter into your own hands: a brief review of the effect of nearby-hands on visual processing. Vis Res 72:74–77. https://doi.org/10.1016/j.visres.2012.09.005
Van der Velden J, Zheng Y, Patullo BW, Macmillan DL (2008) Crayfish recognize the faces of fight opponents. PLoS One 3(2):e1695. https://doi.org/10.1371/journal.pone.0001695

Vangeneugden J, Vancleef K, Jaeggli T, Van Gool L, Vogels R (2010) Discrimination of locomotion direction in impoverished displays of walkers by macaque monkeys. J Vis 10(4):22–22. https://doi.org/10.1167/10.4.22

Vestner T, Tipper SP, Hartley T, Over H, Rueschemeyer S-A (2019) Bound together: social binding leads to faster processing, spatial distortion, and enhanced memory of interacting partners. J Exp Psychol Gen 148(7):1251–1268. https://doi.org/10.1037/xge0000545

Vick S-J, Waller BM, Parr LA, Pasqualini MCS, Bard KA (2007) A cross-species comparison of facial morphology and movement in humans and chimpanzees using the facial action coding system (FACS). J Nonverbal Behav 31(1):1–20. https://doi.org/10.1007/s10919-006-0017-z

Ward E, Ganis G, Bach P (2019) Spontaneous vicarious perception of the content of another's visual perspective. Curr Biol 29(5):874–880. https://doi.org/10.1016/j.cub.2019.01.046

Wimmer H, Perner J (1983) Representation and constraining function of wrong beliefs in young children's understanding of deception. Cognition 13:103–128. https://doi.org/10.1016/0010-0277(83)90004-5

Zuberbühler K (2003) Referential signaling in non-human primates: cognitive precursors and limitations for the evolution of language. Adv Stud Behav 33:265–307. https://doi.org/10.1016/S0065-3454(03)33006-2

Part III
Social Cognition

Chapter 7
Attending to Others' Visual Attention

James R. Anderson and Yuko Hattori

Abstract This chapter selectively reviews research on nonhuman primates' responses to cues that indicate the focus of other individuals' attention, in particular, visual attention. The chapter begins with an overview of primates' abilities to use various gestural and facial cues in object-choice tasks, in which the aim is to select one of two or more baited containers based on the cueing behavior of an informant, who might be cooperative or competitive, conspecific or human. This is followed by coverage of a range of species' tendencies to visually co-orient (or gaze follow), and factors that influence this response. Throughout the chapter, consideration is given to species differences in sensitivity to, and ability to make inferences based on, other individuals' visual behavior. The chapter closes with a discussion and examples of "checking back" as a possible indicator of greater perspective-taking skills in primates, and a brief look at complementary approaches that have contributed to our current understanding of other species' understanding of attentional cues.

Keywords Object choice · Gaze following · Visual co-orientation · Eyes · Joint attention · Checking back

Introduction

In this chapter we summarize our studies in one area of social cognition—responses to attentional cues from other individuals—in nonhuman primates (hereafter: primates). For visually dominant species that live in social groups, the focus of other individuals' attention is important. Conspecifics frequently look at each other as well

J. R. Anderson (✉)
Department of Psychology, Kyoto University Graduate School of Letters, Kyoto, Japan
e-mail: j.r.anderson@psy.bun.kyoto-u.ac.jp

Y. Hattori
Primate Research Institute, Kyoto University, Inuyama, Japan
e-mail: hattori.yuko.3w@kyoto-u.ac.jp

J. R. Anderson, H. Kuroshima (eds.), *Comparative Cognition*,
https://doi.org/10.1007/978-981-16-2028-7_7

as at their surroundings, and this monitoring of other individuals' focus and changes of attention is a valuable behavior. Gaze following—or visual co-orientation—can facilitate detection of potentially important information, for example, about what others are doing, or what is happening elsewhere in the environment, including possible sources of danger. Given the very large number of studies about following other individuals' gaze, we make no attempt to present them all in detail here. Other reviews adequately cover multiple inter-related aspects of sensitivity and responses to gaze; they also discuss research on more species (including non-primates) and procedures, and include developmental, evolutionary, sensory and cognitive perspectives (e.g., Emery 2000; Itakura 2004; Anderson and Vick 2008; Rosati and Hare 2009; Shepherd 2010; Davidson et al. 2014). Topics addressed in our own studies that are presented here include, among others: use of human attentional cues by primates in object-choice tasks, visual co-orientation or gaze following, understanding of gaze as a predictor to action, and how other cues might influence object-choice performance and gaze following.

The Object-Choice Task

Around 30 years ago, one of us (JRA) was starting to study the effects of aging on cognition in monkeys. To run experiments on discrimination and reversal learning, delayed response, matching-to-sample, etc., we considered constructing a version of the venerable Wisconsin General Test Apparatus (WGTA), first developed by Harlow and Bromer (1938). One feature of the original WGTA is a window or screen that allows the experimenter to unobtrusively watch the subject during the choice phase of the task, eliminating the possibility of inadvertent behavioral cues from the experimenter. This concern arose in part from the case of Clever Hans, the horse that "counted" and "answered questions." Experiments eventually revealed that Hans' impressive performances were based not on arithmetical skills or general knowledge, but instead on the behaviors (often subtle) of people who were present as he "solved" the problems. Slight shifts in posture, facial expression or gaze direction were cues for Hans to stop tapping the ground with his hoof (which is how he "answered") (Pfungst et al. 1999).

In the end, the apparatus we made for our studies of learning and memory was much simpler than the typical WGTA (it more closely resembled what is sometimes called a Klüver tray or formboard; see Kluver 1937). It consisted of a wheelable table with a tray containing three shallow, circular depressions ("wells") in which food rewards could be placed and then covered with objects (various small junk objects) (Anderson et al. 1996a). An opaque screen could be lowered by the experimenter to hide the baiting phase of the trial and then raised for the choice phase when the subject reached through the bars of the test cage to touch or displace one of the objects. As a precaution against inadvertent cuing, during the choice phase the experimenter always remained still, hands by her sides, and visually fixating the central well (which was never baited or covered). We were confident that this experimenter neutrality made such cuing unlikely, and in fact we had shown this in studies of monkeys' use of explicit experimenter-given cues, as described below.

Investigating Use of Experimenter-Given Cues

Despite psychologists' awareness of Clever Hans effects, until the mid-1990s there was a dearth of systematic research on animals' use of "experimenter-given cues" in learning tasks. Our first study to address this issue involved an experimenter presenting various behavioral cues to individually tested adult capuchin monkeys (*Sapajus apella*) in an attempt to help them as they performed an object-choice task (Anderson et al. 1995). After a piece of food had been placed under one of two junk objects behind the opaque screen and so out of view of the monkey, the screen was raised to reveal the tray with the two objects and the experimenter (a familiar human) standing behind the tray, facing the monkey. In Experiment 1, the experimenter stood with hands by his side, staring only at the midpoint between the two objects (control condition) or with his head and eyes clearly oriented toward the baited object (cue condition), and he then pushed the tray forward so the monkey could touch the baited or the non-baited object (leading to reward or no reward). To our surprise, even after more than 500 control and cue trials in alternation, the monkeys continued to choose at around chance level, with no evidence that they made use of the experimenter's gaze cue to select the correct object.

Experiment 2 used the same two conditions but with a visible baiting procedure and then a short delay, making it a kind of delayed response task. Although the monkeys' overall performance was better than in Experiment 1, again there was no difference between the cue and control conditions: no advantage of the gaze cue. In Experiment 3, the experimenter combined gaze with pointing: his arm and index finger were extended to within about 15 cm from the baited object. This combined gaze + point cue led to significantly higher correct scores in the cue condition for all three monkeys, whereas their choices in the control condition remained at chance. Next, to see whether the monkeys would now use the gaze cue alone, the procedure in Experiment 4 replicated that of Experiment 2. So did the outcome: the monkeys' performances fell to chance level again, indicating no effective use of gaze without pointing. Indeed, when pointing was used as the sole cue while the experimenter's head and eyes remained fixed as in the control condition (Experiment 5), the monkeys again chose the correct object significantly above chance, indicating that they used the experimenter's hand but not his gaze direction when choosing. In summary, this first study showed that capuchin monkeys were unexpectedly poor at using the gaze direction of a human "helper" as a cue to find a hidden reward.

In a follow-up study, we presented three adult female rhesus monkeys (*Macaca mulatta*) with the same control and cue conditions as in the capuchin study, along with two new cue conditions: point + move, in which the pointing index finger moved repeatedly up and down, and point + tap, in which the index finger tapped on the table just behind the baited object (Anderson 1996; Anderson et al. 1996b). The monkeys showed marked individual differences in their overall use of the pointing cues, but like in the first study, they all failed to use the experimenter's gaze direction as a cue. When the experimenter presented another gaze cue in which his head and eyes were much closer to the baited object (around 15 cm) than before (around 1 m), one monkey showed significantly better than baseline performance, and another showed a similar though non-significant trend. However, when only the experimenter's eyes were oriented toward the correct object (with head facing forward), no monkey scored above chance. This study thus provided further evidence that monkeys do not readily follow a human's line of gaze to select the correct object for reward.

Monkeys are not incapable of exploiting gaze cues; with experience they can learn to do so. Shoji Itakura first showed this by training a young capuchin monkey to use a head + eyes gaze cue starting at a reduced distance from the target object of around 15 cm, before increasing it to around 60 cm (Itakura and Anderson 1996). However, when the cue consisted of an eyes-only movement (head oriented toward the monkey, eyes-only shift toward the target), the trained capuchin's performance fell to baseline. Relatively poor use of experimenter-given gaze cues, particularly eyes-only gaze, by chimpanzees on object-choice tasks was also reported by Call et al. (1998) and by Barth et al. (2005). Young chimpanzees started to use close-pointing cues before gaze cues; the latter remained a difficult task for the three subjects (Okamoto-Barth et al. 2008), although one of them was earlier reported to use an eyes-only glance cue (Okamoto et al. 2002). Itakura and Tanaka (1998) did report that chimpanzees, an orangutan, and human infants were more apt than monkeys to use an eyes-only glance cue to choose the correct object. In a study with capuchins (Vick and Anderson 2000), two of three monkeys did eventually learn to use eye gaze alone as a cue, but when the experimenter's eye direction conflicted with her head direction the monkeys prioritized the latter cue. Furthermore, their use of gaze cues deteriorated when the original experimenter was replaced by a new one.

The early object-choice studies mentioned above have been followed by dozens of others in a wide range of species including other primates (see, e.g., Vick et al. 2001; Byrnit 2004, 2008; Peignot and Anderson 1999; Burkart and Heschl 2006; Tan et al. 2014; Caspar et al. 2018), birds, and various domesticated and non-domesticated mammals. Among the various factors investigated in these studies are the subjects' biological, social, and psychological characteristics, the setting, experience of interacting with humans, the model (e.g., human or conspecific), and various parameters of the task itself. Mulcahy and Hedge (2012) discuss procedural aspects of gaze and object-choice studies in primates.

In the first study from Fujita's laboratory to focus on monkeys' response to human gaze cues, Kuroshima et al. (2002) tested capuchins with a "knower-guesser" test, adapting procedures used by Povinelli et al. (1990, 1991) to study chimpanzees' and rhesus monkeys' abilities to understand that seeing leads to knowing. In those latter studies, the apes but not the monkeys learned to preferentially choose an object indicated by someone who had seen a critical event (baiting of the object) vs. someone who merely guessed. In their first experiment, Kuroshima et al. (2002) showed that all four capuchin monkeys eventually learned to choose the one of three containers proposed (by placing a hand on it) by the knower instead of one proposed by a guesser. Just before this choice phase of each trial the knower had slightly tilted each container and looked under it (and so saw the food hidden under one of them), whereas the guesser did not manipulate any container. As a control, in the second experiment both knower and guesser touched the containers, but only the knower looked under them. In the final experiment, the order in which knower and guesser arrived at the test area was randomized. The authors cautiously concluded that overall, their results were consistent with capuchin monkeys being able to understand that seeing leads to knowing.

"Cooperative" vs. "Competitive" Contexts

One important development in experimenter-given cues/object-choice research was the realization that better results might be obtained by changing from a "cooperative" human model to a "competitive" one. Hare (2001) pointed out that in monkeys and apes gaze is not generally used to communicate the presence of hidden food to others. Instead, gaze is more likely to signal possession, and other individuals are more likely to avoid conflict by approaching or reaching for food that another individual is looking at. Stronger evidence for perspective-taking abilities in chimpanzees compared to monkeys has been found in experimental feeding competition studies. In these experiments, subjects are allowed to monitor whether a conspecific might be able to see a desirable food item before the two individuals are allowed access to it (Hare et al. 2000, 2001, 2003; Canteloup et al. 2017). Hare and Tomasello (2004) reported that captive chimpanzees performed better to pointing and reaching gestures from a competitive human or chimpanzee than from cooperative partners. The first study to focus on primates' use of human gaze cues in a competitive context was done with Guinea baboons (*Papio papio*) (Vick and Anderson 2003). In that study, if the subjects reached for the item the experimenter was looking at, instead of giving that item as a reward the experimenter rapidly withdrew the tray, picked up the item and pretended to eat it. In this situation baboons quickly learned to take the food item that the experimenter was *not* looking at. Again, head + eyes was found to be a more salient cue than eye gaze on its own. In another competition-based study, rhesus monkeys preferentially "stole" food from a human who could see the food that was between him and the monkey than from a human who could not see the food (Flombaum and Santos 2005).

Competitive or "contest" situations were subsequently used to test lemurs. In one study, ring-tailed lemurs (*Lemur catta*) (but not three other lemur species) were more likely to approach food that was "uncontested" (a nearby human paid no attention to the food) than "contested" food (watched by a nearby human). Furthermore, only the ring-tailed lemurs showed a tendency to look upward when a human did so. The authors interpreted the ring-tails' use of gaze cues in terms of their relatively larger and more complex social groups (Sandel et al. 2011; see also MacLean et al. 2013).

Further evidence that ring-tailed lemurs were sensitive to attention direction of a human competitor in a food-getting situation was presented by Bray et al. (2014). However, some ring-tailed and brown lemurs (*Lemur fulvus*) with extensive experience of being fed by hand actually preferentially approached food that was just in front of a face-on human (Botting et al. 2011).

Visual Co-orientation with Humans

The overall poor abilities of primates to use experimenter-given gaze cues in cooperative object-choice tasks, as described above, appeared to conflict with multiple reports of spontaneous gaze shifting to follow others' direction of gaze. Such reports blossomed after Itakura (1996) described great apes' tendency to visually re-orient to follow a human's changing direction of attention (head + eyes cue). In that study, some monkeys followed gaze if it was accompanied by a pointing gesture. Anderson and Mitchell (1999) reported visual co-orientation in stumptailed macaques (*Macaca arctoides*): all four adults tested reliably turned to look in the same direction as a human (combined head + eye shift), but a juvenile did not

Fig. 7.1 Visual co-orientation with a human by an adult stumptailed macaque, illustrated in four frames from an original video. (Top left) Monkey and human are sitting calmly. (Top right) Human changes gaze direction; monkey notices. (Bottom right) Monkey turns to look in the same direction as the human. (Bottom left) Human's gaze returns to the monkey; monkey looks back at the human

(Fig. 7.1). In the same study, none of four adult black lemurs (*Eulemur macaco*) showed reliable evidence of visual co-orientation with the human model. However, Sandel et al. (2011) have since presented positive evidence of visual co-orientation with a human in ring-tailed lemurs (*Lemur catta*).

Two- to six-year-old pigtailed macaques (*Macaca nemestrina*) correctly visually re-oriented in response to a head + eyes shift by a human, but only fully adult macaques did so when the change of attention was signaled by eye gaze alone (Ferrari et al. 2000). Young rhesus monkeys and chimpanzees were reported to co-orient with a human who switched gaze direction (with head + eyes), and they were also less likely than adults to stop following repeated, uninformative attentional shifts (Tomasello et al. 2001). A later study of a large sample of semi-free-ranging rhesus macaques confirmed that following human gaze started in infancy, peaked in the juvenile period, and then declined in adulthood (Rosati et al. 2016). Unlike older adult rhesus monkeys, however, older Barbary macaques showed no decline in the tendency to follow a human's shift in gaze direction (Rosati and Santos 2017), a difference that the authors related to a species difference in social tolerance. Rhesus monkeys were also more likely to look up when a human looked up to the sky than if the human looked into an overhead box. Further experimentation supported the authors' view that monkeys might not be strongly motivated to gaze follow in noncompetitive contexts (Bettle and Rosati 2019). In the only study to date of a colobine monkey species, zoo-housed golden snub-nosed monkeys visually co-oriented with a human, but they did not use gaze as a cue to solve an object-choice task (Tan et al. 2014).

Earlier studies of laboratory chimpanzees described visual co-orientation when a human appeared "distracted," even when the human's gaze shift involved only the eyes (Povinelli and Eddy 1996a, b, 1997). In a comparative study, Bräuer et al. (2005) reported similarities in following a human's gaze in all four species of great apes. In a study comparing visual co-orientation by capuchin and squirrel monkeys, (Anderson et al. 2005) included attention shifts not only by one human model, but simultaneous or near-simultaneous attention shifts by two models; for example, two people might switch direction of attention simultaneously or one shortly after the other. There was little overall effect of such different combinations on gaze following by the monkeys, but capuchin monkeys responded more than squirrel monkeys, and both species showed a stronger tendency than reported in Itakura (1996). Comparative studies of social gaze generally align with Fujita's (2001) view that species differences in visual perception are to be expected based on ecology and evolution.

Inferring Intentions and Attention from Gaze Cues

Human children do not simply follow other individuals' gaze, they also use it to predict what someone's desire or aim is. For example, by around school age, children identify the object that somebody looks at for longer as the one that the

Fig. 7.2 Illustration of "hand situation" (left) and "table situation" (right) in experiment described in Hattori et al. (2010a, b)

person "wants to have" (Montgomery et al. 1998). Three- to four-year-old children even read eye direction in a simple cartoon face as a cue signaling which of four possible objects the cartoon character desires (Baron-Cohen et al. 1995).

Do other primate species predict that an object looked at is more likely to be picked up by the looker than an object that is not looked at? In one study, captive cotton-top tamarins (*Saguinus oedipus*, a small New World monkey) saw a human looking at and then manipulating one of two objects. After three such demonstrations a test trial was run. When the test trial consisted of the human looking at the object again but then manipulating the other (i.e., previously visually ignored) object, the monkeys' looking response was slightly but significantly longer (by less than 1 s), which the authors suggested reflected violation of the monkeys' expectation of another repeat of the look-and-touch sequence (Santos and Hauser 1999). However, when squirrel monkeys (*Saimiri sciureus*) and capuchin monkeys in Fujita's laboratory were tested using similar methods, there was no compelling evidence of such an expectancy violation effect. Capuchin monkeys responded slightly more positively when the human actor was replaced by a puppet (Anderson et al. 2004), but we concluded that extent to which monkeys (and indeed other species) predict others' actions based on their visual behaviors needed further study; this is still the case.

In another study from Fujita's laboratory, individually tested capuchin monkeys begged an experimenter to give food that was under one of two containers (the second one was empty), by reaching or "pointing" toward the baited one (Hattori et al. 2007). The experimenter either looked at the monkey during the first 5 s of the trial, or ignored the monkey and looked up at the ceiling instead. In a second experiment the experimenter either looked at the midpoint between the two containers, or assumed the same posture but with her eyes closed. Although the monkeys did not point differentially in these conditions, they looked longer at the experimenter's face when she looked at the monkey (Experiment 1) and when her eyes were open (Experiment 2). Thus, the monkeys were clearly sensitive to attentional cues, but the authors expressed caution about equating this to "mind-reading."

To better distinguish between responding to others' referential gestures and understanding their attentional states, Hattori et al. (2010a, b, 2011) developed a new "cooperative paradigm," in which the subjects (capuchins and chimpanzees, respectively) could request food held in the experimenter's hand ("Hand situation," Fig. 7.2). In this situation, subjects need only attract the experimenter's attention to be able to successfully request food. Their behavior in this situation was compared with trials in which they could request food that was simply lying on a table ("Table situation," Fig. 7.2). It was predicted that the cooperative paradigm might more accurately reveal primates' understanding of attentional states.

Six capuchin monkeys and five chimpanzees were tested. As predicted, both species showed sensitivity to the experimenter's different attentional states and to eye gaze (eyes open vs. eyes closed) only in the former task (i.e., Hand situation). These results suggest that an "unfamiliar" situation (i.e., Table situation), that requires subject to produce referential gestures (pointing) to direct the experimenter's attention to food, may limit primates' abilities to express understanding of other's attentional states, even though they are sensitive to them.

Conspecific and Allospecific Cues

A common feature of the studies discussed so far is that the various cues presented to the nonhuman primate subjects are from humans. Of course, primates have evolved to respond primarily to interact with members of their own species, not humans. It is therefore important to also investigate sensitivity and responses to attentional cues of conspecifics. The power of conspecific cues was shown in a simple but clever experiment that involved the sudden appearance of a food item, visible to one of two monkeys or apes in their familiar group environment. Individuals of five species tested (chimpanzees and four Old World monkey species) quickly visually co-oriented when the partner shifted attention to look at the newly appeared food, thus showing an "ecologically valid" co-orientation response (Tomasello et al. 1998).

In a study of visual co-orientation in cotton-top tamarins tested with their familiar pair-mate, the monkeys showed no evidence of co-orienting to a human's gaze cues; however, they frequently co-oriented with their conspecific partner (Neiworth et al. 2002). The greater likelihood of visual co-orientation to conspecific than human gaze stimuli may also be true for lemurs. Although earlier studies found no evidence of human gaze following by lemurs (Itakura 1996; Anderson and Mitchell 1999), ring-tailed lemurs with telemetric tracking devices attached to their heads were reported to follow their companions' gaze (for which head direction was used as a proxy) to objects in the environment (Shepherd and Platt 2008). Itakura et al. (1999) reported similar performances by chimpanzees tested on object-choice with cues

Fig. 7.3 Images of a chimpanzee (left) and a human (right) each looking at one of two identical objects, as used in the comparative eye-tracking study by Hattori et al. (2010a, b)

from a human or from another chimpanzee, especially when the cue involved physically approaching the correct object. In two studies that illustrate the power of conspecific attentional cues and associated emotional signals (facial expressions, vocalizations; see also the next section), Morimoto and Fujita (2011, 2012) found that reaches toward a container by individually tested capuchin monkeys depended on how a familiar conspecific reacted when looking into the container (in which there could be a neutral, a positive, or a negative object).

One study that used photographic images to investigate responses to gaze cues, nicely illustrates different sensitivity and attentiveness to conspecific vs. allospecific social cues. Hattori et al. (2010a) used eye-tracking technology to investigate looking patterns in chimpanzees and humans. Static images of a chimpanzee model or a human model similarly gazing at one of two objects placed one on either side of the model were displayed (Fig. 7.3), and how chimpanzees and humans looked at each image was recorded and analyzed.

The data showed that chimpanzee looking patterns were modulated more by chimpanzee images than human images, whereas human looking patters were modulated equally by images of the two species. In particular, chimpanzees looked at the face of the chimpanzee model more than that of the human model, but no such difference was found in humans. Similar species-specific sensitivity and attending to social cues was also seen in another study, showing that nonhuman apes followed conspecific gaze more reliably than allospecific gaze (Kano and Call 2014). These studies suggest that both the social stimuli and the context are important when studying nonhuman primates' abilities to follow and act upon other individuals' gaze.

Other Attentional Cues

We have already noted that another individual's change of head orientation appears overall more powerful than an eyes-only change in shifting a primates' direction of attention. This tendency became clear in studies of spontaneous visual co-orientation with live human models. However, with experience (i.e., training), primates can learn to respond to human eye gaze only cues. Some studies that combined visual attentional with vocal cues (e.g., chimpanzee-like sounds, or nonsense words) in object-choice tasks reported enhanced performance in some—though not all—chimpanzees (e.g., Itakura et al. 1999; Call et al. 2000). Others have combined changes in attention orientation with changes in facial expression, assessing possible influences of the latter in the co-orientation response. In the first such study, individually tested pigtailed macaques showed no evidence of greater visual co-orientation when the human model changed her facial expression quickly after shifting her head and eyes to a new orientation (Paukner et al. 2007). By contrast, Goossens et al. (2008) reported an enhancing effect of one of three human-modeled facial expressions, the bared-teeth face (associated with fear or submission), on gaze following in long-tailed macaques (*Macaca fascicularis*). No such submissive expression was included in Paukner et al.'s (2007) study. Whether a human mimicked surprise facially and vocally while looking at an object influenced the likelihood of individually tested bonobos and chimpanzees looking toward the target of the human's attention (MacLean and Hare 2012). Studies of responses to conspecific attentional shifts also show the influence of accompanying emotional cues. Teufel et al. (2010) found that one specific type of facial expression in particular, often emitted when observing third-party interactions, was effective for eliciting conspecific gaze following in group-living, semi-free-ranging Barbary macaques (*Macaca sylvanus*). As previously mentioned, Morimoto and Fujita (2011, 2012) also reported that decision-making based on conspecific attentional cues varied with the nature of the associated emotional responses.

Checking Back

When a monkey or ape follows a model's gaze, either to a specific object or if they see nothing of particular interest as a result of their visual co-orientation response, it sometimes looks back at the model's face and then follows up with another co-orientation response. This phenomenon, sometimes referred to as "looking back," "checking back," or "double looking," was first reported in a gaze following study by Call et al. (1998). It was noted to be absent in a young infant chimpanzee tested with human gaze cues (Okamoto et al. 2004), but it is clearly present in chimpanzee and other great ape juveniles, and especially frequent in adults (Bräuer et al. 2005). Checking back may imply some kind of perspective taking in which the gaze follower infers that the model perceives something, which the follower then also wants to perceive (see Carpenter and Call 2015, for a critical discussion of how gaze following in nonhuman primates compares cognitively with that of humans). It was therefore interesting that gibbons and siamangs showed no double looking to a human looking up, which the authors attributed to their relatively less complex social systems (Liebal and Kaminski 2012), and that neither spider monkeys nor capuchin monkeys showed looking back responses to a human model (Amici et al. 2009).

Given our previous finding of stronger visual co-orientation responses in capuchins than squirrel monkeys (Anderson et al. 2005) and the voluminous literature on capuchins impressive cognitive abilities (for summaries see Anderson 1996; Fragaszy et al. 2004), we ran another study to compare co-orientation and checking back in capuchin and squirrel monkeys (Anderson and Hattori, unpublished). We tested the responses of five adults of each species to visual attention switches by a familiar adult human (model), consisting of the latter switching from looking at the monkey (in a familiar, transparent test cage) to looking at one of two computer screens, positioned behind and to either side of the monkey's cage. The model gazed at the screen immediately after the appearance of a geometric figure on that screen (congruent condition), after the appearance of the geometric figure on the non-gazed-at screen (incongruent condition), or when both screens remained blank (control condition). The model held his gaze for 10 s, before a 60-s inter-trial interval (both screens blank) after which the procedure was repeated. Each session consisted of 10 trials of only one condition, with the gazed-at screen pseudo-randomized across trials. All three conditions were presented twice. No rewards were given to the monkeys during these sessions.

Video analyses revealed significantly more first-response co-orientation events (i.e., after the model's attention switch the monkey first turned its head toward the gazed-at screen rather than the other screen) in capuchins than squirrel monkeys (means per session: 5.3 and 3.7, respectively), with no effect of congruence. Furthermore, capuchins but not squirrel monkeys showed a marked decrease in frequency of co-orientation from the first set of presentations to the second. Concerning checking back, we found no species difference; the only significant effect concerned overall frequency, which decreased by more than 50% between the first and second set of presentations (means: 3.9 and 1.9 per session). In summary, these data supported our earlier findings of a greater overall tendency to visually co-orient with a human in capuchins than squirrel monkeys. Therefore, unlike Amici et al. (2009), we did find evidence of checking back in New World monkeys, with no difference in the overall frequency between capuchins and squirrel monkeys. A parsimonious conclusion at this stage is that these two species do not differ in the extent to which perspective taking is implicated in their co-orienting responses, if at all.

Conclusion

In this chapter, we focused particularly on studies that use live models (either humans or conspecifics) to examine gaze following and use of gaze cues in nonhuman primates. As mentioned in our Introduction, however, other approaches to studying attention monitoring in primates and other species have also contributed toward building a better understanding of responses to attentional cues and their underlying mechanisms, influences, development and evolution. For example, some studies have used attentional cues in static pictures or in videos to examine gaze-cue effects (e.g., Vick et al. 2006; Emery et al. 1997; Tomonaga 2007; Ruiz et al. 2009; Kano et al. 2018; Spadacenta et al. 2019). Others have sought to elucidate electrophysiological responses and neural structures involved in responding to gaze cues (e.g., Jellema et al. 2000; Hoffmn et al. 2007; Kamphuis et al. 2009; Pryluk et al. 2020; Ramezanpour and Their 2020). Yet another approach considers primate eye morphology from evolutionary and socioecological perspectives (Kobayashi and Kohshima 1997, 2001). This originally led to the hypothesis of nonhuman primates' "camouflage eyes" vs. humans' "gaze signaling" eyes, but recent research has shown that great ape species in fact have eye morphology that enhances detectability of gaze direction (Perea-García et al. 2019). All of these lines of research, along with those that we have described more fully, have improved our understanding of the biology and psychology of paying attention to others' attention. We now also have a better grasp of the evolutionary emergence of various mechanisms, ranging from an innate attractiveness to eyelike stimuli, to reflexive co-orientation, to mental state attributions based on others' direction of attention. We look forward to further progress in this broad and vibrant field of comparative cognition.

References

Amici F, Aureli F, Visalberghi E, Call J (2009) Spider monkeys (*Ateles geoffroyi*) and capuchin monkeys (*Cebus apella*) follow gaze around barriers: evidence for perspective taking? J Comp Psychol 123:368–374

Anderson JR (1996) Chimpanzees and capuchin monkeys: comparative cognition. In: Russon AE, Bard KA, Parker ST (eds) Reaching into thought: the minds of the great apes. Cambridge University Press, Cambridge, pp 23–56

Anderson JR, Mitchell RW (1999) Macaques but not lemurs co-orient visually with humans. Folia Primatol 70:17–22

Anderson JR, Vick SJ (2008) Primates' use of others' gaze. In: Itakura S, Fujita K (eds) Origins of the social mind. Springer, Tokyo, pp 39–64

Anderson JR, DeMonte M, Kempf J (1996a) Discrimination learning and multiple reversals in young adult and older adult monkeys (*Macaca arctoides*). Q J Exp Psychol 49B:193–200

Anderson JR, Montant M, Schmitt D (1996b) Rhesus monkeys fail to use gaze direction as an experimenter-given cue in an object-choice task. Behav Process 37:47–55

Anderson JR, Sallaberry H, Barbier H (1995) Use of experimenter-given cues during object-choice tasks by capuchin monkeys. Anim Behav 49:201–208

Anderson JR, Kuroshima H, Kuwahata H, Fujita K (2004) Do squirrel monkeys (*Saimiri sciureus*) and capuchin monkey (*Cebus apella*) predict that looking leads to touching? Anim Cogn 7:185–192

Anderson JR, Kuroshima H, Hattori Y, Fujita K (2005) Attention to combined attention in new world monkeys (*Cebus apella, Saimiri sciureus*). J Comp Psychol 119:461–414

Anderson JR, Sallaberry P, Barbier H (2015) Use of experimenter-given cues during object-choice tasks by capuchin monkeys. Anim Behav 49:201–208

Baron-Cohen S, Campbell R, Karmiloff-Smith A, Grant J, Walker J (1995) Are children with autism blind to the mentalistic significance of the eyes? Br J Dev Psychol 13:379–398

Barth J, Reaux JE, Povinelli DJ (2005) Chimpanzees' (*Pan troglodytes*) use of gaze cues in object-choice tasks: different methods yield different results. Anim Cogn 8:84–92

Bettle R, Rosati AG (2019) Flexible gaze-following in rhesus monkeys. Anim Cogn 22:673–686

Botting JL, Wiper ML, Anderson JR (2011) Brown (*Eulemur fulvus*) and ring-tailed lemurs (*Lemur catta*) use human head orientation as a cue to gaze direction in a food choice task. Folia Primatol 82:165–176

Bräuer J, Call J, Tomasello M (2005) All great ape species follow gaze to distant locations and around barriers. J Comp Psychol 119:145–154

Bray J, Krupenye C, Hare B (2014) Ring-tailed lemurs (*Lemur catta*) exploit information about what others can see but not what they can hear. Anim Cogn 17:735–744

Burkart J, Heschl A (2006) Geometrical gaze following in common marmosets (*Callithrix jacchus*). J Comp Psychol 120:120–130

Byrnit J (2004) Nonenculturated orangutans' (*Pongo pygmaeus*) use of experimenter-given manual and facial cues in an object-choice task. J Comp Psychol 118:309–315

Byrnit J (2008) Gorillas' (*Gorilla gorilla*) use of experimenter-given manual and facial cues in an object-choice task. Anim Cogn 12:404–404

Call J, Hare BA, Tomasello M (1998) Chimpanzee gaze following in an object-choice task. Anim Cogn 1:89–99

Call J, Agnetta B, Tomasello M (2000) Cues that chimpanzees do and do not use to find hidden objects. Anim Cogn 3:23–34

Canteloup C, Poitrasson I, Anderson JR, Poulin N, Meunier H (2017) Factors influencing deceptive behaviours in Tonkean macaques (*Macaca tonkeana*). Behaviour 154:765–784

Carpenter M, Call J (2015) How joint is the joint attention of apes and human infants? In: Metcalfe J, Terrace HS (eds) Agency and joint attention. Oxford Scholarship Online, Oxford. https://doi.org/10.1093/acprof:oso/9780199988341.003.0003

Caspar KR, Mader L, Pallasdies F, Lindenmeier M, Begall S (2018) Captive gibbons (Hylobatidae) use different referential cues in an object-choice task: insights into lesser ape cognition and manual laterality. PeerJ 6:e5348. https://doi.org/10.7717/peerj.5348

Davidson GL, Butler S, Juricic F, Thornton A, Clayton NS (2014) Gaze sensitivity: function and mechanisms from sensory and cognitive perspectives. Anim Behav 87:3–15

Emery NJ (2000) The eyes have it: the neuroethology, function and evolution of social gaze. Neurosci Biobehav Rev 24:581–604

Emery NJ, Lorincz EN, Perrett DI, Oram MW, Baker CI (1997) Gaze following and joint attention in rhesus monkeys (*Macaca mulatta*). J Comp Psychol 111:286–293

Ferrari PF, Kohler E, Fogassi L, Gallese V (2000) The ability to follow eye gaze and its emergence during development in macaque monkeys. Proc Natl Acad Sci USA 97:13997–14002

Flombaum JI, Santos LR (2005) Rhesus monkeys attribute perception to others. Curr Biol 15:447–452

Fragaszy DM, Visalberghi E, Fedigan LM (2004) The complete capuchin: the biology of the genus *Cebus*. Cambridge University Press, Cambridge

Fujita K (2001) What you see is different from what I see: species differences in visual perception. In: Matsuzawa T (ed) Primate origins of human cognition and behavior. Springer, Tokyo, pp 29–54

Goossens BMA, Dekleva M, Reader SM, Sterck EHM, Bolhuis JJ (2008) Gaze following in monkeys is modulated by observed facial expressions. Anim Behav 75:1673–1681

Hare B (2001) Can competitive paradigms increase the validity of experiments on primate social cognition? Anim Cogn 4:269–280

Hare B, Tomasello M (2004) Chimpanzees are more skilful in competitive than in cooperative cognitive tasks. Anim Behav 68:571–581

Hare B, Call J, Agnetta B, Tomasello M (2000) Chimpanzees know what conspecifics do and do not see. Anim Behav 59:771–785

Hare B, Call J, Tomasello M (2001) Do chimpanzees know what conspecifics know? Anim Behav 61:139–151

Hare B, Adressi E, Call J, Tomasello M, Visalberghi E (2003) Do capuchin monkeys, *Cebus apella*, know what conspecifics do and do not see? Anim Behav 65:131–142

Harlow HF, Bromer J (1938) A test-apparatus for monkeys. Psychol Rec 2:434–436

Hattori Y, Kuroshima H, Fujita K (2007) I know you are not looking at me: capuchin monkeys'(*Cebus apella*) sensitivity to human attentional states. Anim Cogn 2:141–148

Hattori Y, Kano F, Tomonaga M (2010a) Differential sensitivity to conspecific and allospecificcues in chimpanzees and humans: a comparative eye-tracking study. Biol Lett 6:610–613

Hattori Y, Kurishima H, Fujita K (2010b) Tufted capuchin monkeys (*Cebus apella*) show understanding of human attentional states when requesting food held by a human. Anim Cogn 13:87–92

Hattori Y, Tomonaga M, Fujita K (2011) Chimpanzees (*Pan troglodytes*) show more understanding of human attentional states when they request food in the experimenter's hand than on the table. Interact Stud 12:418–429

Hoffmn KL, Gothard KM, Schmid MC, Logothetis NK (2007) Facial-expression and gaze-selective responses in the monkey amygdala. Curr Biol 17:766–772

Itakura S (1996) An exploratory study of gaze-monitoring in nonhuman primates. Jpn Psychol Res 38:174–180

Itakura S (2004) Gaze-following and joint visual attention in nonhuman animals. Jpn Psychol Res 46:216–226

Itakura S, Anderson JR (1996) Learning to use experimenter-given cues during an object-choice task by a capuchin monkey. Curr Psychol Cogn 15:103–112

Itakura S, Tanaka M (1998) Use of experimenter-given cues during object-choice tasks by chimpanzees (*Pan troglodytes*), an orangutan (*Pongo pygmaeus*), and human infants (*Homo sapiens*). J Comp Psychol 112:119–126

Itakura S, Agnetta B, Hare B, Tomasello M (1999) Chimpanzee use of human and conspecific social cues to locate hidden food. Dev Sci 2:448–456

Jellema T, Baker CI, Wicker B, Perrett DI (2000) Neural representation for the representation of the intentionality of actions. Brain Cogn 44:280–302

Kamphuis S, Dicke PW, Their P (2009) Neuronal substrates of gaze following in monkeys. Eur J Neurosci 29:1732–1738

Kano F, Call J (2014) Cross-species variation in gaze following and conspecific preference among great apes, human infants and adults. Anim Behav 91:137–150

Kano F, Moore R, Krupenye C, Hirata S, Tomonaga M, Call J (2018) Human ostensive signals do not enhance gaze following in chimpanzees, but do enhance object-oriented attention. Anim Cogn 21:1205. https://doi.org/10.1007/s10071-018-1205-z

Kluver H (1937) Certain effects of lesions of the occipital lobes in macaques. J Psychol 4:383–401

Kobayashi H, Kohshima S (1997) Unique morphology of the human eye. Nature 387:767–768

Kobayashi H, Kohshima S (2001) Unique morphology of the human eye and its adaptive meaning: comparative studies on external morphology of the primate eye. J Hum Evol 40:419–435

Kuroshima H, Fujita K, Fuyuki A, Masuda T (2002) Understanding of the relationship between seeing and knowing by tufted capuchin monkeys (*Cebus apella*). Anim Cogn 5:41–48

Liebal K, Kaminski J (2012) Gibbons (*Hylobates pileatus, H. moloch, H. lar, Symphalangus syndactylus*) follow human gaze, but do not take the visual perspective of others. Anim Cogn 15:1211–1216

MacLean EL, Hare B (2012) Bonobos and chimpanzees infer the target of another's attention. Anim Behav 83:345–353

MacLean EL, Sandel AA, Bray J, Oldenkamp RE, Reddy RB, Hare BA (2013) Group size predicts social but not nonsocial cognition in lemurs. PLoS ONE 8(6):e66359. https://doi.org/10.1371/journal.pone.0066359

Montgomery DE, Bach LM, Moran C (1998) Children's use of looking behavior as a cue to detect another's goal. Child Dev 69:692–705

Morimoto Y, Fujita K (2011) Capuchin monkeys (*Cebus apella*) modify their own behaviors according to a conspecific's emotional expressions. Primates 52:279–286

Morimoto Y, Fujita K (2012) Capuchin monkeys (*Cebus apella*) use conspecifics' emotional expressions to evaluate emotional valence of objects. Anim Cogn 15:341–347

Mulcahy NJ, Hedge V (2012) Are great apes tested with an abject object-choice task? Anim Behav 83:313–321

Neiworth JJ, Burman MA, Basile BM, Lickteig MT (2002) Use of experimenter-given cues in visual co-orienting and in an object-choice task by a new world monkey species, cotton top tamarins (*Saguinus oedipus*). J Comp Psychol 116:3–11

Okamoto S, Tomonaga M, Ishii K, Kawai N, Tanaka M, Matsuzawa T (2002) An infant chimpanzee (*Pan troglodytes*) follows human gaze. Anim Cogn 5:107–114

Okamoto S, Tanaka M, Tomonaga M (2004) Looking back: the 'representational mechanism' of joint attention in an infant chimpanzee. Jpn Psychol Res 46:236–245

Okamoto-Barth S, Tomonaga M, Tanaka M, Matsuzawa T (2008) Development of using experimenter-given cues in infant chimpanzees: longitudinal changes in behavior and cognitive development. Dev Sci 11:98–108

Paukner A, Anderson JR, Fogassi L, Ferrari PF (2007) Do facial gestures, visibility or speed of movement influence gaze following responses in pigtail macaques? Primates 48:241–244

Peignot P, Anderson JR (1999) Use of experimenter-given manual and facial cues by gorillas (*Gorilla gorilla*) in an object-choice task. J Comp Psychol 113:253–260

Perea-García JO, Kret ME, Monteiro A, Hobaiter C (2019) Scleral pigmentation leads to conspicuous, not cryptic, eye morphology in chimpanzees. Proc Natl Acad Sci U S A 116:19248–19250

Pfungst O, Povinelli DJ, Bierschwale DT, Cech CG (1999) Comprehension of seeing as a referential act in young children, but not juvenile chimpanzees. Br J Dev Psychol 17:37–60

Povinelli DJ, Eddy TJ (1996a) What young chimpanzees know about seeing. University of Chicago Press, Chicago

Povinelli DJ, Eddy TJ (1996b) Factors influencing young chimpanzees' (*Pan troglodytes*) recognition of attention. J Comp Psychol 110:335–345

Povinelli DJ, Eddy TJ (1997) Specificity of gaze-following in young chimpanzees. Br J Dev Psychol 15:213–222

Povinelli DJ, Nelson KE, Boysen ST (1990) Inferences about guessing and knowing by chimpanzees (*Pan troglodytes*). J Comp Psychol 104:203–210

Povinelli DJ, Parks KA, Novak MA (1991) Do rhesus monkeys (*Macaca mulatta*) attribute knowledge and ignorance to others? J Comp Psychol 105:318–325

Pryluk R, Shohat Y, Morozov A et al (2020) Shared yet dissociable neural codes across eye gaze, valence and expectation. Nature 586:95–100

Ramezanpour H, Their P (2020) Decoding of the other's focus of attention by a temporal cortex module. Proc Natl Acad Sci USA 117:2663–2670

Rosati AG, Hare B (2009) Looking past the model species: diversity in gaze-following skills across primates. Curr Opin Neurobiol 19:45–51. https://doi.org/10.1016/j.conb.2009.03.002

Rosati AG, Santos LR (2017) Tolerant Barbary macaques maintain juvenile levels of social attention in old age, but despotic rhesus macaques do not. Anim Behav 130:199–207

Rosati AG, Arre AM, Platt ML, Santos LR (2016) Rhesus monkeys show human-like changes in gaze following across the lifespan. Proc R Soc B 283:20160376
Ruiz A, Gómez JC, Roeder JJ, Byrne RW (2009) Gaze following and gaze priming in lemurs. Anim Cogn 12:427–434
Sandel AA, MacLean EL, Hare B (2011) Evidence from four lemur species that ringtailed lemur social cognition converges with that of haplorhine primates. Anim Behav 81:925–931
Santos LR, Hauser MD (1999) How monkeys see the eyes: cotton-top tamarins' reaction to changes in visual attention and action. Anim Cogn 2:131–139
Shepherd SV (2010) Following gaze: gaze-following behavior as a window into social cognition. Front Integr Neurosci 4:5. https://doi.org/10.3389/fnint.2010.00005
Shepherd SV, Platt ML (2008) Spontaneous social orienting and gaze following in ring-tailed lemurs (*Lemur catta*). Anim Cogn 11:13–20
Spadacenta S, Dicke PW, Their P (2019) Reflexive gaze following in common marmoset monkeys. Sci Rep 9:15292. https://doi.org/10.1038/s41598-019-51783-9
Tan J, Tao R, Su Y (2014) Testing the cognition of the forgotten colobines: a first look at golden snub-nosed monkeys (*Rhinopithecus roxellana*). Int J Primatol 35:376–393
Teufel C, Gutmann A, Pirow R, Fischer J (2010) Facial expressions modulate the ontogenetic trajectory of gaze-following among monkeys. Dev Sci 13:913–922
Tomasello M, Call J, Hare B (1998) Five primate species follow the visual gaze of conspecifics. Anim Behav 55:1963–1069
Tomasello M, Hare B, Fogleman T (2001) The ontogeny of gaze following in chimpanzees, *Pan troglodytes*, and rhesus macaques, *Macaca mulatta*. Anim Behav 61:335–343
Tomonaga M (2007) Is chimpanzee (*Pan troglodytes*) spatial attention reflexively triggered by the gaze cue? J Comp Psychol 121:156–170
Vick SJ, Anderson JR (2000) Learning and limits of use of eye gaze by capuchin monkeys (*Cebus apella*) in an object-choice task. J Comp Psychol 114:200–207
Vick SJ, Anderson JR (2003) Use of human visual attention cues by olive baboons (*Papio anubis*) in a competitive task. J Comp Psychol 117:209–216
Vick SJ, Bovet D, Anderson JR (2001) Gaze discrimination learning in olive baboons (*Papio anubis*). Anim Cogn 4:1–10
Vick SJ, Toxopeus I, Anderson JR (2006) Pictorial gaze cues do not enhance long-tailed macaques' performance on a computerized object-location task. Behav Process 73:308–314

Chapter 8
Understanding Others' Behavior: Effect of One's Own Experience

Hika Kuroshima

Abstract For social animals, including humans, it is important to be able to understand and predict others' actions. Previous studies have shown that not only human infants but also great apes can detect others' goals or intentions from their actions. The discovery of the mirror neuron system led to the suggestion that one's own action experience affects understanding of others' actions. For example, in one study 3-month-old infants predicted the goal of another individual's reaching behavior after experiencing similar reaching themselves. In this chapter, I review some studies about abilities to understand others' mental states and the effect of one's own action experience in human infants and non-human primates. I then introduce our own research projects on this topic, for which our study species were capuchin monkeys and dogs. The results of these three projects suggest that one's own active experience affects the understanding of others' actions, and that this relationship might have the same underlying mechanisms in diverse species.

Keywords Social cognition · Understanding of others · Action experience · Social learning · Capuchin monkeys · Dogs

Introduction

For social non-human animals, as for humans, it is important to be able to understand and predict others' actions. This ability allows us to engage in smooth communication with each other, to adapt our behavior in cooperative or competitive situations, and to learn new skills from others. Human infants interpret people's actions in terms of their goals or intentions, and they predict future actions, for example, grasping or reaching, by 6–9 months of age (Hamlin et al. 2008; Király et al. 2003; Woodward 1998, 1999, 2005; Woodward et al. 2009). For instance, in Woodward (1998), 5-, 6-, and 9-month-old infants observed as an experimenter grasped one of two toys

H. Kuroshima (✉)
Department of Psychology, Kyoto University Graduate School of Letters, Kyoto, Japan

J. R. Anderson, H. Kuroshima (eds.), *Comparative Cognition*,
https://doi.org/10.1007/978-981-16-2028-7_8

several times (habituation phase), after which the experimenter made the same grasping movement but for the other toy, or else moved to grasp in another direction but for the same toy as during habituation. By the age of 6 months, infants looked longer at the situation where the experimenter grasped the new toy via the old path than when the old toy was grasped via the new path. Later, by the age of 12 months, infants flexibly interpret observed animated movements as goal-directed, according to the context (Csibra 2003; Gergely and Csibra 2003; Gergely et al. 1995; Phillips and Wellman 2005). Although chimpanzees (*Pan troglodytes*) differ from humans in their ratio of looking toward an actor's face to looking elsewhere, they were found to show predictive eye movements toward the goal when a human actor performed a goal-directed behavior; in this respect they resembled 12-month-olds and human adults (Myowa-Yamakoshi et al. 2012). Call et al. (2004) reported that chimpanzees responded differently to a human actor who was unwilling versus one who was physical unable to deliver food to the subjects. Similar results have been obtained with capuchin monkeys (*Sapajus apella*) (Phillips et al. 2009) and Tonkean macaques (*Macaca tonkeana*) (Canteloup and Meunier 2017). These types of studies provide evidence that not only humans, but also non-human primates detect others' goals or intentions from their actions.

In the late 1990s, neurons involved in the shared neural expression of perception and production of actions—the so-called mirror neurons—were discovered in the premotor and posterior parietal cortex of the macaque brain (Gallese et al. 1996; Rizzolatti et al. 1996). The prominent feature of these neurons was their activation both when the subject performed goal-directed actions and when it observed the same actions performed by another individual (human actor or conspecific). After this discovery, researchers studying humans found similar neural activation in the inferior frontal cortex, posterior inferior frontal gyrus (IFG), and adjacent ventral premotor cortex (PMC) (see Fox et al. 2016 for a review). These areas were referred to collectively as the "mirror-neuron systems (MNS)." The discovery of MNS had several implications for understanding other's actions. According to simulation theory (Rizzolatti et al. 2001), others' mental states are represented from one' own perspective. That is, we can understand other people's mental states by tracking or matching their actions with our own mental states. From this perspective, one's own action experience may affect the understanding of others' goals or intentions.

In this chapter I introduce three of our studies about the influence of one's own action experience on the understanding of others' actions and propose possible projects for future research. Taken together, these studies on monkeys and dogs also nicely illustrate the broader comparative perspective that has characterized several of the projects conducted in Kazuo Fujita's lab.

Understanding Other's Action Based on One's Own Action Experience

If one's own action experience influences predictions about another's goal-directed behavior, a participant should react differently depending on whether or not an observed action is present in the participant's own action repertoire. Studies that take this perspective include a pioneering one by Sommerville et al. (2005). One group of 3-month-old infants picked up a pair of toys using a mitten with Velcro ("reach first" group). A second group of infants did not have this experience ("watch first" group). Then, both groups of infants observed an adult experimenter's hand wearing a mitten reach toward one of two toys for several trials (habituation phase). In the test trial, the experimenter reached for the new toy via the old path or for the old toy via a new path (because the position of the toys was switched). Infants in the reach first group were found to look longer when the experimenter reached for the new toy via the old path than the old toy via the new path. Based on this result, the authors concluded that one's own action experience can lead to rapid learning and transfer of this knowledge to an observed event. Also, Kanakogi and Itakura (2011) demonstrated that predicting another's goal-directed behavior was related to development of locomotory ability. Several other studies have reported similar results in infants (Cannon et al. 2012; Loucks and Sommerville 2018; Rosander and von Hofsten 2011; Sommerville and Woodward 2005; Woodward and Guajardo 2002)

Is the link between action experience and goal understanding or prediction unique to humans? If this link involves MNS, we should expect to find similar evidence in non-human primates. To clarify the possible evolutionary origins of this link, we tested New World monkeys, tufted capuchins (*Sapajus apella*) (Kuroshima et al. 2014). The subjects were 8 captive-born, adult males (14–20 years old) housed at the University of Georgia. In Experiment 1, we ran two test sessions consisting of 12 regular and four test trials for each monkey. Regular trials started with an experimenter (hereafter *E*) entering the experimental room and putting a food reward into a clear glass jar and closing the lid in front of the monkey, and then leaving the room. Then an actor (hereafter *A*) entered the room, tried to open the jar for about 3 s, opened it and gave the reward to the monkey. In test trials, which were inserted randomly among the regular trials, after *E* put the reward into the jar and left the room, *A* entered the room and manipulated the jar using one of two actions ("*Familiar action*" or "*Unfamiliar action*") for 30 s (Fig. 8.1). In the familiar action, *A* tried to open the jar in the same manner as in regular trials, but never opened it. In the unfamiliar action, *A* repeatedly touched the sides of the jar (about ten times) with the extended index fingers of both hands. During all trials, the subject monkey was videotaped and the videos were later coded for measures of looking behavior.

Video analysis of the monkeys' behavior revealed that they all looked at *A*'s face or the jar and hands for significantly longer when *A* performed the familiar action than the unfamiliar action. The monkeys had learned the outcome of the familiar action during the regular trials. This result showed that duration of looking toward *A*'s face and the apparatus were reliable behavioral indices of the monkeys' expectation about the outcome of *A*'s action.

In Experiment 2, we introduced a new food container that could be opened in two ways (by lifting the lid or by pulling a drawer) and showed the monkeys *A* opening the container by lifting the lid in regular action trials, and two new attempts using novel actions in test trials (Fig. 8.2). In the test trials, the monkeys never got a reward, because *A* never managed to open the container, the same as in test trials in

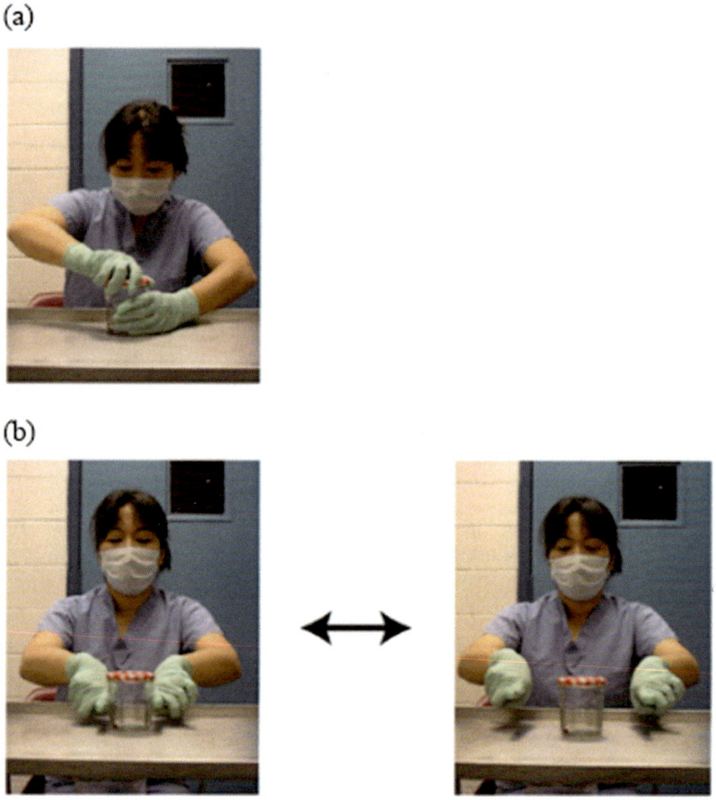

Fig. 8.1 Actor's action in regular trials and familiar test trials (**a**) and unfamiliar test trials (**b**). The figure is from Kuroshima et al. (2014)

Experiment 1. Then we confirmed that the monkeys looked at *A*'s face and the container equally for both unfamiliar actions.

In Experiment 3, we first trained the monkeys to open the container by using one of the unfamiliar actions from Experiment 2 (pulling the drawer) to get a reward. So, this unfamiliar action now became an "experienced action" for the monkeys. Following training, we repeated the same procedure as in Experiment 2. The result was that monkeys looked longer when *A* tried to open the container with their experienced action than with the unexperienced action. In other words, the monkeys anticipated a specific outcome of the other's action based on their own action experience, even if they did not directly observe the link between other's action and the outcome. Their action experience influenced their perception of the other's action.

However, the results described above do not allow us to conclude that the monkeys' own action experiences affect interpretation of the other's *mental state*, such as goal or intention. As Biro and Leslie (2007) and Meltzoff and Brooks (2008) pointed out, the monkeys might simply use the outcome of the their own action to

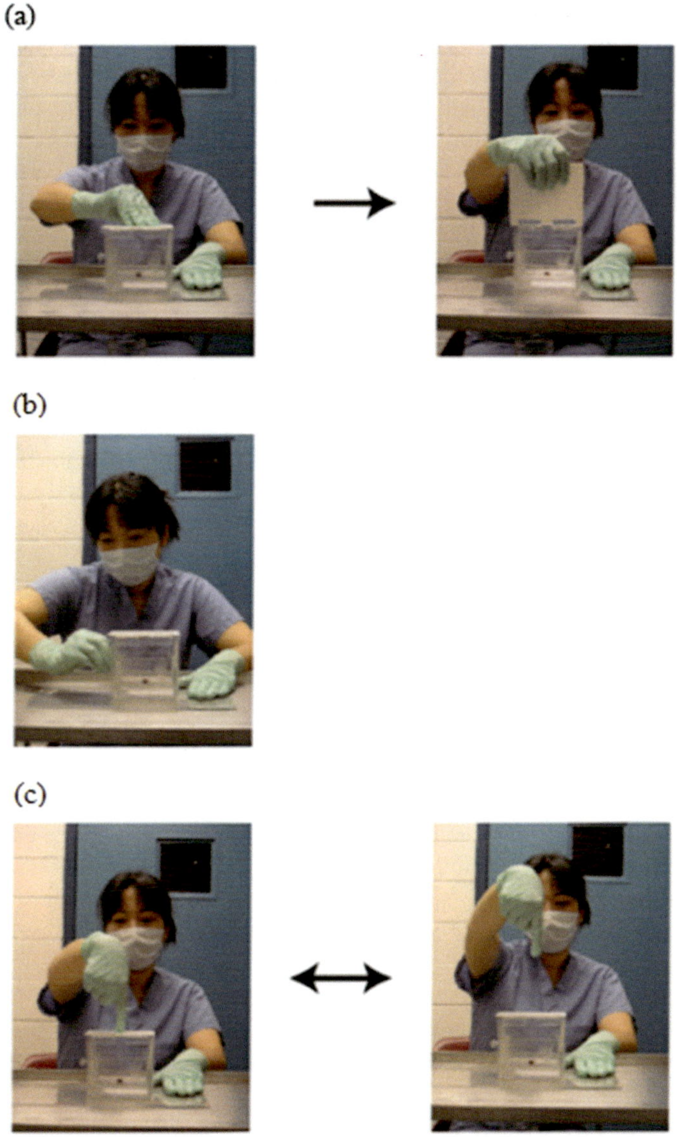

Fig. 8.2 Actor's action in regular trials in Experiment 2 (**a**), unfamiliar action 1 (**b**), and unfamiliar action 2 (**c**) in test trials. Before the test phase in Experiment 3, because the monkeys were trained to open the container by unfamiliar action 1, this became the experienced action and novel action 2 became the unexperienced action in Experiment 3. The figure is from Kuroshima et al. (2014)

expect the same outcome of the other's action. Meltzoff and Brooks (2008) investigated whether first-person experience with visual occluders influenced infants' understanding of others' visual experiences. They demonstrated that 12-month-old infants who had experienced their own view affected by an opaque blindfold were less likely to follow a blindfolded experimenter's head movement toward an object than infants whose blindfold had a window and therefore did not produce the experience of one's view being blocked. Similarly, Myowa-Yamakoshi et al. (2011) reported that 12-month-old infants who had experienced occluded vision by wearing a blindfold looked longer when the blindfolded actor performed successfully than in vain, whereas they looked longer when a non-blindfolded actor failed rather than succeeded. These results suggest that human infants use their own mental state or representation, such as knowledge gained from their own visual experience, to interpret others' actions. Recently, Kano et al. (2019) demonstrated that great apes who experienced visual barriers in their daily life anticipated an actor's action in a false belief task. The overall experimental evidence suggests that human infants and great apes can apply knowledge gained through their own visual experiences to predict others' next actions, and more specifically, they can infer the invisible mental states of others based on their own visual experiences. However, for a stronger conclusion that they do this through simulation, some kinds of additional tests would be needed, to rule out non-mentalistic accounts.

Understanding Environmental Properties: Observing Others' Actions and Deploying One's Own Action Experience

In our daily lives, we use others' actions not only to make inferences about those individuals' intentions or goals, but also about various aspects of the environment, such as physical causality, heaviness, hardness, temperature, and so on. Studies of infants' and preschool children's causal understanding suggest that their imitation of novel target actions is influenced by their own action experiences (Sobel and Sommerville 2010; Sommerville and Woodward 2005; Williamson and Meltzoff 2011; Williamson et al. 2008). For example, Wood et al. (2013) demonstrated that 5-year-olds who had spontaneously learned how to get a reward from a container showed less over-imitation than children whose prior experience of the container consisted simply of observing another individual successfully opening it. This finding suggests that one's own action experiences facilitate causal understanding of the situation, resulting in less redundant imitation of another individual's irrelevant actions.

Does active experience with the environment influence understanding of physical properties in other species in observation learning situations? We approached this question by investigating the effect of active experience in capuchin monkeys (Kuroshima et al. 2008). Although capuchins show some evidence of social learning in tool-using contexts, seen especially nut-cracking behavior in wild and semi-wild groups (Fragaszy et al. 2004; Ottoni et al. 2005; Ottoni and Mannu 2001), numerous laboratory studies have reported failures to imitate novel action sequences used successfully by a conspecific or human to solve a problem (Custance et al. 1999; Fragaszy and Visalberghi 2004; Fragaszy et al. 2011; Fragaszy and Visalberghi 1989; Munkenbeck Fragaszy and Visalberghi 1990; Visalberghi 1987). But although capuchin monkeys do not appear to acquire new behaviors by imitation, acquisition can benefit from processes such as stimulus enhancement and social facilitation. Therefore, we examined whether capuchin monkeys would choose an appropriate action from their own action repertoire to solve a problem, after seeing another individual failing on the same task (i.e., using a wrong action).

In Experiment 1 of Kuroshima et al. (2008), we introduced three identical-looking transparent containers which could be opened in three different ways (top-open, bottom-open, and top- and bottom-open types) (Fig. 8.3). In the training phase, using two one-way-to-open containers, three monkeys housed in Fujita's lab at Kyoto University (two males: Pigmon and Zinnia, one female: Kiki; 3–8 years old during the study) were trained to open their container by observing a human demonstrator's successful action; opening resulted in access to a reward that was in the container. After this training, using a two-ways-to-open container, the demonstrator showed the monkeys a successful or an unsuccessful action (a "mistake" trial): in the latter the actor tried to open the container from either top or bottom but failed. In the mistake trials, the monkey could then open the container by using the alternative action, but none of the monkeys was seen to do so by spontaneously using the alternative to the demonstrator's unsuccessful action.

In Experiment 2, we used new one-way-to-open containers to train the same monkeys to immediately correct their own mistake (Fig. 8.4). Then we conducted the same test as in Experiment 1, using new two-ways-to-open containers. This correction experience made no difference to the monkeys' behavior: they did not spontaneously show the successful action after seeing the demonstrator's mistake.

In Experiment 3, we began by confirming that the monkeys could immediately switch to the alternative action when their first attempt to open the container failed. In the test phase, we positioned two monkeys face to face; now, the demonstrator was a conspecific, not a human, and each monkey served as both model and subject. The second monkey got the chance to open the container only if the first monkey failed to do so. We used the same containers as in Experiment 1. The results showed that the two adults (Kiki and Pigmon) were able to solve a task by taking into account the demonstrator's mistake. This finding suggests that capuchin monkeys can understand another's action—and not just the outcome of that action—when the action is already in their own action repertoire. Because their action experience facilitates understanding of the other's action, the latter's mistake can easily be corrected, as if it were one's own mistake.

However, even when the monkeys had experience of switching to the alternative action after their first attempt failed in Experiments 2 and 3, they persisted with their own preferred action after observing a human's mistake or a mistake by a specific monkey (Zinnia). It is possible that individual relationships, or the other individual's skill level, exert an influence. Ottoni et al. (2005) reported that semi-free-ranging brown capuchins prefer to watch more skilled nut-cracking monkeys, which implies sensitivity to others' skill levels. In our study the two successful monkeys, Kiki and

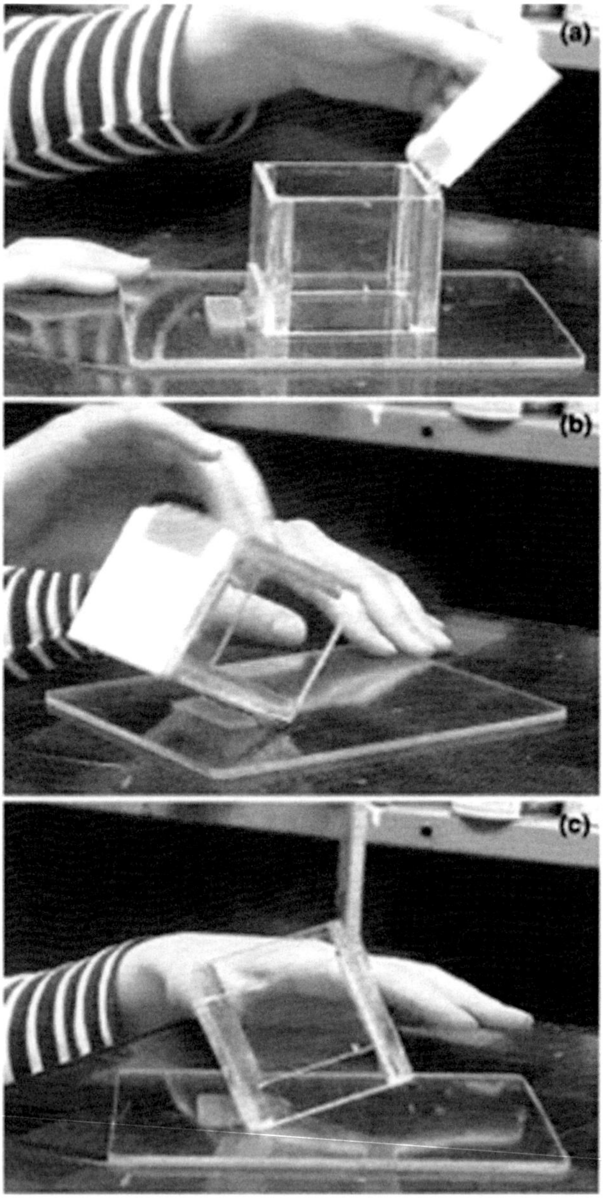

Fig. 8.3 Three types of transparent food containers used in Experiments 1 and 3. They were identical in appearance but required different actions for opening: by the lid (**a**), at the bottom (**b**), or both (**c**). We used two one-way-to-open containers (**a**, **b**) in regular trials and two-ways-to-open container (**c**) in test trials. The figure is from Kuroshima et al. (2008)

Pigmon, were older than Zinnia, who was in fact the youngest male in the social group. Conceivably, the older pair paid little regard to Zinnia's ability. Also, in other studies we have shown that capuchin monkeys evaluate interactions between third parties for traits such as reciprocity and cooperativeness (Anderson et al. 2013a, b, 2017). How different traits of individuals might affect how others interpret their actions will be an interesting topic for future studies.

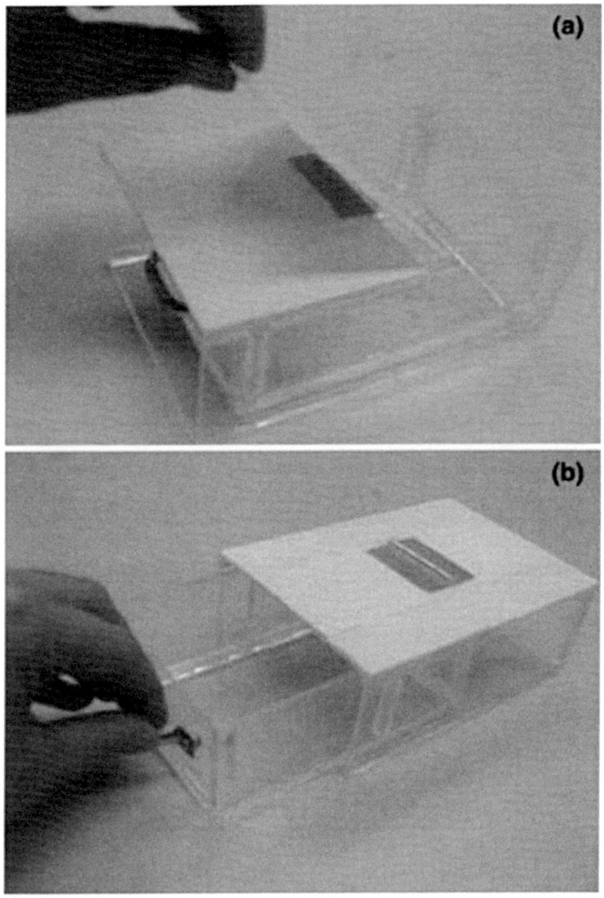

Fig. 8.4 The food containers used in Experiment 2. They looked identical in appearance but could be opened by different actions: on the lid (**a**) or on a drawer (**b**). The third container could be opened using either lid or drawer (two-ways-to-open type). The figure is from Kuroshima et al. (2008)

More recently, we investigated a similar though different ability in dogs. Our question was whether dogs would infer the weight of different doors upon observing a human demonstrator's actions with the doors. It is widely accepted that due to artificial selection, dogs have become high sensitive to various human communicative gestures such as pointing, facial expressions, vocalization, etc. (Albuquerque et al. 2016; Buttelmann and Tomasello 2013; Deputte and Doll 2011; Hare 1998, 1999; Kaminski and Nitzschner 2013; Miklosi et al. 1998; Turcsan et al. 2015). However, it is not clear whether the dogs make inferences about physical properties of the environment based on observed actions by humans or whether such inferences might be affected by the dogs' own active experiences. Kuroshima et al. (2017) addressed this topic by using a situation in which dogs observed differential manipulation of doors by humans.

The study consisted of two experiments. In Experiment 1, dogs saw a human demonstrator pushing open two swing doors; the doors looked the same, but one was "light" and the other was "heavy" (Fig. 8.5). The demonstrator pushed the light door

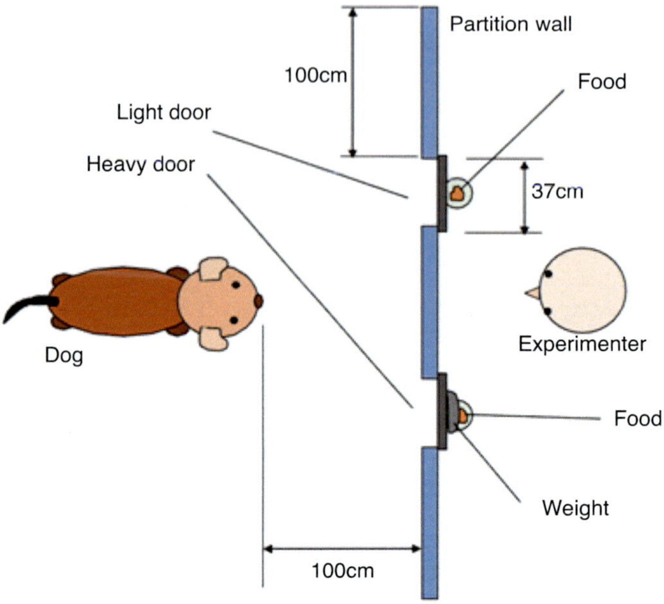

Fig. 8.5 A top view of the experimental setting. This figure is from Kuroshima et al. (2017)

quickly and with little effort, and the heavy door slowly and with some difficulty. A piece of food was placed behind each door. After the demonstration, we allowed the dogs to choose one door, push it open, and therefore get the food. We hypothesized that if the dogs inferred the relative heaviness of the doors based on how the doors moved in response to the human's action, they should prefer the light door, not the heavy one. However, in this situation the dogs chose randomly; they did not use the human's action (indirect clue) to represent the physical properties of the objects.

In Experiment 2, to assess the influence of one's own active experience (direct clue), we gave naïve dogs the opportunity to interact with both doors before observing the human demonstrator's action. The dogs were allowed to push open each door (light and heavy) and retrieve a food reward from behind each door; they did this twice for each door, once on the left and once on the right. We counterbalanced the door's weight and side for each trial. Through this experience, the dogs presumably learned that there was a heavy door and a light one, but without pushing, they could not tell which was which. Following this phase, the dogs watched the demonstrator's action on each door. We hypothesized that if the dog's active experience influenced their inference based on the demonstrator's actions, they should subsequently prefer the light door to the heavy one. In contrast to the results of Experiment 1, now 20 out of the 25 dogs tested (80%) chose the light door, an effect highly above chance ($p = 0.004$, two-tailed binominal test). There was no effect of age, sex, or body size on the dogs' preferences. The contrasting outcomes of Experiments 1 and 2 can be taken as evidence that active experience improved dogs' understanding of the demonstrator's action.

However, it is still unclear whether our dogs actually matched their own body movements and the human demonstrator's. In our study the dogs used their muzzle or head to push against the doors, whereas the human demonstrator used her hands. Also, the dogs opened only one door each time, but the demonstrator was shown to open each door twice. We do not yet know precisely what clues the dogs learn to use through their own action experiences. One possibility is that they pay attention to the movement of the doors, rather than the body movement of the demonstrator. This could be tested by running a "ghost" condition in which the dogs observe only the doors' movements (slow or quick) with no human action visible. Indeed, the same experiment could also be done using conspecific demonstrators.

Conclusion

Our studies about how one's own action experiences influence understanding of others' actions in capuchin monkeys and dogs (Kuroshima et al. 2008, 2014, 2017) show that active experience changes perception or understanding of others' action in these two species. This raises the possibility that such an ability to understand others may exist in other species of non-human animals too. However, there remain some problems that need to be addressed by researchers in this area.

First, as described earlier, subjects may predict others' actions based on the observed outcome (or end state) or an object's movement caused by the action, rather than understanding the goal or intention of the individual observed. Our results suggest that some knowledge acquired through one's own action experience can facilitate understanding of others' actions, but we cannot yet conclude confidently that the subjects understand others' mental states. Also, how do species transcend differences in morphology, for example, between humans and dogs? We need more information about what kinds of experience and what kinds of knowledge are required to understand others' mental states.

Second, for an individual to infer another's mental state through a kind of projection of self (as proposed by simulation theory), the individual should be aware of its own mental state. To what extent do non-human animals have awareness of their own mental states? How consciously can they access them? Recent studies have presented evidence for some metacognitive abilities in several species of non-human animals (apes: Call 2010; rhesus monkeys (*Macaca mulatta*): Brown et al. 2019; Hampton 2001; Templer and Hampton 2012; rats (*Rattus norvegicus*): Foote and Crystal 2007; Templer et al. 2017). Further studies could usefully consider a wider range of "self-understandings," including, for example, emotions, knowledge, memory, and the ability for introspection. And of course, testing with more animal species will enrich discussion of the evolutionary origins of these abilities. This line of research may eventually strengthen the argument that understanding of self is the foundation for understanding of others.

References

Albuquerque N, Guo K, Wilkinson A, Savalli C, Otta E, Mills D (2016) Dogs recognize dog and human emotions. Biol Lett 12(1):20150883. https://doi.org/10.1098/rsbl.2015.0883

Anderson JR, Kuroshima H, Takimoto A, Fujita K (2013a) Third-party social evaluation of humans by monkeys. Nat Commun 4:1561. https://doi.org/10.1038/ncomms2495

Anderson JR, Takimoto A, Kuroshima H, Fujita K (2013b) Capuchin monkeys judge third-party reciprocity. Cognition 127(1):140–146. https://doi.org/10.1016/j.cognition.2012.12.007

Anderson JR, Bucher B, Chijiiwa H, Kuroshima H, Takimoto A, Fujita K (2017) Third-party social evaluations of humans by monkeys and dogs. Neurosci Biobehav Rev. https://doi.org/10.1016/j.neubiorev.2017.01.003

Biro S, Leslie AM (2007) Infants' perception of goal-directed actions: development through cue-based bootstrapping. Dev Sci 10(3):379–398. https://doi.org/10.1111/j.1467-7687.2006.00544.x

Brown EK, Basile BM, Templer VL, Hampton RR (2019) Dissociation of memory signals for metamemory in rhesus monkeys (*Macaca mulatta*). Anim Cogn 22(3):331–341. https://doi.org/10.1007/s10071-019-01246-5

Buttelmann D, Tomasello M (2013) Can domestic dogs (*Canis familiaris*) use referential emotional expressions to locate hidden food? Anim Cogn 16(1):137–145. https://doi.org/10.1007/s10071-012-0560-4

Call J (2010) Do apes know that they could be wrong? Anim Cogn 13(5):689–700. https://doi.org/10.1007/s10071-010-0317-x

Call J, Hare B, Carpenter M, Tomasello M (2004) 'Unwilling' versus 'unable': chimpanzees' understanding of human intentional action. Dev Sci 7(4):488–498. https://doi.org/10.1111/j.1467-7687.2004.00368.x

Cannon EN, Woodward AL, Gredeback G, von Hofsten C, Turek C (2012) Action production influences 12-month-old infants' attention to others' actions. Dev Sci 15(1):35–42. https://doi.org/10.1111/j.1467-7687.2011.01095.x

Canteloup C, Meunier H (2017) 'Unwilling' versus 'unable': Tonkean macaques' understanding of human goal-directed actions. PeerJ 5:e3227. https://doi.org/10.7717/peerj.3227

Csibra G (2003) Teleological and referential understanding of action in infancy. Phil Trans R Soc Lond Ser B Biol Sci 358(1431):447–458. https://doi.org/10.1098/rstb.2002.1235

Custance D, Whiten A, Fredman T (1999) Social learning of an artificial fruit task in capuchin monkeys (*Cebus apella*). J Comp Psychol 113(1):13–23. https://doi.org/10.1037/0735-7036.113.1.13

Deputte BL, Doll A (2011) Do dogs understand human facial expressions? J Vet Behav 6(1):78–79. https://doi.org/10.1016/j.jveb.2010.09.013

Foote AL, Crystal JD (2007) Metacognition in the rat. Curr Biol 17(6):551–555. https://doi.org/10.1016/j.cub.2007.01.061

Fox NA, Bakermans-Kranenburg MJ, Yoo KH, Bowman LC, Cannon EN, Vanderwert RE (2016) Assessing human mirror activity with EEG mu rhythm: a meta-analysis. Psychol Bull 142(3):291–313. https://doi.org/10.1037/bul0000031

Fragaszy DM, Visalberghi E (1989) Social influences on the acquisition of tool-using behaviors in tufted capuchin monkeys (*Cebus apella*). J Comp Psychol 103(2):159–170. https://doi.org/10.1037/0735-7036.103.2.159

Fragaszy D, Visalberghi E (2004) Socially biased learning in monkeys. Learn Behav 32(1):24–35. https://doi.org/10.3758/bf03196004

Fragaszy D, Izar P, Visalberghi E, Ottoni EB, de Oliveira MG (2004) Wild capuchin monkeys (*Cebus libidinosus*) use anvils and stone pounding tools. Am J Primatol 64(4):359–366. https://doi.org/10.1002/ajp.20085

Fragaszy DM, Deputte B, Cooper EJ, Colbert-White EN, Hemery C (2011) When and how well can human-socialized capuchins match actions demonstrated by a familiar human? Am J Primatol 73(7):643–654. https://doi.org/10.1002/ajp.20941

Gallese V, Fadiga L, Fogassi L, Rizzolatti G (1996) Action recognition in the premotor cortex. Brain 119(2):593–609. https://doi.org/10.1093/brain/119.2.593

Gergely G, Csibra G (2003) Teleological reasoning in infancy: the naive theory of rational action. Trends Cogn Sci 7(7):287–292. https://doi.org/10.1016/s1364-6613(03)00128-1

Gergely G, Nádasdy Z, Csibra G, Bíró S (1995) Taking the intentional stance at 12 months of age. Cognition 56(2):165–193. https://doi.org/10.1016/0010-0277(95)00661-H

Hamlin JK, Hallinan EV, Woodward AL (2008) Do as I do: 7-month-old infants selectively reproduce others' goals. Dev Sci 11(4):487–494. https://doi.org/10.1111/j.1467-7687.2008.00694.x

Hampton RR (2001) Rhesus monkeys know when they remember. Proc Natl Acad Sci 98(9):5359. https://doi.org/10.1073/pnas.071600998

Hare B (1998) Communication of food location between human and dog (*Canis familiaris*). Evol Commun 2(1):137–159. https://doi.org/10.1075/eoc.2.1.06har

Hare B (1999) Domestic dogs (*Canis familiaris*) use human and conspecific social cues to locate hidden food. J Comp Psychol 113(2):173–177. https://doi.org/10.1037/0735-7036.113.2.173

Kaminski J, Nitzschner M (2013) Do dogs get the point? A review of dog–human communication ability. Learn Motiv 44(4):294–302. https://doi.org/10.1016/j.lmot.2013.05.001

Kanakogi Y, Itakura S (2011) Developmental correspondence between action prediction and motor ability in early infancy. Nat Commun 2:341. https://doi.org/10.1038/ncomms1342

Kano F, Krupenye C, Hirata S, Tomonaga M, Call J (2019) Great apes use self-experience to anticipate an agent's action in a false-belief test. Proc Natl Acad Sci USA 116(42): 20904–20909. https://doi.org/10.1073/pnas.1910095116

Király I, Jovanovic B, Prinz W, Aschersleben G, Gergely G (2003) The early origins of goal attribution in infancy. Conscious Cogn 12(4):752–769. https://doi.org/10.1016/s1053-8100(03)00084-9

Kuroshima H, Kuwahata H, Fujita K (2008) Learning from others' mistakes in capuchin monkeys (*Cebus apella*). Anim Cogn 11(4):599–609. https://doi.org/10.1007/s10071-008-0150-7

Kuroshima H, Kaiser I, Fragaszy DM (2014) Does own experience affect perception of others' actions in capuchin monkeys (*Cebus apella*)? Anim Cogn 17(6):1269–1279. https://doi.org/10.1007/s10071-014-0760-1

Kuroshima H, Nabeoka Y, Hori Y, Chijiiwa H, Fujita K (2017) Experience matters: dogs (*Canis familiaris*) infer physical properties of objects from movement clues. Behav Process 136:54–58. https://doi.org/10.1016/j.beproc.2017.01.013

Loucks J, Sommerville J (2018) Developmental change in action perception: is motor experience the cause? Infancy 23(4):519–537. https://doi.org/10.1111/infa.12231

Meltzoff AN, Brooks R (2008) Self-experience as a mechanism for learning about others: a training study in social cognition. Dev Psychol 44(5):1257–1265

Miklosi A, Polgardi R, Topal J, Csanyi V (1998) Use of experimenter-given cues in dogs. Anim Cogn 1(2):113–121. https://doi.org/10.1007/s100710050016

Munkenbeck Fragaszy DM, Visalberghi E (1990) Social processes affecting the appearance of innovative behaviors in capuchin monkeys. Folia Primatol 54(3-4):155–165. https://doi.org/10.1159/000156439

Myowa-Yamakoshi M, Kawakita Y, Okanda M, Takeshita H (2011) Visual experience influences 12-month-old infants' perception of goal-directed actions of others. Dev Psychol 47(4):1042–1049. https://doi.org/10.1037/a0023765

Myowa-Yamakoshi M, Scola C, Hirata S (2012) Humans and chimpanzees attend differently to goal-directed actions. Nat Commun 3:693. https://doi.org/10.1038/ncomms1695

Ottoni EB, Mannu M (2001) Semifree-ranging tufted capuchins (*Cebus apella*) spontaneously use tools to crack open nuts. Int J Primatol 22(3):347–358. https://doi.org/10.1023/A:1010747426841

Ottoni EB, de Resende BD, Izar P (2005) Watching the best nutcrackers: what capuchin monkeys (*Cebus apella*) know about others' tool-using skills. Anim Cogn 8(4):215–219. https://doi.org/10.1007/s10071-004-0245-8

Phillips AT, Wellman HM (2005) Infants' understanding of object-directed action. Cognition 98 (2):137–155. https://doi.org/10.1016/j.cognition.2004.11.005

Phillips W, Barnes JL, Mahajan N, Yamaguchi M, Santos LR (2009) 'Unwilling' versus 'unable': capuchin monkeys' (*Cebus apella*) understanding of human intentional action. Dev Sci 12 (6):938–945. https://doi.org/10.1111/j.1467-7687.2009.00840.x

Rizzolatti G, Fadiga L, Gallese V, Fogassi L (1996) Premotor cortex and the recognition of motor actions. Cogn Brain Res 3(2):131–141. https://doi.org/10.1016/0926-6410(95)00038-0

Rizzolatti G, Fogassi L, Gallese V (2001) Neurophysiological mechanisms underlying the understanding and imitation of action. Nat Rev Neurosci 2(9):661–670. https://doi.org/10.1038/35090060

Rosander K, von Hofsten C (2011) Predictive gaze shifts elicited during observed and performed actions in 10-month-old infants and adults. Neuropsychologia 49(10):2911–2917. https://doi.org/10.1016/j.neuropsychologia.2011.06.018

Sobel DM, Sommerville JA (2010) The importance of discovery in children's causal learning from interventions. Front Psychol 1:176. https://doi.org/10.3389/fpsyg.2010.00176

Sommerville JA, Woodward AL (2005) Pulling out the intentional structure of action: the relation between action processing and action production in infancy. Cognition 95(1):1–30. https://doi.org/10.1016/j.cognition.2003.12.004

Sommerville JA, Woodward AL, Needham A (2005) Action experience alters 3-month-old infants' perception of others' actions. Cognition 96(1):B1–B11. https://doi.org/10.1016/j.cognition.2004.07.004

Templer VL, Hampton RR (2012) Rhesus monkeys (*Macaca mulatta*) show robust evidence for memory awareness across multiple generalization tests. Anim Cogn 15(3):409–419. https://doi.org/10.1007/s10071-011-0468-4

Templer VL, Lee KA, Preston AJ (2017) Rats know when they remember: transfer of metacognitive responding across odor-based delayed match-to-sample tests. Anim Cogn 20(5):891–906. https://doi.org/10.1007/s10071-017-1109-3

Turcsan B, Szantho F, Miklosi A, Kubinyi E (2015) Fetching what the owner prefers? Dogs recognize disgust and happiness in human behaviour. Anim Cogn 18(1):83–94. https://doi.org/10.1007/s10071-014-0779-3

Visalberghi E (1987) Acquisition of nut-cracking behaviour by 2 capuchin monkeys (*Cebus apella*). Folia Primatol 49(3-4):168–181. https://doi.org/10.1159/000156320

Williamson RA, Meltzoff AN (2011) Own and others' prior experiences influence children's imitation of causal acts. Cogn Dev 26(3):260–268. https://doi.org/10.1016/j.cogdev.2011.04.002

Williamson RA, Meltzoff AN, Markman EM (2008) Prior experiences and perceived efficacy influence 3-year-olds' imitation. Dev Psychol 44(1):275–285

Wood LA, Kendal RL, Flynn EG (2013) Copy me or copy you? The effect of prior experience on social learning. Cognition 127(2):203–213. https://doi.org/10.1016/j.cognition.2013.01.002

Woodward AL (1998) Infants selectively encode the goal object of an actor's reach. Cognition 69 (1):1–34. https://doi.org/10.1016/s0010-0277(98)00058-4

Woodward AL (1999) Infants' ability to distinguish between purposeful and nonpurposeful behaviors. Infant Behav Dev 22(2):145–160. https://doi.org/10.1016/s0163-6383(99)00007-7

Woodward AL (2005) The infant origins of intentional understanding. Adv Child Dev Behav 33:229–262

Woodward AL, Guajardo JJ (2002) Infants' understanding of the point gesture as an object-directed action. Cogn Dev 17(1):1061–1084. https://doi.org/10.1016/s0885-2014(02)00074-6

Woodward AL, Sommerville JA, Gerson S, Henderson AME, Buresh J (2009) The emergence of intention attribution in infancy. In: The psychology of learning and motivation, vol 51. Elsevier, London, pp 187–222. https://doi.org/10.1016/s0079-7421(09)51006-7

Chapter 9
Behavioral Coordination and Synchronization in Non-human Primates

Yuko Hattori

Abstract Primate species are highly social animals, capable of collaborating and coordinating with others. Such activities are used not only for purposes such as hunting, territorial defense and inter-group competition, but also for social bonding. In humans, across many cultures behavioral coordination and synchrony occur in activities such as marching, dancing and singing. Advanced rhythmic ability underlies such sophisticated activities, but the evolutionary pathway of this ability is unclear. In this chapter, I review coordination and collaboration in non-human primates, focusing on timing and rhythm, which have recently been attracting increased attention from researchers. Based on comparisons of rhythmic ability in humans and non-humans, possible evolutionary pathways of coordination and synchronization for bonding in humans are also discussed.

Keywords Behavioral coordination · Behavioral synchrony · Social cognition · Rhythm · Communication · Entrainment

Behavioral Coordination in Non-human Primates

Highly social primates often flexibly collaborate and coordinate their movements for various purposes. Functional purposes include hunting, territorial defense, and inter-group competition (e.g., Boesch 2003; Langergraber et al. 2017), observed in many primate species. Cooperation generally refers to the behavioral coordination of two or more individuals for a common goal. However, not many studies have investigated to what extent primates coordinate with each other when a complex sequence of actions is necessary for solving a problem. In one example of such research, Hattori et al. (2005) asked whether capuchin monkeys would spontaneously cooperate and divide sequential actions by coordinating their movements. Six monkeys

Y. Hattori (✉)
Primate Research Institute, Kyoto University, Inuyama, Japan
e-mail: hattori.yuko.3w@kyoto-u.ac.jp

J. R. Anderson, H. Kuroshima (eds.), *Comparative Cognition*,
https://doi.org/10.1007/978-981-16-2028-7_9

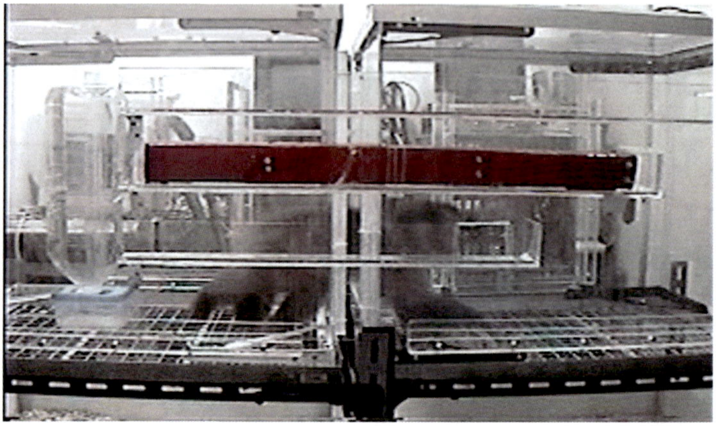

Fig. 9.1 A capuchin monkey, Pigmon, operates a cooperative problem-solving task by himself

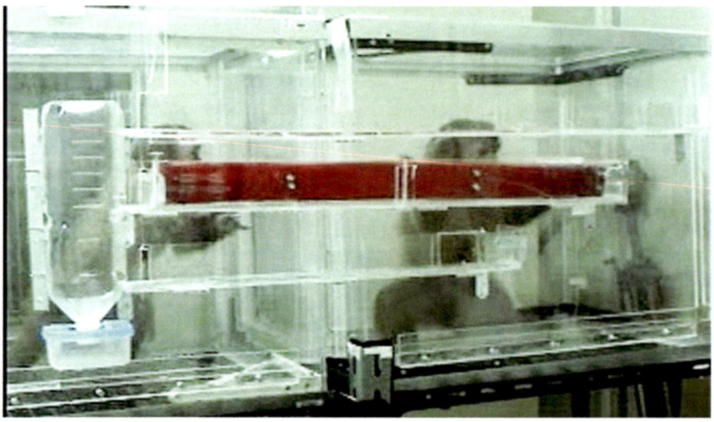

Fig. 9.2 Two capuchin monkeys, (Kiki (left) and Pigmon (right), participate in a cooperative problem-solving task

were trained individually to perform the necessary sequence of actions for rewards, and then tested in pairs to see if they could solve the task by spontaneously dividing the sequence of actions (Figs. 9.1 and 9.2). Although they knew how to solve the task when they were alone, when they were in the same situation with a partner, the monkeys quickly learned to collaborate with each other by coordinating their actions. After just one or two trials, all three pairs solved the task quickly (within 10 s), showing the monkeys' ability to coordinate their movements flexibly even in an artificial, complex cooperative task.

In addition to situations in which they are pursuing a common goal, humans and other primates also coordinate and synchronize their movements for social bonding purposes. During conversation, for example, people spontaneously move in synchrony with each other's speech rhythms (Newtson 1994) and match one another's postures (Shockley et al. 2003). Moreover, humans actively coordinate and synchronize their movement in music-related activities such as dancing and singing. In many cultures across the world behavioral coordination and synchrony occur in activities such as marching, dancing, and singing (McNeill 1995). We can capture and interpret a beat in a rhythmic pattern in infants as young as 5–24 months (Winkler et al. 2009); this early ability later allows us to sing and dance in time to music. Such advanced synchronous movement even occurs unconsciously when we hear rhythmically complex music. Humans, frequently coordinate nonconsciously with each other, and this has many positive social consequences: When others mimic us, we like them more, empathize with them more, and are more helpful to them (e.g., Chartrand and van Baaren 2009). Although imitation is relatively rare in NHPs (NHPs), they recognize when their actions are being imitated (Haun and Call 2008; Paukner et al. 2005; Nielsen et al. 2005) and there is some evidence that they show more affiliative responses to those who recently imitated them (Paukner et al. 2009).

In monkeys and apes, social play is one of the most frequently observed phenomena involving movement coordination between individuals (Palagi and Paoli 2007). Burghardt (2005) defines play as any behavior that meets all five of the following criteria: (1) it must have elements that do not contribute toward the organism's immediate survival; (2) it must be spontaneous or rewarding to the organism; (3) it must differ from strictly functional expressions of behavior either structurally or temporally; (4) it must be repeated in a similar form during at least a portion of the organism's ontogeny; and (5) it occurs when the organism is in a "relaxed field," meaning that it is healthy, sated, and not under stress from competing systems (e.g., feeding, mating, fear). Although many NHPs play with other group members in both simple and complex ways, great apes (i.e., bonobos and chimpanzees, Palagi 2006) play with greater frequency than many other primates. Furthermore, sometimes during play facial mimicry of the play face, which represents positive emotion, occurs between individuals (i.e., open-mouth faces (OMFs), Davila Ross et al. 2008). This suggests that playful mood and positive emotion are "contagious" between individuals.

Compared to other primate species, humans seem to coordinate and collaborate with others in more sophisticated ways. This might be because humans have an advanced ability to synchronize their movements to external rhythms. Recently, chimpanzees were reported to also spontaneously coordinate with a partner's movements (Yu and Tomonaga 2016). However, little is yet known about the extent to which such rhythmic ability is shared among non-human primates (NHPs). In this chapter, I review coordination and collaboration in NHPs, focusing on timing and rhythm, which have been largely neglected by researchers. I will also introduce recent findings about sensitivity to, and influence of, auditory rhythms on movements in NHPs, comparing them with humans. Finally, I discuss how similar rhythmic ability is, between humans and NHPs, given its importance as a foundation for collaborating and coordinating with others, and the possible evolutionary emergence of human advanced rhythmic abilities.

Coordination in Primate Vocal Communication

Vocal duetting by adult pairs has been found in monogamous old-world species: *Indri, Tarsius*, *Presbytis*, and *Hylobates* (Haimoff 1986). Although organizational features differ across these species, the duets are produced by coordinating of vocalizations by mated pairs, and the duetting in the latter three species is sequentially organized (Haimoff 1986). These songs last from 5 to 30 min, which is far longer than vocalizations in other primate species. The main function of the duetting behavior is thought to be maintenance of social cohesion within the bonded pair and spatial organization among neighboring family groups.

With much simpler and fewer sequences, other primates also coordinate or exchange vocalizations among bonded individuals. Some primates also modify characteristics of their vocalizations to be similar to other individuals' vocalizations. For example, in infant marmosets, turn-taking with parents in early development shapes their phee call (Takahashi et al. 2016). Even adult pygmy marmosets (*Cebuella pygmaea*) and common marmosets (*Callithrix jacchus*) modify their call structures when bonded (Snowdon and Elowson 1999). Another study using playback found that Japanese macaques matched some acoustic features of their coo calls in reply to calls that they heard (Sugiura 1998). Chimpanzees (*Pan troglodytes*) were also reported to adjust the temporal structure of acoustic vocalizations during chorusing with other individuals (Fedurek et al. 2013). These studies indicate that, although NHP vocalizations are quite stereotyped compared to humans, to some extent they adjust duration, frequency or timing of their vocalizations to exchange or coordinate them, and that this phenomenon might function to strengthen social relationships.

Sensitivity to Auditory Rhythms in Primates

Although vocal control in NHPs is less flexible than in humans, the former are sensitive to complex sound patterns that they themselves cannot produce. For example, one study measured event-related potentials (ERPs) in an awake chimpanzee and found that she processed the sound of her name differently from other sounds. However, to what extent NHPs perceive and recognize auditory rhythms is not well known. In humans, newborn babies perceive and anticipate timing of an eight-beat pattern. Winkler et al. (2009) presented sleeping newborns with rock drum rhythms, with occasional omissions at different metrical positions. They used ERPs to measure the mismatch negativity (MMN) as a response to expectancy violation. They found that omitting sound at the most salient metrical position elicited a larger MMN than an omission at the lowest level of metrical salience, suggesting that newborn babies (even when asleep) perceive beat and anticipate the arrival of the next sound. Honing et al. (2012, 2018) tested macaque monkeys using the same procedure, but they found no compelling evidence that monkeys perceive beat. However, monkeys did perceive isochronous rhythms, showing larger MMNs when sound failed to arrive when it was supposed to than when it did. So, monkeys can perceive simple rhythm and anticipate the timing of the next sound, but they do not perceive metrical structure. Anticipating future sounds based on past events was also shown in chimpanzees testing with an auditory oddball paradigm (Ueno et al. 2008). Therefore, NHPs perceive isochronous rhythms, but so far there is no evidence that they perceive complex beat in the same way as humans.

Intrinsic Motivation for Synchronizing Movements to External Rhythms

The extent to which NHPs can match their movements to external rhythms has been studied only in the most recent decade. Most studies were conducted with macaque monkeys and great apes (chimpanzees and bonobos). Overall, their ability is rudimentary compared with humans, but there are also some differences in response to external rhythms between monkeys and apes. Zarco et al. (2009) first systematically trained rhesus monkeys to tap with whole hand on a button in synchrony with auditory and visual rhythms and compared them with humans. It took several years to train monkeys to tap multiple times reliably synchronized to a metronome (i.e., more than 70% trials correct), which suggests that monkeys do not have intrinsic motivation for matching movement to external rhythms, and that this task is difficult for them.

By contrast, humans have a clear tendency to synchronize their movements to auditory rhythms such as music, even in early development. Infants who showed longer rhythmic engagement also smiled for longer, suggesting that coordinating and synchronizing their movements to musical beat was rewarding to them (Zentner and Eerola 2010). Zarco et al. (2009) also reported that bias toward auditory as opposed to visual cues for accurately maintaining time intervals was observed in humans but not rhesus monkeys. Instead, monkeys were shown to accurately synchronize eye movements to visual rhythms (Takeya et al. 2018), suggesting that monkeys may be better at coordinating movements to visual rhythms than auditory rhythms. Another study that used a button-pressing behavioral measure between face-to-face pairs of monkeys showed that visual information from the partner induced a higher degree of synchronization, whereas auditory information had no such effect (Nagasaka et al. 2013), consistent with a visual modality bias for coordinating their movements in monkeys.

To explore this issue in great apes, I introduced an electric keyboard to three chimpanzees, and trained them to tap two keys alternately 30 times (Hattori et al. 2013, 2015).

A metronome sound at different tempi was played as distractor stimuli while they were tapping, and we investigated whether the chimpanzees would spontaneously synchronize their tapping to the onset of the sound stimulus. Overall, the

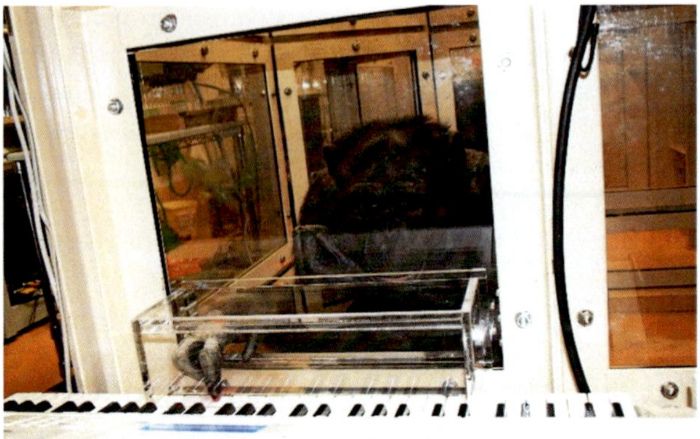

Fig. 9.3 A female chimpanzee, Ai, is tapping a keyboard

chimpanzees' tapping was entrained intermittently to the metronome when the tempo was close to natural tapping tempo of each chimpanzee. For example, when no stimulus sound was presented, the natural tapping rate of chimpanzee Ai was 578.5 ms (median tapping rate in No stimulus condition). Only when hearing the 600-ms beat interval did Ai spontaneously and accurately aligne her tapping to the beat, even though she had never been required to pay attention to auditory stimuli. This indicates that individual chimpanzees have intrinsic motivation for aligning movements to auditory rhythms, a phenomenon not observed in macaque monkeys. Interestingly, chimpanzees showed not only synchronized tapping (i.e., matching their tapping to the onset of a sound stimulus), but also another type of coordination. Additional analysis revealed that chimpanzee Ayumu, for example, coordinated his tapping with sound in antiphase or 60° (Fig. 9.3) in some sessions. However, several differences from synchronization in human tapping were also observed. First, none of the chimpanzees showed flexible alignment of tapping to the metronome other than the one close to their spontaneous tapping rate. By contrast, humans can intentionally synchronize their tapping to various rates in a range between 200 ms and 1800 ms (Repp 2005). Additionally, the chimpanzees' tapping accuracy was relatively weak and lacking evidence of negative asynchrony, which commonly appears when humans try to tap in synchrony with auditory rhythms (Repp 2006). Therefore, this spontaneous synchronization or entrainment is considered unintentional, which makes sense as the chimpanzees were only trained to tap two keys and never to pay attention to auditory stimuli. A similar type of synchronization or entrainment has also been reported in a bonobo (Large and Gray 2015), Kuni, who entrained and synchronized her drum strikes within a range around her spontaneous motor tempo. Kuni had never been trained to match her movement to auditory rhythm, so what emerged was her spontaneous response. Although the stimulus sound was produced by a human experimenter's drumming movements and the effect might not be due to only auditory stimuli, it appears that bonobos and chimpanzees have some intrinsic motivation for synchronizing movements to external rhythms.

Concerning modality effects on movement timing, one recent study showed that there is no clear modality bias in chimpanzees. Pairs of chimpanzees participated in a finger-tapping task on a touch-panel while they either sat next to each other (Yu and Tomonaga 2015) or sat face-to-face (Yu and Tomonaga 2016). In both situations, one chimpanzee spontaneously adjusted her tapping so that the difference in tapping onset between her and her partner (i.e., her mother) became smaller. This indicates that, although not all chimpanzees will adjust their movement with each other, some chimpanzee do so when they perceive other's rhythms either auditory or visually.

Effect of Auditory Rhythms on Movement: Sound-Induced Rhythmic Movement

In humans, listening to music induces rhythmic movement, suggesting a close connection between the auditory and motor areas in the brain. From an early stage of development, human infants spontaneously engage in movement to music (Fujii et al. 2014; Zentner and Eerola 2010). Later, this movement becomes more accurately synchronized with the pulse of music (Eerola et al. 2006). Although previous studies have largely focused on finger tapping to reveal rhythmic ability in humans, music induces a variety of body movements. Research suggests that when humans dance to music, the whole body and body parts are entrained differently at each metrical level (Burger et al. 2013, 2014; Toiviainen et al. 2010). Neurobiological studies have shown that this advanced rhythmic ability depends on close connections between auditory and motor areas in the brain (for reviews see Zatorre et al. 2007; Patel and Iversen 2014). Motor areas in the brain are recruited even when humans listen passively to beat-based rhythms, suggesting that the motor system plays an important role in processing such rhythms. For example, in a study using functional magnetic resonance imaging participants were instructed not to move when hearing the beat (Grahn and Rowe 2009). Despite the lack of movement that study found increased levels of activity in the putamen (basal ganglia) and supplementary motor area when participants listened to beat-based versus non beat-based rhythms (Grahn and Rowe 2009).

Among NHPs, chimpanzees have been described as performing "rain dances" and "waterfall dances" in the wild. At the start of a heavy downpour, adult male chimpanzees show rhythmic swaying that differs from normal charging displays in that it omits any intimidation of other chimpanzees. There are also some differences in the rain dances performed between sites (Whiten et al. 1999). For example, at Tai, the rain dance was reported to be totally silent and similar to a slow-motion version of normal display (Whiten et al. 2001), whereas at Gombe, it was described as a wild social display with noisy pant-hoots that occurred at the onset of heavy storms (Goodall 1986; Hashimoto 1998). These displays are considered as some kind of collective response to the sound and fury of nature, and have been linked to human rituals in response to natural forces that are sensorially overwhelming. However,

Fig. 9.4 A male chimpanzee, Akira, swaying while sitting with his feet on the floor and hearing the auditory stimulus (Hattori and Tomonaga 2020)

despite its existence being known for more than half a century, the rain dance has not yet been analyzed in detail (McGrew 2011). Chimpanzees also show vigorous displays at waterfalls. According to Goodall (2005): "...Deep in the forest are some spectacular waterfalls. Sometimes as a chimpanzee—most often an adult male—approaches one of these falls his hair bristles slightly, a sign of heightened arousal. As he gets closer, and the roar of falling water gets louder, his pace quickens, his hair becomes fully erect, and upon reaching the stream he may perform a magnificent display close to the foot of the falls. Standing upright, he sways rhythmically from foot to foot, stamping in the shallow, rushing water, picking up and hurling great rocks...This 'waterfall dance' may last for ten or fifteen minutes." This waterfall dance is more interesting than the rain dance because the site of the waterfall is constant and should be familiar, but the chimpanzees approach it to engage in such activity (McGrew 2011).

Hattori and Tomonaga (2020) experimentally investigated whether sound actually induces any type of movement in chimpanzees. In animal welfare research, rhythmic movement such as swaying or rocking induced is often described as "stereotyped behavior," which may be induced by sound such as white noise (Berkson and Mason 1964). These responses are reported mostly in individuals reared in inadequate captive environments, often socially deprived. However, the chimpanzees we tested are group-reared and housed in a large, environmentally enriched compound with other group members (Matsuzawa et al. 2006). We conducted playback experiment with seven adults (three males, four females) using a typical 2-measure rock drum accompaniment pattern as previously used in studies with infants and macaque monkeys (Honing et al. 2012, 2018). We found that all seven chimpanzees engaged with an auditory beat, but there was clear sex difference: the beat induced more rhythmic movement and vocalization in male than female chimpanzees (Fig. 9.4).

The longer and more vigorous whole-body swaying, and more frequent and longer vocalizations by the male chimpanzees, suggest that males may be more sensitive and responsive to auditory stimuli than females. Actually, this is consistent with reports on wild chimpanzees; for example, acoustic communication occurs more frequently in males than females, including drumming (Arcadi et al. 1998,

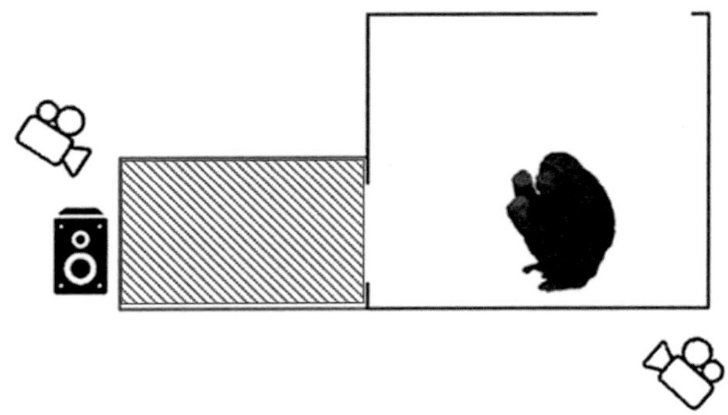

Fig. 9.5 Experimental setting in which preference for sound was tested. The hatched area represents the "sound source area"

2004) and chorusing (Fedurek et al. 2013). Higher sensitivity and hence larger response to sounds in males may be related to chimpanzees' patriarchal society, in which males frequently use acoustic communication to confirm their social relationships and protect their territory and community members.

We continued to explore the effect of the type of rhythm on movement in one male chimpanzee, Akira, who responded more than 50% of the time the sound was played. Specifically, we tested how rhythm tempo affected his movement periodicity and whether only regular rhythmic patterns induced rhythmic movement. Akira showed two types of swaying movements: horizontally while in a quadrupedal posture and vertically while in a bipedal posture. Therefore, we analyzed the effect of beat tempo on movement periodicity separately for each posture. Akira responded flexibly to different sound tempi in both postures. In bipedal posture, movement periodicity was positively correlated with beat tempo, meaning that he swayed faster when the beat sound was faster. But while standing quadrupedal, he swayed more stably to the middle-tempo beats than to the other two tempi, possibly because the former was his preferred tempo and he was less distracted (Hattori and Tomonaga 2020). These responses are markedly different from distressed stereotyped responses to aversive stimuli reported in the animal welfare literature. However, rhythmic swaying was also induced by random rhythms, from which we do not normally extract regular pulse. Humans infants showed less rhythmic movement when they heard speech than music (Zentner and Eerola 2010). Therefore, different responses to regular and random rhythms might be a uniquely human trait.

Another important aspect of chimpanzees' sound-induced rhythmic swaying is the possible positive emotion during those activities. We measured Akira's distance from the sound source and found that he stayed near it for significantly longer when sounds were played that when no sounds were played (Hattori and Tomonaga 2020). This suggests that the sound was an attractive stimulus for him. This tendency was replicated when the sound was played only when he was already in proximity to the sound source (Fig. 9.5).

Following the procedure in the previous study (Experiment 1 in Hattori and Tomonaga 2020), a trial took 120 s (2 min). We ran only 1 trial per day, and 10 trials in total were conducted. We played the sound stimulus (inter-onset interval: 180 ms = 83 bpm) only when Akira was in the sound source area (Fig. 9.5). The experiment

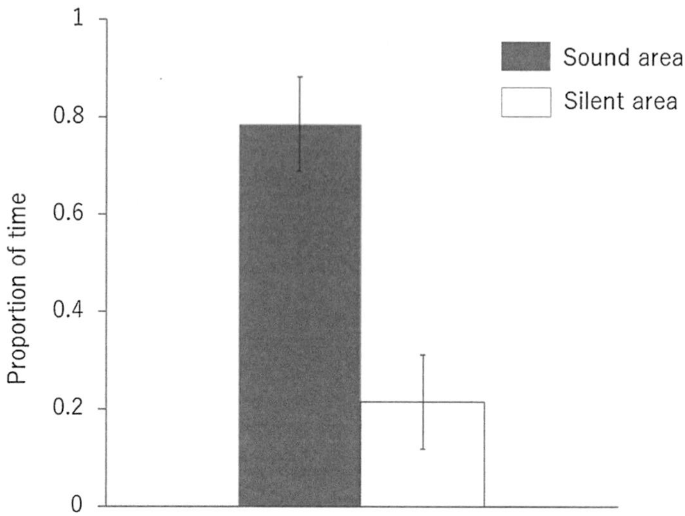

Fig. 9.6 Proportion of time spent in each area with 95% CI

was recorded by a digital video camera and Akira's location was analyzed with Adobe Premier Pro CC2020.

I conducted a one-sample two-tailed t-test looking whether the proportion of time spent in the sound source area was different from chance (0.5) and found significant difference [$t(9) = 6.660$, $P < 0.001$]. This indicates that Akira preferred to stay close to where he could hear the sound, indicating that his rhythmic swaying was not an aversive reaction.

Evolution of Coordinating and Synchronizing Movements for Social Bonding

Humans and other primates use social touching or grooming for reinforcing social relationships (Dunbar and Shultz 2010). Several studies have suggested that the μ-opioid-receptor (MOR) system might be involved in this social bonding mechanism (Meller et al. 1980; Fabre-Nys et al. 1982). However, because the size of human social networks exceeds what can be maintained by dyadic social touching (Dunbar 2012), it has been proposed that other mechanisms such as social laughter, singing, and dancing have evolved to release endogenous opioids, as grooming does. In humans, social laughter is known to trigger endogenous opioid release (Manninen et al. 2017). It is possible social play-induced laughter in NHPs activates similar neurochemical mechanisms for social bonding.

Since human ancestors diverged from the common ancestor with chimpanzees and bonobos ca 7 million years ago, Hominin group size has increased progressively to a community size of around 150 (Dunbar 2008, 2012). In the course of human evolution, limitations on time for maintaining close social relationship by grooming required a change in behaviors used in bonding. Sophisticated rhythmic coordination and collaboration such as dancing and singing allow humans to engage in a form of grooming-at-a-distance, sometimes involving several individuals simultaneously. Modern humans' uniquely advanced rhythmic abilities enable us to engage in such activities with ease. But although the exact same activities are not observed in NHPs, some underlying mechanisms are shared to various degrees with other species.

References

Arcadi AD, Robert C, Boesch (1998) Buttress drumming by wild chimpanzees: temporal patterning, phrase integration into loud calls, and preliminary evidence for individual distinctiveness. Primates 39:505–518

Arcadi AC, Robert F, Mugurusi (2004) A comparison of buttress drumming by male chimpanzees from two populations. Primates 45:135–139

Berkson G, Mason WA (1964) Stereotyped behaviors of chimpanzees: relation to general arousal and alternative activities. Percept Mot Skills 19:635–662

Boesch C (2003) Complex cooperation among Taï chimpanzees. In: de Waal FBM, Tyack PL (eds) Animal social complexity: intelligence, culture, and individualized societies. Harvard University Press, Cambridge, pp 93–110. https://doi.org/10.4159/harvard.9780674419131.c7

Burger B, Thompson MR, Luck G, Saarikallio S, Toiviainen P (2013) Influences of rhythm- and timbre-related musical features on characteristics of music-induced movement. Front Psychol 4:183

Burger B, Thompson MR, Luck G, Saarikallio SH, Toiviainen P (2014) Hunting for the beat in the body: on period and phase locking in music-induced movement. Front Hum Neurosci 8:903

Burghardt GM (2005) The genesis of animal play: testing the limits. MIT Press, Cambridge

Chartrand TL, van Baaren R (2009) Human mimicry. In: Zanna M (ed) Advances in experimental social psychology, vol 41. Academic, San Diego, pp 219–274

Davila Ross M, Menzler S, Zimmermann E (2008) Rapid facial mimicry in orangutan play. Biol Lett 4(1):27–30

Dunbar RI (2008) Mind the gap: or why humans aren't just great apes. Proc Br Acad 154:403–423

Dunbar RI (2012) Bridging the bonding gap: the transition from primates to humans. Phil Trans R Soc Lond B 367:1837–1846

Dunbar RIM, Shultz S (2010) Bondedness and sociality. Behaviour 147(7):775–803

Eerola T, Luck G, Toiviainen P (2006) An investigation of pre-schoolers' corporeal synchronization with music. In: Baroni M, Addessi AR, Caterina R, Costa M (eds) Proceedings of the 9th international conference on music perception and cognition. Alma Mater Studiorium University of Bologna, Bologna, pp 472–476

Fabre-Nys C, Meller RE, Keverne EB (1982) Opiate antagonists stimulate affiliative behaviour in monkeys. Pharmacol Biochem Behav 16:653–659

Fedurek P, Schel AM, Slocombe KE (2013) The acoustic structure of chimpanzee panthooting facilitates chorusing. Behav Ecol Sociobiol 67:1781–1789

Fujii S et al (2014) Precursors of dancing and singing to music in three- to four-months-old infants. PLoS One 9:e97680

Goodall J (1986) The chimpanzees of gombe. In: Pattern of behavior. The Belknap Press of Harvard University Press, Cambridge

Goodall J (2005) Primate spirituality. Encyclopedia of religion and nature. Continuum, London

Grahn JA, Rowe JB (2009) Feeling the beat: premotor and striatal interactions in musicians and nonmusicians during beat perception. J Neurosci 29:7540–7548
Haimoff EH (1986) Convergence in the duetting of monogamous old world primates. J Hum Evol 15:51–59. https://doi.org/10.1016/S0047-2484(86)80065-3
Hashimoto C (1998) Chimpanzees of the Kalinzu forest, Uganda. Pan Afr News 5:6–8
Hattori Y, Tomonaga M (2020) Rhythmic swaying induced by sound in chimpanzees (*Pan troglodytes*). Proc Natl Acad Sci 117(2):936–942
Hattori Y, Kuroshima H, Fujita K (2005) Cooperative problem solving by tufted capuchin monkeys (*Cebus apella*): spontaneous division of labor, communication, and reciprocal altruism. J Comp Psychol 119(3):335
Hattori Y, Tomonaga M, Matsuzawa T (2013) Spontaneous synchronized tapping to an auditory rhythm in a chimpanzee. Sci Rep 3:1566. https://doi.org/10.1038/srep01566
Hattori Y, Tomonaga M, Matsuzawa T (2015) Distractor effect of auditory rhythms on self-paced tapping in chimpanzees and humans. PLoS One 10(7):e0130682. https://doi.org/10.1371/journal.pone.0130682
Haun DBM, Call J (2008) Imitation recognition in great apes. Curr Biol 18(7):R288–R290
Honing H, Merchant H, Háden GP, Prado L, Bartolo R (2012) Rhesus monkeys (*Macaca mulatta*) detect rhythmic groups in music, but not the beat. PLoS ONE 7:e51369. https://doi.org/10.1371/journal.pone.0051369
Honing H, Bouwer FL, Prado L, Merchant H (2018) Rhesus monkeys (*Macaca mulatta*) sense isochrony in rhythm, but not the beat: additional support for the gradual audiomotor evolution hypothesis. Front Neurosci 12:475. https://doi.org/10.3389/fnins.2018.00475
Langergraber KE, Watts DP, Vigilant L, Mitani JC (2017) Group augmentation, collective action, and territorial boundary patrols by male chimpanzees. Proc Natl Acad Sci 114(28):7337–7342. https://doi.org/10.1073/pnas.1701582114
Large EW, Gray PM (2015) Spontaneous tempo and rhythmic entrainment in a bonobo (*Pan paniscus*). J Comp Psychol 129:317–328
Matsuzawa T, Tomonaga M, Tanaka M (2006) Cognitive development in chimpanzees. Springer, Tokyo
Manninen S, Tuominen L, Dunbar RI et al (2017) Social laughter triggers endogenous opioid release in humans. J Neurosci 37(25):6125–6131. https://doi.org/10.1523/JNEUROSCI.0688-16.(2017)
McGrew WC (2011) Pan symbolicus. In: Henshilwood CS, D'Errico F (eds) Homo symbolicus. John Benjamins Publishing Company, Amsterdam, pp 1–12
McNeill WH (1995) Keeping together in time: dance and drill in human history. Harvard University Press, Cambridge
Meller RE, Keverne EB, Herbert J (1980) Behavioral and endocrine effects of naltrexone in male talapoin monkeys. Pharmacol Biochem Behav 13:663–672
Nagasaka Y, Chao ZC, Hasegawa N, Notoya T, Fujii N (2013) Spontaneous synchronization of arm motion between Japanese macaques. Sci Rep 3:1151. https://doi.org/10.1038/srep01151
Newtson D (1994) The perception and coupling of behavior waves. In: Vallacher RR, Nowak A (eds) Dynamical systems in social psychology. Academic Press, San Diego, pp 139–167
Nielsen M, Collier-Baker E, Davis JM, Suddendorf T (2005) Imitation recognition in a captive chimpanzee (*Pan troglodytes*). Anim Cogn 8:31–36. https://doi.org/10.1007/s10071-004-0232-0
Palagi E (2006) Social play in bonobos (*Pan paniscus*) and chimpanzees (*Pan troglodytes*): implications for natural social systems and inter-individual relationships. Am J Phys Anthropol 129(3):418–426
Palagi E, Paoli T (2007) Play in adult bonobos (*Pan paniscus*): modality and potential meaning. Am J Phys Anthropol 134(2):219–225
Patel AD, Iversen JR (2014) The evolutionary neuroscience of musical beat perception: the action simulation for auditory prediction (ASAP) hypothesis. Front Syst Neurosci 8:57
Paukner A, Anderson JR, Borelli E, Visalberghi E, Ferrari PE (2005) Macaques (*Macaca nemestrina*) recognize when they are being imitated. Biol Lett 1:219

Paukner A, Suomi SJ, Visalberghi E, Ferrari PF (2009) Capuchin monkeys display affiliation toward humans who imitate them. Science 325:880–883. https://doi.org/10.1126/science.1176269
Repp BH (2005) Sensorimotor synchronization: a review of the tapping literature. Psychon Bull Rev 12(6):969–992
Repp BH (2006) Does an auditory distractor sequence affect self-paced tapping? Acta Psychol 121 (1):81–107
Shockley K, Santana MV, Fowler CA (2003) Mutual interpersonal postural constraints are involved in cooperative conversation. J Exp Psychol 29:326–332
Snowdon CT, Elowson AM (1999) Pygmy marmosets modify call structure when paired. Ethology 105(10):893–908
Sugiura H (1998) Matching of acoustic features during the vocal exchange of coo calls by Japanese macaques. Anim Behav 55:673–687
Takahashi DY, Fenley AR, Ghazanfar AA (2016) Early development of turn-taking with parents shapes vocal acoustics in infant marmoset monkeys. Phil Trans R Soc Lond B 371:20150370
Takeya R, Patel AD, Tanaka M (2018) Temporal generalization of synchronized saccades beyond the trained range in monkeys. Front Psychol 9:2172
Toiviainen P, Luck G, Thompson MR (2010) Embodied meter: hierarchical eigenmodes in music-induced movement. Music Percept 28:59–70
Ueno A, Hirata S, Fuwa K, Sugama K, Kusunoki K, Matsuda G et al (2008) Auditory ERPs to stimulus deviance in an awake chimpanzee (*Pan troglodytes*): towards hominid cognitive neurosciences. PLoS ONE 3:e1442. https://doi.org/10.1371/journal.pone.0001442
Whiten A et al (1999) Cultures in chimpanzees. Nature 399:682–685
Whiten A et al (2001) Charting cultural variation in chimpanzees. Behaviour 138:1481–1516
Winkler I, Háden GP, Ladinig O, Sziller I, Honing H (2009) Newborn infants detect the beat in music. Proc Natl Acad Sci USA 106:2468–2471
Yu L, Tomonaga M (2015) Interactional synchrony in chimpanzees: examination through a finger-tapping experiment. Sci Rep 5(10218)
Yu L, Tomonaga M (2016) Unidirectional adaptation in tempo in pairs of chimpanzees during simultaneous tapping movement: an examination under face-to-face setup. Primates 57 (2):181–185
Zarco W, Merchant H, Prado L, Mendez JC (2009) Subsecond timing in primates: comparison of interval production between human subjects and rhesus monkeys. J Neurophysiol 102:3191–3202. https://doi.org/10.1152/jn.00066.2009
Zatorre RJ, Chen JL, Penhune VB (2007) When the brain plays music: auditory-motor interactions in music perception and production. Nat Rev Neurosci 8:547–558
Zentner M, Eerola T (2010) Rhythmic engagement with music in infancy. Proc Natl Acad Sci USA 107:5768–5773

Chapter 10
The Lasting and the Passing: Behavioural Traditions and Opportunities for Social Learning in Wild Tufted Capuchin Monkeys

Eduardo B. Ottoni

Abstract There is abundant evidence that allows us to consider wild tufted capuchin monkeys' toolkits as behavioural traditions. Developmental studies show that infants' interest in nutcracking and adults' tolerance of scrounging enhance opportunities for social learning. Field experiments have examined the socially mediated diffusion of new behaviours. The difference between forest populations' lack of customary tool use and the typical savannah toolkit—including stone "hammers" for nutcracking—seems sufficiently explained by terrestriality. By contrast, the narrower distribution of customary use of tools for probing cannot be accounted for by distinct diets or environmental affordances. Opportunities for social learning may be framed in Niche Construction theory, as social diffusion may depend on the conspicuousness and permanence of tools and leftovers. This is the case for nutcracking, which is highly conspicuous, leaves lasting environmental changes, and frequently allows delayed scrounging (enabling direct observation and delayed stimulus enhancement). The use of stick probes, however, creates fewer opportunities for social learning: the events are quick and less conspicuous, scrounging opportunities are minimal, and there are no lasting "tool use sites". This may explain the observed distribution of probe use: the lesser the role of environmental niche construction, the greater the role of social dynamics in the diffusion of innovations and the establishment of traditions.

Keywords Tufted capuchin monkeys · Tool use · Socially biased learning · Behavioural traditions · Cultural niche construction

---

E. B. Ottoni (✉)
Department of Experimental Psychology, University of São Paulo, São Paulo, Brazil

J. R. Anderson, H. Kuroshima (eds.), *Comparative Cognition*,
https://doi.org/10.1007/978-981-16-2028-7_10

Introduction

One afternoon, a quarter of century ago, Angela Perondi, an undergraduate student, stood at my office door. I had agreed to supervise an experiment on tool-aided problem-solving by semi-captive tufted capuchin monkeys—basically, a "naturalistic" replication of a laboratory experiment with captive capuchins, conducted by Westergaard and Fragaszy (1987). The group of tufted capuchins in Angela's experiment lived on a small river island in an urban park (Tietê Ecological Park, São Paulo, Brazil), and had just started to explore our molasses' filled problem-box when, one evening on her way back from the island, while crossing the Park's "Conservation Area" Angela heard some banging sounds from the woods. "Ah—these are the capuchins cracking palm nuts with stones. . ."—casually explained the park keeper who transported her to and from the island by boat.

Though tool-aided problem-solving had long been observed in captive primates, no cases of spontaneous use of tools by wild animals beyond single, stereotyped "species-typical" instances or anecdotal reports were known at the time, except in chimpanzees. After Goodall's (1971) descriptions of probe tools' use by the chimpanzee population at Gombe (Tanzania), the growing number of long-term study sites across Africa started revealing the diversity and variation of forms of tool use by wild chimpanzees, mostly in foraging contexts, but also in bodily care and communication. McGrew (1992) examined variation in the toolkits of chimpanzee populations, referring to chimpanzees' "material culture". Whiten et al. (1999), in a broader meta-analysis, dropped the "material", as they included activities other than tool use (e.g. social behaviours, use of "medicinal plants"). From an evolutionary perspective, "cultured" chimpanzees (McGrew 2004)—our closest living relatives—were not an unexpected idea. But only in the last few decades, a broader understanding of "culture" as including any behavioural traditions whose individual acquisition relies on socially learned and transmitted information (Laland and Hoppitt 2003) took hold—along with newer perspectives on the relationships between genetic and cultural evolution (Richerson and Boyd 1985), within a Darwinian framework.

Comparing behavioural and cognitive features in chimpanzees' and orangutans' tool use to those inferred from hominin Oldowan tools, Wynn and McGrew (1989) talked about an "ape adaptive grade". After those features were reported in at least one Old World monkey species (*Macaca fascicularis*, Malaivijitnond et al. 2007) and one New World genus (*Sapajus* sp.), it had to be expanded to a "simian adaptive grade" (McGrew et al. 2019).

I remember Angela standing by my door and telling me about the tool-using capuchins, 25 years ago, as if it was yesterday. It changed our labs' history.

Fig. 10.1 Stone-aided nutcracking at Fazenda Boa Vista (FBV) (Photo by Tiago Falótico)

Spontaneous Tool Use by Wild Tufted Capuchin Monkeys

Tufted, or "robust" capuchins (formerly, *Cebus apella*—now, the *Sapajus* genus) are highly encephalized New World monkeys inhabiting a variety of forest and savannah habitats—from the Amazon forest to southern Brazil. Although tool use in captive capuchins was already well known and studied (see, for instance, Antinucci and Visalberghi 1986; Westergaard and Fragaszy 1987; Chevalier-Skolnikoff 1989; Fragaszy and Visalberghi 1989; Anderson 1990; Visalberghi 1990; Visalberghi and Limongelli 1994), long-term studies in forest environments had never produced any evidence of spontaneous, customary tool use (Ottoni and Izar 2008). Only some anecdotes were available, about the use of hard substrates to crack open nuts (Izawa and Mizuno 1977) or stone-aided oyster predation (Fernandes 1991). When we first discovered capuchins' stone-aided nutcracking of palm nuts in the Tietê Park (TEP; Ottoni and Mannu 2001), some suggested it might be an artefact of captivity, since many of our semi-free subjects were former captives. The findings from our first long-term field sites showed that was not the case.

In Fazenda Boa Vista (FBV; Fragaszy et al. 2004a), the monkeys are specialized in stone-aided cracking of the extremely hard palm nuts found in the *Cerrado* (a kind of savanna[1]). This requires the skilful handling of very heavy "hammer" stones

[1]The "cerrado" is a tropical, wet savanna covering an extensive area in the plateaus of central Brazil, with semi-humid climate, exhibiting a mosaic of vegetation types, from grassy "campos" with a few trees, to areas with more extensive canopy cover ("cerradão") and denser "gallery forests" along rivers.

Fig. 10.2 Stick probe tool use at Serra da Capivara National Park (SCNP) (Photo by Tiago Falótico)

(Fig. 10.1). In Serra da Capivara National Park (SCNP), in the drier *Caatinga*,[2] there are no palm trees, but stone tools are used for many different purposes, including cracking seeds, and digging for tubers or trapdoor spiders' burrows (Mannu and Ottoni 2009; Falótico and Ottoni 2016; Falótico et al. 2017). As we started surveying wild populations in the savanna, we found that in such environments, percussive stone tool use was the rule rather than the exception, its occurrence being confirmed in all ten surveyed groups[3] (Mendes et al. 2015). More recently, we found archaeological evidence that the use of lithic tools to pound open encased food in SCNP dates back at least 3000 years (Haslam et al. 2016; Falótico et al. 2019).

Customary use of *probe tools*, by contrast, has only been observed in the Serra da Capivara National Park population—in all three groups systematically observed there (plus a couple of episodes observed in other groups; Mannu and Ottoni 2009; Falótico and Ottoni 2014). There are very few reports of probing from elsewhere, and they are single, anecdotal observations. Sticks in SCNP are used for a variety of purposes, such as probing insect nests (Fig. 10.2) or spiders' burrows, but mostly for dislodging vertebrate prey—lizards—from rock crevices. Differently

[2]The "caatinga" is a dryer, desert-like, and very seasonal kind of savanna in interior north-eastern Brazil, with dry winters and wet summers, which vegetation includes xeric shrubland, cacti and small, thorny trees that lose all leaves during the dry season, and semi-deciduous forests in hill slopes and near rivers. There are extensive transition areas ("ecotones") between "cerrado' and "caatinga" areas.

[3]Selected from 20 localities we got reports of nutcracking from local inhabitants (the other ten localities were not surveyed due to logistic issues, such as authorizations or availability of local guides).

from stone tools, probe tools are usually modified before use: leaves and side branches are trimmed, and the bark may be removed.

As it is much less conspicuous than nutcracking, probing is likely to be underreported to some extent, but its verified absence in FBV is hard to explain by environmental affordances alone: lizards are hunted in both sites, and the availability of raw tool material is not an issue. A peculiar—and still poorly understood—aspect of probe use is that it is almost exclusively performed by males (Moura and Lee 2010; Falótico and Ottoni 2014).

Tool use is not restricted to contexts such as foraging, self-defence, or bodily care: in the Pedra Furada group (SCNP), we followed the spread of a new "fad" among adult female capuchins—the "communicative" use of stone projectiles as an apparent "enhancement" to their oestrus displays towards preferred males who were "taking their time" to copulate (Falótico and Ottoni 2013).

The Ontogeny of Stone Tool Use for Nutcracking

Since the beginning of our research on the semi-free capuchin group in TEP, we have been examining the ontogeny of nutcracking behaviour. Stone tool use is slow to develop. From the "inept manipulation" of nuts and rocks by infants, performance improves gradually across the first years of life (Ottoni and Mannu 2001). Simple manipulation of objects can be observed around 8–12 weeks of age; manipulation combining objects and substrates appears almost at the same time and is highly diversified by 16–24 weeks. Percussive behaviours involving stones and nuts were observed around 32–80 weeks. The first successful episodes of nutcracking were performed by a few juveniles with a little over 2 years of age (Resende et al. 2008), but it takes at least 3 years to reach adultlike proficiency. This is roughly similar to what is observed in the development of percussive tool use in chimpanzees (whose life expectancies are just slightly greater than capuchins'), which do not perform nutcracking before 3.5 years of age (Inoue-Nakamura and Matsuzawa 1997).

Proper percussive tools have to be hard (not friable), and the heavier, the better, requiring fewer strikes to crack a nut—though some skilful nutcrackers may prefer smaller stones that suite their particular techniques. The selection of tools improves with the experience: when exposed to potential "hammer" stones of different weights, the adults in Tietê Park showed a significant preference for one of them (the second heaviest, at 1300 g), while juveniles tended to select stones closer to the "anvil" (Falótico and Ottoni 2005). When exposed to (visually similar) hard and soft flat surfaces to use as "anvils"—with "hammer" stones lying between them, the six adult monkeys which visited the experimental cracking site never tried to use the soft surface, suggesting some understanding of the physical affordances of the available materials (Fujita et al. 2007). The selection of appropriate percussive tools has also been examined in the lab (Schrauf et al. 2008) and in the wild—both in naturalistic (Ferreira et al. 2009) and experimental situations (Visalberghi et al. 2009). Tool transport may be costly (when "hammer" stones are not already close by the "anvils"), but the monkeys are capable of minimizing the cost, for instance, by first collecting the nuts, then choosing stones away from the nuts´ location, but closer to the "anvil" (Corat et al. 2016).

As we focused on the development of stone tool use, we quickly noticed the importance of both *infants' curiosity*, and *adult males' tolerance* to being closely observed and scrounged from. Furthermore, although the best strategy is usually just to "watch the dominant males", when there were a lot of less experienced tool manipulators around, including adult males (as was the case in that group at the time), the youngsters easily learned who were the best nutcrackers to follow, watch, and scrounge from (Ottoni et al. 2005; Coelho et al. 2015). The same patterns were later observed in the wild (Silva 2008).

In this case, scrounging and social learning are not conflicting tactics (as suggested by Giraldeau and Lefebrve 1987, but see Giraldeau and Beauchamp 1999), as a "watch-the-most-successful" strategy (Laland 2004) favours both. The underlying decision process might be guided by an actual understanding of the relative proficiency ranking of potential targets—in a sort of "prestige" hierarchy, distinct and independent of "dominance" (Henrich and Gil-White 2001), but it could also just be the result of differential associative or reinforcement histories with each individual observational and scrounging "target". Whatever the underlying cognitive mechanism, this "strategy" can play a key role in the establishment of tool use as a behavioural tradition, as the preferred scrounging "targets"—the most efficient nutcrackers—are also the best models to learn the technique from.

Field Experiments on Tool Use and Social Learning

Toolkit Complexity, Problem-Solving, and Generalization

Because we had such a favourable situation for comparisons—two long-studied populations (in Fazenda Boa Vista and in Serra da Capivara) with different toolkits, and in FBV we could be sure about the absence of probe tool use—we examined the effects of those distinct "cultural repertoires" on their performances in an experimental probing task (Cardoso and Ottoni 2016). We exposed a group in each site to the same novel task of probing for molasses in a Plexiglas box with a slit opening at the top. This task is quite similar to the one used by Gruber et al. (2009) with the Sonso and Kanyawara chimpanzee communities, and it yielded comparable results. In Serra da Capivara, all adult and juvenile males in the Pedra Furada group quickly learned to extract the molasses with probes that they fashioned for the job. In FBV, by contrast, no monkey ever solved the task, even after more than twice the time of exposure to the box. After "sampling" some molasses from the top of the box in a Habituation Phase, despite showing interest in the contents, they soon stopped visiting the box. During an added "Facilitation" phase—with pre-inserted sticks available, they did just what females did in SCNP when they found abandoned probes inserted and left there by males: they extracted them, licked, and then dropped them, but never re-inserted them. Even more curious—and contrary to our expectations—no monkey in FBV ever tried to crack the Plexiglas box with stones. SCNP monkeys, though, did this sometimes (before discovering the most effective technique), which we attributed to their already more diversified use of the more abundantly available stones.

Experimental Approaches to Social Diffusion

Although some insights into social learning mechanisms in customary tool use behaviours can be gained from developmental studies, actual observation of spontaneous diffusion of novel behaviours in the wild only happens in lucky and rare cases (as in the stone-throwing females' oestrus display). However, we can engineer such observations by experimental means. We examined the cultural capacities of wild capuchins in two SCNP groups (Pedra Furada and Jurubeba) by following the open diffusion of an innovation—the extraction of food from a two-action problem-box, comparing socially biased and asocial diffusion models with the aid of the statistical tools of Network-Based Diffusion Analysis (NBDA). We found strong support for observational social learning of the task solution (Ottoni et al. 2016; Kendal et al., in preparation).

Another experiment on the social diffusion of probe tool use (in the semi-free TEP group), suggests that, in this case at least, social influences on learning may be limited to "stimulus enhancement", where the observer's attention is drawn toward the task and the tools, promoting interaction with the apparatus and thus expediting individual learning. Witnessing proficient probing did not result in any immediate copy of the relevant behaviours by observers, which apparently rules out "imitation" processes. This might be related to a difficulty of capuchin monkeys in understanding causality involving three items, as suggested by results of experiments by Visalberghi and Limongelli (1994) and Fujita et al. (2003) (Fujita et al. 2011, though, observed a better performance in another three-item tool task, suggesting at least a rudimentary understanding of causal relations; also, it is important to remember that nutcracking involves the coordination of two loose objects—nut and "hammer"—plus the substrate—which can be terrestrial or, sometimes, arboreal).

Although these findings might restrict possibilities of a "cumulative culture" in capuchins, any form of socially biased learning—not only imitation—can, in principle, lead to the establishment and maintenance of behavioural traditions (see Fragaszy and Visalberghi 2001). For our six successful molasses-probers, success came after a long time and much "useless" stick manipulation around the box. However, their generalization of probe use to solve a second, different probe task (*pushing* solid food items out of another problem-box) was extremely fast (Ottoni and Rufo 2019; Rufo & Ottoni, in preparation), and comparable to the generalization capacities exhibited by Fujita et al. (2011) subjects in their three-factors' tasks.

Explaining Differences in Tool Use Between Populations

The *Sapajus* genus includes all former *Cebus apella* subspecies (Fragaszy et al. 2004b). Given their broad distribution, genetic differences could, in principle, at least partly explain the differences between toolkits of distinct populations. Though most data reported here come from *Sapajus libidinosus* in the savanna, there are reports of tool use by other *Sapajus* species. In *S. xanthosternos*, at least, nutcracking seems restricted to dry environments, being apparently absent in humid Atlantic forest (Canale et al. 2009). There are no reports, so far, about tool-using *S. libidinosus* in forest environments, but mangrove populations do use lithic tools (Santos et al. 2019). Also, the semi-free group in Tietê Ecological Park (TEP) is mixed, and some of the individuals are probably hybrids (as in many laboratory groups). So, there is not much reason to expect critical genetic differences between *Sapajus* species, although there is research currently being done on this issue.

Tool Use in the Wild: Ecological Explanations

Moura and Lee (2004) suggested that tool use by tufted capuchins was a strategy to overcome food scarcity in the dry season—the "Necessity Hypothesis" (Fox et al. 1999). Visalberghi et al. (2005) argued instead that the degree of *terrestriality* was likely a better predictor of stone tool use, rather than food availability. Forest environments are frequently poorer than savannas in food resources, and, at least for FBV monkeys, palm nuts are no "fallback food"; rather, they are processed and consumed when available—which is usually when other fruits are also available (Spagnoletti et al. 2012).

Another ecological explanation—based on the relative abundance of suitable lithic materials—may apply to the differences in the diversification of stone tools use—greater in the SCNP, where suitable stones are abundant, and more "specialized" in FBV, where they are rare. Greater opportunities for the use of lithic tools may have fostered greater behavioural generalization.

"Terrestriality" explanations, as well as those based on availability of raw tool materials, can be included in the so-called Opportunity Hypothesis (Fox et al. 1999), which proposes that encounter rates with tool materials, and resources whose exploitation requires tools, affect the likelihood of tool invention and frequency of tool use, thus explaining observed tool use patterns. But it can be asked whether there are any plausible ecological explanations for the difference in the use of *stick probes*. Raw material is equally available in both sites, and there are no big differences in rates of predation on lizards or spiders. The terrestriality argument clearly does not hold in the case of probe use (think about orangutans! van Schaik et al. 1996). So, why is customary use of tools for probing so rare? Considering the evidence that tool use learning depends to some extent on social influences—by direct interactions or through Niche Construction—the causes for these behavioural differences may be found in *different opportunities for socially biased learning*.

Capuchin Tool Use: Evidence of Behavioural Traditions?

Differences in behavioural repertoires of different populations of a given species have been used as evidence of cultural traditions, after excluding variation that can be explained by genetic or ecological differences (Whiten et al. 1999). An important criticism of the "group-comparison" method[4] is that it relies on characteristics unrelated to the essential feature of traditions—their dependence on the social context for acquisition of the behaviours in question by new practitioners (Fragaszy 2003). Although it can identify interesting targets for further study (when genetic or ecological explanations for behavioural differences do not suffice), it is prone to "false negatives" and "positives" (see, for instance, Humle and Matsuzawa 2002 on chimpanzees' ant-dipping), as it does not examine the role of social interactions and Niche Construction in the individual acquisition of the behaviours. One way to improve this situation is through longitudinal and developmental studies, as well as diffusion experiments (see above).

Furthermore, while environmental settings are obviously causally relevant to most behaviour, ecological explanations do not rule out cultural processes—nor vice-versa. The environment defines the *opportunities* of encountering various potential food resources and raw tool materials, affecting not only the expression of any "innate" behaviour or opportunities for individual learning, but also the likelihood of both *innovation* and *socially mediated learning*:

> Ecological explanations may constitute alternative explanations, but they do not rule out cultural differences (...). The 'method of exclusion' neglects ecological influences on culture, which, ironically, may be critical for understanding technology and thus material culture (Koops et al. 2014).

The difference between the toolkits of capuchin savanna and forest populations—the "terrestriality" factor notwithstanding—may be also related to different opportunities for observing conspecifics' behaviour. Compared to the savanna, in the forest there are *visibility* issues, and monkeys may forage in smaller parties (fission–fusion), depending on food availability (Izar et al. 2011). The greater complexity of SCNP capuchins' toolkit may also be related to group sizes: SCNP groups tend to be big, sometimes more than 50 individuals, whereas at the time of our initial study (Fragaszy et al. 2004a), the group observed in FBV had only 10 individuals. In theory, the smaller the social group, the fewer opportunities for innovation and the maintenance of learned techniques (for discussions on human cases, see, for instance, Henrich 2004; Derex et al. 2013).

---

[4]Also known as "Exclusion", "Comparative", "Geographic", "Ethnographic", or "Regional Contrast" method. See Whiten et al. (1999), Fragaszy (2003), and Schuppli and Van Schaik (2019).

Can observability biases also explain the sex bias in probe use? We have no indication of a "cognitive" sex bias in capuchin tool use: the observed sex differences in percussive stone tool use seem to be primarily related to strength and body mass constrains. They are pronounced in FBV, where hard palm nuts are cracked with heavy "hammer" stones, but less so where the processed fruits are smaller and softer, as in SCNP. Besides, females in a semi-free park population did learn to use probes as efficiently as males after facilitation in an experimental task (Cardoso & Ottoni, unpublished data), and there are anecdotal observations of probe use by females in bodily care (Haslam and Falótico 2015).

More than stone tool use, socially biased learning of probe use in most natural conditions would rely on direct observation of brief, frequently unpredictable behaviours. This means that the groups' social dynamics might bias the opportunities of males and females: sex-biased social networks and social proximity, for instance, might explain sex-biased intragroup diffusion (a hypothesis still being examined). Sex biases in tool use may also explain intergroup diffusion patterns. Among tufted capuchin monkeys, adult males tend to migrate, while females are philopatric. This might create different opportunities for intergroup diffusion, and explain why customary probe tool use—male-biased but generally rare (reported only in SCNP), is present in all three systematically observed SCNP groups, while females' stone-enhanced oestrus display has so far been observed in only one group.

Tool Use and Innovation

The criteria proposed by Reader and Laland (2003) to identify an innovation include the actual observation of its first occurrence in a population subject to long-term research—something rare outside experimental settings. According to their definition, which refers to "innovation" at the population level, only the SCNP females' stone throwing could be classified as an innovation (which then spread through social learning).

Ramsey et al. (2007) proposed a different approach, similar to the "group-comparison" method used to identify cultural traits (Whiten et al. 1999), based on the degree of ecological or genetic determination, and distinguishing between "innovations" at the population and individual levels. They classified novel learned behaviours in an individual's repertoire according to the processes that generate them, on a continuum between "innovation", "social learning", and "environmental induction."[5] Applying their criteria to wild *Sapajus* tools (see Fig. 10.3), stone-aided nutcracking should be considered a "[very] weak innovation" (or as not being an "innovation"), as its learning depends to some extent on both social mediation and

[5]Ramsey et al. (2007) define "environmentally induced novel behaviours" as new behaviours that are expected of all or most individuals of that type, given some novel environmental element. "Environmental induction" is, thus, a concept close to Gibson (2015) "environmental affordances".

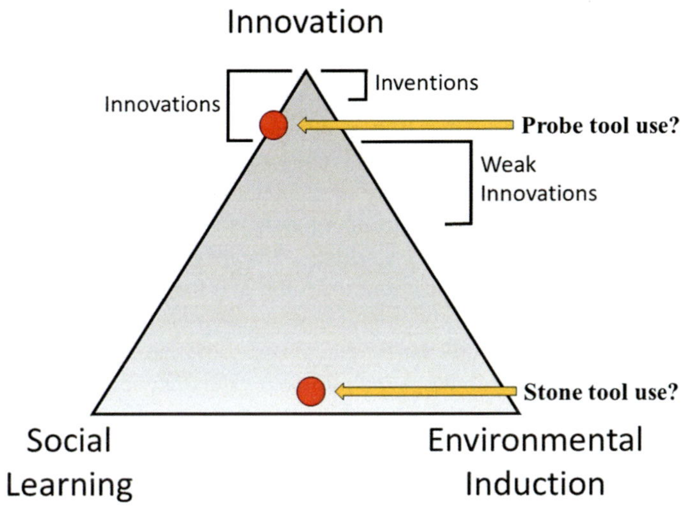

Fig. 10.3 The place of capuchin monkeys' tool use traditions in the continuum between innovation, social learning, and environmental induction (based on Ramsey et al. 2007)

environmental induction (the more terrestrial habits of savanna populations, associated with the availability of lithic material and encased fruit), and some potentially inherited predispositions for percussive behaviours may be involved (capuchins' palm-fruit cracking by percussion against a substrate, without the use of tools, was reported by Izawa and Mizuno 1977). The more diversified use of stone tools by SCNP monkeys may also be considered a set of "minor innovations", a generalization possibly fostered by the greater availability of tool material.

The customary use of probes by SCNP males is different in many respects. On the one hand, it is not induced by any local environment peculiarity. On the other hand, opportunities for social learning are restricted to direct observation of brief episodes, whereas nutcrackers' behaviour influences naïve individuals' learning not only while it happens, but also by establishing long-lasting nutcracking sites. Probe use, being less conspicuous, seems to involve individual innovations that in only one known case (in the SCNP population) spread socially to become a population-level tradition.

"Environmental induction" is usually conceived in terms of ecological constraints or affordances, clearly distinct from "social mediation". In some cases, however, environmental induction and social mediation can be directly connected, when the latter consists of or includes some form of "cultural Niche Construction" (Laland et al. 2000).

Niche Construction and Opportunities for Social Learning

A key factor underlying the radically distinct patterns of occurrence of stone and probe tool use in wild populations may be related to distinct degrees of Niche Construction. Just observing other individuals using tools may not be enough to learn to use a tool skilfully. Fragaszy et al. (2013) proposed a complementary perspective to Socially Biased Learning models, framed in Niche Construction theory. They suggest that enduring artefacts—like stone tools left at nutcracking sites—scaffold individuals' practice and learning, and thereby promote the maintenance of technical traditions. This sounds fine when it comes to nutcracking. But they go on to affirm that "All investigated cases of habitual tool use in wild chimpanzees and capuchin monkeys include youngsters encountering durable artefacts, most often in a supportive social context". However, this may not always be the case.

The Lasting and the Passing: Observability Biases?

There are some serious differences in opportunities for observing nutcracking and of the use of probes by capuchin monkeys: differences in *duration* and *predictability* of the behaviours, in the *permanence of artefacts*, and in *opportunities for scrounging*. Nutcracking is noisy, it requires some time, and it generates enduring residual artefacts, the "nutcracking sites": suitable surfaces, hard and level (the "anvils"), not far from the palm trees (or other encased fruit plants), where previously used "hammer" stones are usually left, along with husks and often edible endosperm leftovers, which can be consumed later ("delayed scrounging"). So, nutcracking sites provide "stimulus enhancement" that lasts longer than the behaviour itself, as well as some "scaffolding" for naive individuals, constituting a case of environmental Niche Construction.

This contrasts sharply with the case of capuchins' probe use, where opportunities for *direct social learning* are usually much rarer and those for interacting with a *modified environment*, almost null. Differently from chimpanzees' termite nests probing (Pascual-Garrido 2017), were there are plenty of opportunities for observation and reuse of tools, probe use events by capuchins are generally much less predictable, opportunistic (as when predating a lizard that tries to hide in a rock crevice), fast, silent, and there is usually nothing left to scrounge. Differently from termite fishing chimpanzees (Musgrave et al. 2016), probe tool transfers were never observed in SCNP capuchins, and probe reuse was very rare (Falótico, personal communication).[6] If on trees or cliffs, these episodes may be less visible than terrestrial activity, and they seldom promote any lasting "stimulus enhancement"

[6] In our field experiments on probe tool use (Cardoso and Ottoni 2016), on the other hand, there were indeed opportunities for tool reuse.

(as when bee nests' are targeted), so social learning opportunities may rely exclusively on direct observation, which depends on luck, and being in proximity to potential models. Furthermore, since probing is virtually absent among capuchin females, their youngsters do not enjoy the same learning opportunities as young chimpanzees.

Conclusions

Tufted capuchin monkeys' technological traditions seem to involve different degrees of innovation and diffusion, from the widespread use of percussive stone tools—facilitated by innate predispositions and environmental affordances, to the rarer use of probe tools, which, apparently, spread by means of socially biased learning in only one population. Beyond direct observation, behavioural traditions may be transmitted by means of the environmental changes they create, which amounts to Cultural Niche Construction. The distinct degrees of Niche Construction and Observability associated with different forms of tool use may help to explain the difference between the widespread stone tool use traditions and the rarer cases of customary probe use. In the latter case, individual innovations may occur, but seldom spread by socially mediated learning to become behavioural traditions.

References

Anderson JR (1990) Use of objects as hammers to open nuts by capuchin monkeys (*Cebus apella*). Folia Primatol 54:138–145. https://doi.org/10.1159/000156437

Antinucci R, Visalberghi E (1986) Tool use in *Cebus apella*: a case study. Int J Primatol 7:351–363. https://doi.org/10.1007/BF02693700

Canale GR, Guidorizzi CE, Kierulff MCM, Gatto CAFR (2009) First record of tool use by wild populations of the yellow-breasted capuchin monkey (*Cebus xanthosternos*) and new records for the bearded capuchin (*Cebus libidinosus*). Am J Primatol 71:366–372. https://doi.org/10.1002/ajp.20648

Cardoso RM, Ottoni EB (2016) The effects of tradition on problem solving by two wild populations of bearded capuchin monkeys in a probing task. Biol Lett 12:20160604. https://doi.org/10.1098/rsbl.2016.0604

Chevalier-Skolnikoff S (1989) Spontaneous tool use and sensorimotor intelligence in *Cebus* compared with other monkeys and apes. Behav Brain Sci 12:561–627. https://doi.org/10.1017/S0140525X00057678

Coelho CG, Falótico T, Izar P, Mannu M, Resende BD, Siqueira JO, Ottoni EB (2015) Social learning strategies for nut-cracking by tufted capuchin monkeys (*Sapajus* spp.). Anim Cogn 18:911–919. https://doi.org/10.1007/s10071-015-0861-5

Corat C, Siqueira J, Ottoni EB (2016) Sequential organization and optimization of the nut-cracking behavior of semi-free tufted capuchin monkeys (*Sapajus* sp.). Primates 57:113–121. https://doi.org/10.1007/s10329-015-0491-1

Derex M, Beugin M, Godelle B, Raymond M (2013) Experimental evidence for the influence of group size on cultural complexity. Nature 503:389–391. https://doi.org/10.1038/nature12774

Falótico T, Ottoni EB (2005) Experimento de escolha de ferramentas para a quebra de cocos por macacos-prego (*Cebus apella*) em semi-liberdade. Anais do XI Congresso Brasileiro de Primatologia (Porto Alegre, Brazil), p 98

Falótico T, Ottoni EB (2013) Stone throwing as a sexual display in wild female bearded capuchin monkeys, *Sapajus libidinosus*. PLoS ONE 8:e79535. https://doi.org/10.1371/journal.pone.0079535

Falótico T, Ottoni EB (2014) Sexual bias in probe tool manufacture and use by wild bearded capuchin monkeys. Behav Process 108:117–122. https://doi.org/10.1016/j.beproc.2014.09.036

Falótico T, Ottoni EB (2016) The manifold use of pounding stone tools by wild capuchin monkeys of Serra da Capivara National Park, Brazil. Behaviour 153:421–442. https://doi.org/10.1163/1568539X-00003357

Falótico T, Siqueira JO, Ottoni EB (2017) Digging up food: excavation stone tool use by wild capuchin monkeys. Sci Rep 7:6278. https://doi.org/10.1038/s41598-017-06541-0

Falótico T, Proffitt T, Ottoni EB, Staff RA, Haslam M (2019) Three thousand years of wild capuchin stone tool use. Nat Ecol Evol 3:1034–1038. https://doi.org/10.1038/s41559-019-0904-4

Fernandes M (1991) Tool use and predation of oysters (*Crassostrea rhizophorae*) by the tufted capuchin, *Cebus apella apella*, in brackish water mangrove swamp. Primates 32:529–531. https://doi.org/10.1007/BF02381944

Ferreira RG, Emidio RA, Jerusalinsky L (2009) Three stones for three seeds: natural occurrence of selective tool use by capuchins (*Cebus libidinosus*) based on an analysis of the weight of stones found at nutting sites. Am J Primatol 71:1–6. https://doi.org/10.1002/ajp.20771

Fox EA, Sitompul AF, Van Schaik CP (1999) Intelligent tool use in wild Sumatran orangutans. In: Parker ST, Mitchell RW, Miles HL (eds) The mentalities of gorillas and orangutans: comparative perspectives. Cambridge University Press, Cambridge, pp 99–116. https://doi.org/10.1017/CBO9780511542305.005

Fragaszy D (2003) Making space for traditions. Evol Anthropol 12:61–70. https://doi.org/10.1002/evan.10104

Fragaszy DM, Visalberghi E (1989) Social influences on the acquisition of tool-using behaviors in tufted capuchin monkeys (*Cebus apella*). J Comp Psychol 103:159–170. https://doi.org/10.1037/0735-7036.103.2.159

Fragaszy DM, Visalberghi E (2001) Recognizing a swan: socially biased learning. Psychologia 44:82–98. https://doi.org/10.3758/BF03196004

Fragaszy DM, Izar P, Visalberghi E, Ottoni EB, Oliveira M (2004a) Wild capuchin monkeys use anvils and stone pounding tools. Am J Primatol 64:359–366. https://doi.org/10.1002/ajp.20085

Fragaszy DM, Visalberghi E, Fedigan LM (2004b) The complete capuchin: the biology of the genus Cebus. Cambridge University Press, Cambridge

Fragaszy DM, Biro D, Eshchar Y, Humle T, Izar P, Resende B, Visalberghi E (2013) The fourth dimension of tool use: temporally enduring artefacts aid primates learning to use tools. Phil Trans R Soc B 368:410. https://doi.org/10.1098/rstb.2012.0410

Fujita K, Kuroshima H, Asai S (2003) How do tufted capuchin monkeys (*Cebus apella*) understand causality involved in tool use? J Exp Psychol 29:233–242. https://doi.org/10.1037/0097-7403.29.3.233

Fujita K, Tsutsumi S, Morimoto Y, Coelho CG, Falótico T, Ottoni EB (2007) Substrate choice in nut-cracking behavior of semi-wild tufted capuchin monkeys (*Cebus apella*). In: Abstracts of the XXVIII International Ethology Conference (Florianopolis, Brazil), pp 8–10

Fujita K, Sato Y, Kuroshima H (2011) Learning and generalization of tool use by tufted capuchin monkeys (*Cebus apella*) in tasks involving three factors: reward, tool, and hindrance. J Exp Psychol 37:10–19. https://doi.org/10.1037/a0020274

Gibson JJ (2015) The ecological approach to visual perception. Psychology Press, New York

Giraldeau LA, Beauchamp G (1999) Food exploitation: searching for the optimal joining policy. Trends Ecol Evol 14:102–106. https://doi.org/10.1016/s0169-5347(98)01542-0

Giraldeau LA, Lefebrve L (1987) Scrounging prevents cultural transmission of food-finding behaviour in pigeons. Anim Behav 35:387–394. https://doi.org/10.1016/S0003-3472(87)80262-2

Goodall J (1971) Tool-using in primates and other vertebrates. Adv Study Behav 3:195–249. https://doi.org/10.1016/S0065-3454(08)60157-6

Gruber T, Muller MN, Strimling P, Wrangham R, Zuberbühler K (2009) Wild chimpanzees rely on cultural knowledge to solve an experimental honey acquisition task. Curr Biol 19:1806–1810. https://doi.org/10.1016/j.cub.2009.08.060

Haslam M, Falótico T (2015) Nasal probe and toothpick tool use by a wild female bearded capuchin (*Sapajus libidinosus*). Primates 56:211. https://doi.org/10.1007/s10329-015-0470-6

Haslam M, Luncz LV, Staff RA, Bradshaw F, Ottoni EB, Falótico T (2016) Pre-Columbian monkey tools. Curr Biol 26:R515–R522. https://doi.org/10.1016/j.cub.2016.05.046

Henrich J (2004) Demography and cultural evolution: how adaptive cultural processes can produce maladaptive losses: the Tasmanian case. Am Antiq 69:197–214. https://doi.org/10.2307/4128416

Henrich J, Gil-White FJ (2001) The evolution of prestige: freely conferred deference as a mechanism for enhancing the benefits of cultural transmission. Evol Hum Behav 22(3):165–196. https://doi.org/10.1016/S1090-5138(00)00071-4

Humle T, Matsuzawa T (2002) Ant-dipping among the chimpanzees of Bossou, Guinea, and some comparisons with other sites. Am J Primatol 58:133–148. https://doi.org/10.1002/ajp.10055

Inoue-Nakamura N, Matsuzawa T (1997) Development of stone tool use by wild chimpanzees (*Pan troglodytes*). J Comp Psychol 111:159–173. https://doi.org/10.1037/0735-7036.111.2.159

Izar P, Verderane MP, Peternelli-dos-Santos L, Mendonça-Furtado O, Presotto A, Tokuda M, Visalberghi E, Fragaszy D (2011) Flexible and conservative features of social systems in tufted capuchin monkeys: comparing the socioecology of *Sapajus libidinosus* and *Sapajus nigritus*. Am J Primatol 73:1–17. https://doi.org/10.1002/ajp.20968

Izawa K, Mizuno A (1977) Palm-fruit cracking behavior of wild black-capped capuchin (*Cebus apella*). Primates 18:773–792. https://doi.org/10.1007/BF02382930

Koops K, Visalberghi E, van Schaik CP (2014) The ecology of primate material culture. Biol Lett 10:20140508. https://doi.org/10.1098/rsbl.2014.0508

Laland KN (2004) Social learning strategies. Learn Behav 32:4–14. https://doi.org/10.3758/BF03196002

Laland KN, Hoppitt W (2003) Do animals have culture? Evol Anthropol 12:150–159. https://doi.org/10.1002/evan.10111

Laland KN, Odling-Smee J, Feldman MW (2000) Niche construction, biological evolution, and cultural change. Behav Brain Sci 23:131–175. https://doi.org/10.1017/S0140525X00002417

Malaivijitnond S, Lekprayoon C, Tandavanittj N, Panha S, Cheewatham C, Hamada Y (2007) Stone-tool usage by Thai long-tailed macaques (*Macaca fascicularis*). Am J Primatol 69:227–233. https://doi.org/10.1002/ajp.20342

Mannu M, Ottoni EB (2009) The enhanced tool-kit of two groups of wild bearded capuchin monkeys in the Caatinga: tool making, associative use, and secondary tools. Am J Primatol 71:242–251. https://doi.org/10.1002/ajp.20642

McGrew WC (1992) Chimpanzee material culture: implications for human evolution. Cambridge University Press, Cambridge

McGrew WC (2004) The cultured chimpanzee: reflections on cultural primatology. Cambridge University Press, Cambridge

McGrew WC, Falotico T, Gumert MD, Ottoni EB (2019) A simian view of the Oldowan: reconstructing the evolutionary origins of human technology. Ch.1 (pp.13–41) in Overmann KA and Coolidge FL (Eds.), Squeezing Minds from Stones: Cognitive Archaeology and the Evolution of the Human Mind [ISBN:9780190854614]. Oxford University Press, UK

Mendes FDC, Cardoso RM, Ottoni EB, Izar P, Villar DNA, Marquezan RF (2015) Diversity of nutcracking tool sites used by *Sapajus libidinosus* in Brazilian cerrado. Am J Primatol 77:535–546. https://doi.org/10.1002/ajp.22373

Moura ACA, Lee P (2004) Capuchin stone tool use in caatinga dry forest. Science 306:9. https://doi.org/10.1126/science.1102558
Moura ACA, Lee P (2010) Wild capuchins show male-biased feeding tool use. Int J Primatol 31:457–470. https://doi.org/10.1007/s10764-010-9406-6
Musgrave S, Morgan D, Lonsdorf E, Mundry R, Sanz C (2016) Tool transfers are a form of teaching among chimpanzees. Sci Rep 6:34783. https://doi.org/10.1038/srep34783
Ottoni EB, Izar P (2008) Capuchin monkey tool use: overview and implications. Evol Anthropol 17:171–178. https://doi.org/10.1002/evan.20185
Ottoni EB, Mannu M (2001) Semifree-ranging tufted capuchins (*Cebus apella*) spontaneously use tools to crack open nuts. Int J Primatol 22:347–358. https://doi.org/10.1023/A%3A1010747426841
Ottoni EB, Rufo HP (2019) Experimentally induced diffusion of probe tool use in a group of tufted capuchin monkeys (*Sapajus* sp). In: Abstracts of the behaviour 2019 conference, Chicago, IL, USA
Ottoni EB, Resende BD, Izar P (2005) Watching the best nutcrackers: what capuchin monkeys (*Cebus apella*) know about others' tool using skills. Anim Cogn 8:215–219. https://doi.org/10.1007/s10071-004-0245-8
Ottoni EB, Coelho CG, Kendal R (2016) Experimental evidence for social learning in wild bearded capuchin monkeys (*Sapajus libidinosus*) inhabiting the Serra da Capivara National Park, Brazil. In: Abstracts of the joint meeting of the International Primatological Society and the American Society of Primatologists, Chicago, IL, USA
Pascual-Garrido A (2017) Termite fishing by Mahale chimpanzees: revisited, decades later. Pan Afr News 24:15–19
Ramsey G, Bastian ML, van Schaik C (2007) Animal innovation defined and operationalized. Behav Brain Sci 30:393–437. https://doi.org/10.1017/S0140525X07002373
Reader SM, Laland KN (2003) Animal innovation: an introduction. In: Reader SM, Laland KN (eds) Animal innovation. Oxford University Press, Oxford, pp 3–35
Resende BD, Ottoni EB, Fragaszy DM (2008) Ontogeny of manipulative behavior and nut-cracking in young tufted capuchin monkeys (*Cebus apella*): a perception–action perspective. Dev Sci 11:812–824. https://doi.org/10.1111/j.1467-7687.2008.00731.x
Richerson PJ, Boyd R (1985) Culture and the evolutionary process. The University of Chicago Press, Chicago
Santos R, Sousa A, Fragaszy D, Ferreira R (2019) The role of tools in the feeding ecology of bearded capuchins living in mangroves. In: Nowak K, Barnett A, Matsuda I (eds) Primates in flooded habitats: ecology and conservation. Cambridge University Press, Cambridge, pp 59–63
Schrauf C, Huber L, Visalberghi E (2008) Do capuchin monkeys use weight to select hammer tools? Anim Cogn 11:413–422. https://doi.org/10.1007/s10071-007-0131-2
Schuppli C, Van Schaik C (2019) Animal cultures: how we've only seen the tip of the iceberg. Evol Hum Sci 1:E2. https://doi.org/10.1017/ehs.2019.1
Silva EDR (2008) Escolha de alvos coespecíficos na observação do uso de ferramentas por macacos-prego (*Cebus libidinosus*) selvagens. Unpublished master's dissertation, University of São Paulo
Spagnoletti N, Visalberghi E, Verderane MP, Ottoni EB, Izar P, Fragaszy D (2012) Stone tool use in wild bearded capuchin monkeys (*Cebus libidinosus*): is it a strategy to overcome food scarcity? Anim Behav 83:1285–1294. https://doi.org/10.1016/j.anbehav.2012.03.002
van Schaik CP, Fox EA, Sitompul AE (1996) Manufacture and use of tools in wild Sumatran orangutans: implications for human evolution. Naturwissenschaften 83:186–188. https://doi.org/10.1007/BF01143062
Visalberghi E (1990) Tool use in *Cebus*. Folia Primatol 54:146–154. https://doi.org/10.1159/000156438
Visalberghi E, Limongelli L (1994) Lack of comprehension of cause-effect relations in tool-using capuchin monkeys (*Cebus apella*). J Comp Psychol 108:15–22. https://doi.org/10.1037/0735-7036.108.1.15

Visalberghi E, Fragaszy DM, Izar P, Ottoni EB (2005) Terrestriality and tool use. Science 308:951. https://doi.org/10.1126/science.308.5724.951c
Visalberghi E, Addessi E, Truppa V, Spagnoletti N, Ottoni EB, Izar P, Fragaszy D (2009) Selection of effective stone tools by wild bearded capuchin monkeys. Curr Biol 19:213–217. https://doi.org/10.1016/j.cub.2008.11.064
Westergaard GC, Fragaszy DM (1987) The manufacture and use of tools by capuchin monkeys (*Cebus apella*). J Comp Psychol 101:159–168. https://doi.org/10.1037/0735-7036.101.2.159
Whiten A, Goodall J, McGrew WC, Nishida T, Reynolds V, Sugiyama T, Wrangham RW, Boesch C (1999) Cultures in chimpanzees. Nature 399:682–685. https://doi.org/10.1038/21415
Wynn T, McGrew WC (1989) An ape's view of the Oldowan. Man 24(3):383–398

Part IV
Social Emotions

Chapter 11
Capuchins (*Sapajus apella*) and their Aversion to Inequity

Manon K. Schweinfurth and Josep Call

Abstract Humans have a strong sense of fairness and are usually averse to unequal treatment for the same action. Ever since Brosnan and de Waal showed a similar effect in capuchin monkeys (*Sapajus apella*), numerous studies using different experimental methods have been conducted to investigate whether animals show inequity aversion like humans do. Capuchin monkeys have become one of the best-studied animals in this area. Our first aim in this chapter was to synthesise the findings in this literature. We found that there is mixed evidence for inequity aversion in capuchin monkeys. Our second aim was to understand this variation by focusing on the following factors: the type of task used, the feeding regime outside the experiment and the monkeys' social environment. To obtain data on some of these factors, as they are not always reported in published studies, we contacted researchers in the main laboratories conducting this work. We found that responses to inequity systematically varied as a function of the task demands and the feeding regime, but not the social environment. Tasks, in particular pulling tasks, that required participants to expend effort to get the food were more likely to detect evidence of inequity aversion. Moreover, monkeys with access to food before or after testing, were more likely to show inequity aversion than those whose access to food was temporarily restricted. We note that our survey is an explorative approach to investigate the variation in reports on inequity aversion in capuchin monkeys. We hope this chapter raises awareness of the complexity of the concept and generates new testable hypotheses, which might advance our understanding of the theoretical foundations of inequity aversion.

Keywords Capuchin monkeys · Inequity · Inequity aversion · Fairness · Food sharing

---

M. K. Schweinfurth (✉) · J. Call
School of Psychology and Neuroscience, University of St Andrews, St Andrews, UK
e-mail: ms397@st-andrews.ac.uk

J. R. Anderson, H. Kuroshima (eds.), *Comparative Cognition*,
https://doi.org/10.1007/978-981-16-2028-7_11

Introduction

You are not you when you are hungry – Snickers®.

Humans across cultures have a strong sense of fairness (Debove, Baumard and André 2017; Fehr and Schmidt 1999). Fairness is usually studied in resource allocation tasks, in which individuals receive equal or unequal treatments for the same or different actions (e.g. Cowell et al. 2019; Schmidt et al. 2016). Humans are generally averse to unequal treatment for the same action or equal treatment of different actions, i.e. they protest, reject unfair offers and even punish those who act unfairly (Henrich et al. 2006; Oosterbeek et al. 2004). Interestingly, they respond to both disadvantageous, i.e. receiving less than others, and advantageous, i.e. receiving more than others, situations (Fehr et al. 2008). Thereby, we act unlike *Homo economicus*, who tries to maximize her own gains and who should accept unfair offers as long as the offers are greater than zero (Henrich et al. 2005). Instead, however, we pay close attention to what others get in relation to what we get (see above). This behaviour is deeply rooted and develops early in middle childhood across different societies (Blake et al. 2015).

Despite its ubiquity, the evolution of inequity aversion is difficult to explain because it involves costly acts, e.g. punishment, and often leads to benefits being lost, even if they are smaller than that of others. Thus, this behaviour is not only costly, but also seems to provide no benefits to the actor and consequently should not evolve. Although inequity aversion serves no obvious immediate benefits, the behaviour might have evolved because of its delayed benefits. First, although individuals should accept any offer that is providing them with at least some benefits, selection acts on relative levels. This means that individuals compete with others and aim at receiving more benefits than others to outcompete them. Therefore, comparing outcomes and responding negatively to different outcomes is beneficial to maximize relative outcomes (Fehr and Schmidt 1999). Second, most individuals can choose to interact with partners from a pool of several (Noë and Hammerstein 1995). Responding negatively to unfair partners can thus be in the interest of the actor by either ensuring this partner will behave fairly in future interactions or by withdrawing from interacting with this unfair partner in the future (André and Baumard 2011). Third, such partner choice based on fairness has important consequences for the evolution of cooperation, which provides additional benefits (e.g. Melis and Semmann 2010). This is because cooperation can be favoured and stabilized through fairness (Brosnan 2011; Debove et al. 2017).

Given that inequity aversion has adaptive significance, is widespread in humans and emerges early in our ontogeny, it is surprising how little we know about its evolutionary origins. By studying inequity aversion in several species, we can make inferences, for instance, about when and under which conditions this behaviour might have evolved and whether the evolution has happened in steps (cf. Brosnan 2013). One of the first studies on inequity aversion in animals investigated the behaviour of the brown or tufted capuchin monkey, *Sapajus [Cebus] apella* (Brosnan and de Waal 2003), which is a highly social New World primate from

Fig. 11.1 Inti (male) eating two spring onions while being observed by three females. The monkeys live in the RZSS Edinburgh Zoo's Living Links Research Center. Picture taken and kindly provided by Kate Grounds

South America (Fig. 11.1). In this study, the monkeys could exchange a token for a food reward. However, while one monkey received a cucumber for the token, a neighbouring monkey received for the same action a much more valued reward, in this case a grape. Although the cucumber was better than nothing, the monkeys either refused to eat the cucumber or refused to hand back the token in approximately half of the trials. This effect was only shown in females and was even more pronounced, when the partner received the grape for free while the other monkey had to work for its cucumber piece.

Almost 20 years and dozens of publications later, capuchin monkeys have become one of the best-studied species in this area. (Note that this study has also stimulated research in several other species, which are reviewed by McGetrick and Range 2018). However, the results in capuchins have not always been consistent, with a number of studies failing to replicate the original findings. Variation in results between studies is commonly attributed to methods, and additional studies have shown that some of this variation can indeed be attributed to methods. For instance, evidence of inequity aversion is dependent on the effort required from participants (Brosnan and de Waal 2014). In effort-free tasks, responses to inequity do not occur despite unequal food distribution. In other cases, assigning variability to methods is more problematic because some studies have been conducted in different labs using the same methods, but they have produced discrepant results.

The aim of this chapter is to synthesize the available evidence in an attempt to explain this variation by asking does the variation simply reflect noise or are differences related to the conditions under which monkeys are studied? Ultimately, by synthesizing what has been learned from studying these monkeys, including the factors that affect their responses, we are hoping to draw the wider implications for the theoretical foundations of inequity aversion. The chapter is organized as follows.

First, we provide an up-to-date overview of the literature with an emphasis on the different tasks that have been employed. Second, we investigate the variation in the outcome of such studies by comparing the feeding regime outside the experiment and the monkeys' social environment. If necessary, we complemented the information listed in each of the studies with contacting the main researchers who conducted the studies to share information on three areas: feeding regime, subject relationships and social housing. We close by drawing some implications that we hope will contribute to advance our understanding of inequity aversion in animals.

Inequity Aversion in Capuchin Monkeys: The State of the Art

Our survey of the literature returned 17 studies on inequity aversion in capuchin monkeys (see Table 11.1). Fifteen of the studies correspond to one of the following three tasks: token exchange, bar/drawer pulling and free-food distribution. Next, we focus in more detail on each of the tasks by presenting their basic procedure and main results. Because some studies included multiple experiments and included more than one task, we will use the word 'dataset' to refer to a particular experiment or task within a published paper and will reserve the word 'study' to refer to the published study itself.

Token Exchange

Researchers have used this task primarily to study inequity aversion and prosociality (cf. Marshall-Pescini et al. 2016), including the motivational basis for rejecting food items. The basic procedure consists of pairs of capuchins exchanging a token for food with an experimenter. Thus, this task requires some effort (give the token) to obtain the food item. While one of the capuchins obtains a high-quality food item for her effort (e.g. grape), her partner obtains a low-quality item (e.g. cucumber) for the same effort. Researchers measure the willingness to exchange the token and whether the low-quality food is eaten. Although the quality of the reward received by the subject is the most common manipulation, researchers have also varied inequality in other ways including effort (extra cost for the same reward) and the quantity (less food for equal effort). Most studies have used transferring high-quality food to an empty cage (no social comparison possible) as a control condition but some studies have included no token exchange (see free-food distribution section) and varied whether high-quality food is visible or not. Five datasets have yielded evidence of inequity aversion and five have not. Researchers have reported evidence for inequity aversion for quality and effort but not quantity. Moreover, some of the positive results concern only females, and most studies used only interactions with familiar partners.

Table 11.1 Overview of studies investigating inequity aversion in capuchin monkeys

Evidence	Inequality	Benefit –/+	Benefit +/–	Control	Task	Participants	Training	N	Sex	Age	Reference	Study motivation
Yes (in ♀)	Quality	x		No exchange, empty cage	Token exchange	Side-by-side, separated	Familiar with task	10	5♂, 5♀	Adult and subadult	Brosnan and de Waal (2003)	Inequity aversion
Yes	Quality	x	x	NA	Bar pulling	Side-by-side, separated	Familiar with task	10	2♂, 8♀	Adult and subadult	Brosnan et al. (2006)	Equity for cooperation
No	Quality	x		Partner cannot reach food	Effort-free food distribution	Side-by-side, separated	No training	12	12♀	4–30	Dindo and de Waal (2007)	Equity and effort
Yes	Quality, effort	x		High-value reward visible or invisible	Token exchange	Side-by-side, separated	Familiar with task or trained for 6 months	13	4♂, 9f	>4	van Wolkenten et al. (2007)	Greed, frustration alternative
Yes (when familiar)	Quality		x	Covered experimenter's face	Token exchange	Side-by-side, separated	Familiar with task	8	8♀	Adult and subadult	de Waal et al. (2008)	Equity in different relationships
Yes (in quality condition)	Quality, quantity	x		High-value reward visible or invisible	Token exchange	Side-by-side, separated or together	Familiar with task	6	5♂, 1♀	7–22	Talbot et al. (2018)	Violation expectation, separation
								7	7♂	5–10		
Yes	Quantity	x		Empty cage	Pull drawer	Opposite, separated	Familiar with task	8	8♂	13–21	Fletcher (2008)	Expectancy violation
Yes	Quantity	x	x	Empty cage, blocked visual contact	Pull drawer	Opposite, separated	10 sessions per 10 trials	6	3♂, 3♀	6–13	Takimoto et al. (2010)	Control partner's

(continued)

Table 11.1 (continued)

Evidence	Inequality	Benefit −/+	Benefit +/−	Control	Task	Participants	Training	N	Sex	Age	Reference	Study motivation
												reward distribution
Yes	Effort	x	x	Empty cage	Pull drawer	Opposite, separated	12–18 sessions per 12 trials	6	3♂, 3♀	8–15	Takimoto and Fujita (2011)	Effect of dominance
No	Quantity	x		NA	Effort-free food distribution	Side-by-side, together	No training	7	4♂, 3♀	Adult and subadult	Amici et al. (2012)	Social tolerance and dominance
No	Quantity	x	x	Empty cage	Token exchange, bar pulling	Side-by-side, separated	Training varied between tasks and subjects	10	?	?	Amici et al. (2014)	Species comparison
No	Quality	x		Empty cage	Effort-free food distribution	Side-by-side, separated	No training	6	2♂, 4♀	4–18	Dubreuil et al. (2006)	Equity and effort
No	Quality	x		NA	Token exchange	Side-by-side, separated	?	11	11♀	6–26	Silberberg, et al. (2009)	Frustration alternative
No	Quality	x		NA	Effort-free food distribution	Side-by-side, separated	No training	8	8♀	Adult (mean: 9.75)	Roma et al. (2006)	Frustration alternative
No	Quality	x	x	Partner cannot reach food	Rotating tray	Opposite, separated	3 phases, 13–20 sessions per 16 trials	8	2♂, 6♀	4–17	McAuliffe et al. (2015)	Costly task
No	Quality	x	x	Empty cage	Token exchange	Side-by-side, separated	3 training and 5 preference	4	1♂, 3♀	adult	Sheskin et al. (2014)	No-cost task

							phases, 20 trials each					
No	Quality, effort	x		Empty cage	Token-exchange, food distribution	Opposite, separated	Familiar with task	5	5♀	3.5–21	Fontenot et al. (2007)	Equity and effort

We describe for each study whether it resulted in evidence for inequity aversion and whether the study used advantageous (+/−) or disadvantageous (−/+) situations for the actor. We also distinguish in which commodity the situations were unequal, i.e. quality (high- vs. low-quality rewards), quantity (more or less of the same reward) and effort (same or different action for the same rewards). Further, we describe the task and control situations and whether the participants were tested facing each other or sitting side-by-side and whether they were separated from each other or together in the testing room. Finally, we specify how many subjects were tested, their age and sex

Bar/Drawer Pulling

This task has been primarily used to investigate inequity aversion and the factors that may modulate its appearance. Moreover, this task has been used to validate the finding of the token-exchange task by adding a different paradigm that eliminates some of the limitations of the original task. The bar/drawer pulling task, just like the token-exchange task, incorporates effort in the form of pulling a bar or drawer. Unlike the token-exchange task, the experimenter does not play an integral role in the task. She simply sets it up and conspecifics solely interact with each other. Pairs of capuchins sit side-by-side separated by a partition or face each other with the apparatus placed between them. Depending on the type of apparatus used, the subject can pull a bar or drawer to deliver food to herself and/or the partner. It is important to note that studies using this design measure whether subjects make fair or unfair offers, while the token-exchange and free-food distribution tasks measure the response to fair or unfair offers. Typically, the subjects can choose between equality (e.g. one grape for each partner) and advantageous inequality (e.g. one grape for the subject and none for the partner). Some studies have also used disadvantageous inequality (e.g. no grape for the subject and one for the partner). Alternatively, researchers have manipulated effort instead of quality by keeping the quality between the two options equal but making one of them harder to pull. Four of the five datasets have produced evidence of inequity aversion in quantity or effort. Owing to the diverse food item combinations used, next we provide a more detailed presentation of the results.

Fletcher (2008) reported that capuchin operators preferred equity over disadvantageous inequity, which is consistent with token-exchange studies in which subjects were typically the recipients. Takimoto et al. (2010) also used this method but contrasted equity and advantageous inequity. They found that capuchins facing subordinate recipients preferred equity over advantageous inequity, further reinforcing the notion of a preference for equity. However, Takimoto and Fujita (2011) could not replicate this result except when both partners played a role in obtaining the food, but not when the operator alone provided the food. Takimoto et al. (2010) also found that when the subordinate recipient was behind an opaque occluder, subjects no longer preferred equity over advantageous inequity.

Monkeys facing a dominant individual responded somehow differently. They showed indifference between equity and advantageous inequity when the dominant was visible, which could be construed as a weak preference for advantageous inequity. However, they displayed an overt preference for advantageous inequity when the dominant recipient was behind the opaque occluder. Takimoto and Fujita (2011) did replicate their original result (indifference) in the case of a dominant recipient and just like it was the case for subordinate recipients, subjects shifted to a preference for the equitable option when both partners contributed to obtaining the food. Brosnan et al. (2006) also used a bar-pulling cooperative task that required both monkeys to simultaneously pull a bar to bring food within reach. In some cases, both monkeys received the same type of food but in other cases one of the monkeys got better food than the other. Brosnan et al. (2006) reported that capuchins solved the task regardless of the food that each of the partners obtained, seemingly indicating a lack of inequity aversion. However, partners who alternated more often which food they obtained in the unequal condition were more successful than those who did not.

Takimoto et al. (2010) also contrasted disadvantageous and advantageous inequity when facing a subordinate or a dominant recipient. Here the equity option was not available and they found that capuchins preferred disadvantageous inequity (they gave more to others than what they got themselves) but they showed indifference between the options, which in this case can be construed as a form of equity, when they faced the dominant recipient. In summary, most datasets, in some form or another, seem to indicate a preference for equity although the use of different choice options and the small sample size makes drawing robust conclusions difficult.

Free-Food Distribution

Researchers have used this task primarily to investigate the effect of effort on inequity aversion responses and less prominently to assess the modulating effect of social factors such as tolerance and dominance as well as the contribution of frustration in determining responses. The basic arrangement for this task is the same of the token-exchange task except that the experimenter provides food to the subject and the partner without requiring the transfer of a token. Thus, unlike the previous two tasks, this task does not require any effort from any of the participants to obtain food.

Just like in the token-exchange task, subjects and partners in the experimental condition receive food items that differ in quality, or in the case of one study, quantity. Several studies did not include any control conditions other than an equality condition, partly because this task is often used as a control for the token-exchange task. Those studies that included control conditions varied whether the partner was absent or present but unable to reach the food given to her. All five studies using this setup have found no evidence of inequity aversion. This means that capuchins are indifferent to the gains of others when they have spent no effort in obtaining the food.

Effect of Different Tasks on Likelihood to Find Inequity Aversion

Table 11.2 summarizes the results of these tasks and also includes two studies using other paradigms that have provided no evidence of inequity aversion in capuchin

Table 11.2 Overview of different task designs to study inequity aversion in capuchin monkeys

	Tasks				
Laboratory	Token exchange	Bar/drawer pulling	Free-food distribution	Other	Total
Yerkes/ LRC	++++	+	-		+++++ / --
Rome	--	-	--		----
Kyoto		++			++
NIH	+		-		+ / -
Yale	-			-	--
New Iberia	-		-		--
UGA		+			+
TOTAL	+++++ / ----	++++ / -	/ -----	/ -	+9 / -12

monkeys. Two findings are particularly noticeable. First, the bar/drawer pulling task has produced mainly positive results, the free-food distribution task uniformly negative results and the token-exchange task has produced mixed results. Second, the three laboratories with the higher number of studies have produced opposing results. Researchers working with the capuchins from the 'Language Research Center at Georgia State University' and the 'Yerkes National Primate Research Center of the Emory University' have reported positive results in the token-exchange task. Researchers working with the 'Primate Center of the Institute of Cognitive Sciences and Technologies' capuchins have reported uniformly negative results in three of the tasks. The rest of the laboratories present a mixture of positive and negative results, which taken in isolation are hard to interpret due to the small sample size.

Different methods, different populations and different laboratories have produced a mix of positive and negative results. This resulted in a lively and still ongoing debate about whether the studies test inequity aversion or whether confounding effects can explain the findings (e.g. Bräuer and Hanus 2012; Henrich 2004; McAuliffe et al. 2015; Wynne 2004). The initial findings were challenged on the grounds that they may indicate frustration at not getting a more valuable visible food rather than reflect any sort of social comparison (Dubreuil et al. 2006; Roma et al. 2006). However, inequity responses disappear when effort is eliminated in the effort-free food distribution task (e.g. Dindo and de Waal 2007; reviewed in Brosnan and de Waal 2014). This means that even though a discrepancy in food allocation, and hence frustration between partners is still present, capuchins do not reject food of lower quality than their partners. Other researchers questioned these findings by arguing that rejecting lower quality exacerbates the effect that caused the response in the first place because the partner is unaffected by the subject rejecting food of a lower quality (Henrich 2004; McAuliffe et al. 2015). In fact, by rejecting low-quality food the subject *increases* rather than *decreases* the inequity with her partner. While this is the case for token-exchange and free-distribution tasks, bar/drawer pulling tasks investigate whether subjects choose a fair or unfair distribution and hence avoid this issue.

Some of the failures to replicate have been attributed to the physical arrangement of the participants. In some studies partners sit side-by-side separated by some partition (e.g. mesh) while in other studies they face each other. De Waal (pers. comm., 15 April 2020) mentioned that the spatial disposition of the individuals might be important, with side-by-side arrangements more likely to yield evidence of inequity aversion. Indeed, most positive findings were obtained using side-by-side designs. However, three studies used an arrangement where partners faced each other in a drawer pulling task but they still found evidence of inequity aversion (cf. Table 11.1). This suggests that the spatial disposition cannot explain all the variation in the data.

As effective as some of these counterarguments can be in explaining some of the negative results, some tasks do replicate the methods of the original studies, including effort present and the right spatial disposition and still find no evidence of inequity aversion in the token exchange or the bar/drawer pulling tasks (cf. Tables 11.1 and 11.2). At the same time, some of the proposed confounding effects have been addressed but without fully explaining why studies resulted in positive findings. So, the mixed evidence remains. Confronted with mixed results, readers have two basic options: dismiss a subset of them on methodological grounds or consider the possibility that both subsets may be valid as they reflect the natural variation in the prosocial behaviour of capuchin monkeys. Our intention in writing this chapter was not to add just another piece to the long list of articles pointing out the weaknesses in this literature (e.g. Bräuer et al. 2006). Our goal here was to explore the second option by taking a more holistic approach that we hope will bring some balance and clarity. To do so, we turn our attention to the between-study factors that might account for this currently unexplained variability.

Factors That Could Influence Behaviours in Unequal Situations

When we disregard datasets that involve effort-free situations (n = 5, Table 11.2; cf. Brosnan and de Waal 2014), nine datasets find evidence for inequity aversion, while six do not (see Table 11.2). Interestingly, our review of the literature revealed some consistent results within laboratories (Table 11.2). As we have outlined above, the studies differ in various aspects but investigate the same concept. Thus, the studies can be considered as conceptual replications (Nosek and Errington 2020).

Mixed findings suggest that inequity aversion may not be a reliable finding. Still, it is surprising to find so many positive findings in this case. One explanation is that different conclusions might be explained by a confirmation bias, i.e. unconsciously biased experimental design, data collection, data analyses or publication (Ioannidis et al. 2014; Stevens 2017). Another explanation might be that studies based on small sample sizes are more likely to result in wrong conclusions (Colquhoun 2014) and are less likely to be replicable (Farrar and Ostojić 2019). Given that the median sample size is only eight (Table 11.1), this might be a source of concern. Furthermore, there is marked variation in the response between participants (Price and Brosnan 2012). Hence the small sample size and noisy measurements make the finding vulnerable to different conclusions and hence reduced replicability.

However, more than half of the studies have resulted in positive findings, obtained in several laboratories, which make confirmatory biases and false positive explanations unlikely. Variation in results may represent systematic differences that can inform the theoretical foundation and generalizability of inequity aversion. Here, we argue that capuchin monkeys might show inequity aversion, but only under some conditions. Different research groups seem to find almost exclusively either positive or negative evidence, which might point towards conditional-dependent evidence (Table 11.2). Methodological approaches of different research groups vary inevitably from each other, as the study subjects and keeping conditions differ greatly from each other. Although the differences seem minor, they might represent systematic differences that could affect findings, leading to this mixed picture. Here, we argue that moving beyond the question 'whether inequity aversion in capuchins exists or not' to asking 'under which conditions is it expressed' is more informative. We hope that by identifying those sources of variation and their impacts on inequity responses, we might be able to advance the theory of inequity aversion.

To investigate such sources of variation, we proposed several factors and predicted their influence on inequity aversion responses in capuchin monkeys. We then discussed those factors with authors of the studies, listed in Table 11.1, to compliment our literature review. We asked them about the feeding and housing conditions of their monkeys and the relationship between the participants (Box 11.1). Almost all researchers shared such information with us, covering the main research facilities which included the Graduate School of Letters of Kyoto University [JP], Laboratory of Comparative Ethology of the National Institutes of Health Animal Center [USA], Language Research Center at Georgia State University [USA], Primate Center of the Institute of Cognitive Sciences and Technologies [IT] and Yerkes National Primate Research Center of the Emory University [USA]. Thus, our chapter coveres information about 13 of the 17 published studies. We summarize the information in Table 11.3. Below we discuss the factors that may influence inequity responses. We distinguish in our predictions whether the factors might decrease the motivation to engage with the task and whether the factors might bias the response of the monkeys for the low- or high-value food option.

Table 11.3 Results of the survey

	Laboratory	1	2	3	4	5*	6	7	8*	9*
	Evidence for inequity aversion	Yes	Yes	Yes	Yes	Yes	No	No	No	No
Food	Fed/or food available before testing	No	Yes	Yes	Yes	?	No	No	No	?
	Ad libitum feeding	No	Yes	Yes	No	?	No	No	No	?
	Rewards were part of diet	Yes	Yes	Yes (quality) and no (quantity)	Yes	?	Yes	Yes	?	?
	Which reward was part of the diet	One or both	Both	Both	Both	?	Low	High	?	?
Partner	Partner familiar/ tolerant	Yes	Yes	Yes	Yes	Yes	Yes	Yes	?	Yes
	Partner bonded	Yes	Yes	Yes	Yes	?	Yes	Yes	?	?
	Dominance difference	Yes	Yes	Yes	Yes	?	Yes	Yes	Yes	?
Housing	Housed with more than one partner	Yes	Yes	Some	Yes	No	Yes	Yes	No	Yes
	Housed with more than five partners	Yes	Yes	No	Yes	No	No	Yes	No	Yes
	Mixed-sex housing groups	Yes	Yes	No	Yes	?	Yes	Yes	Yes	Yes
	Stable housing group	Yes	Yes	Yes	Yes	?	Yes	Yes	NA	?
	Stable dominance hierarchy	Yes	Yes	Yes	Yes	?	Yes	Yes	NA	?

We complemented our literature review with information about the feeding schemes, relationship of participants with each other and housing conditions of the capuchin monkeys. The labs were assigned numbers to provide anonymised data. For labs marked with an asterisk, we had either no contact details or we had not received a reply. Two labs (#3 and #7) were part of the same facility, but the colonies were kept very differently, which is why we split up the labs. We tried to fill in the information for those studies based on the information given in the articles. Question marks indicate missing information

Box 11.1 Additional methodological details

- Did the monkeys have *ad libitum* food or were there short times when there was no food available for them? yes/no.
- Were the food rewards, used for the experiment, part of their normal diet? yes/no.
    - If yes, which one? high-quality/low-quality/both.
- Were the monkeys fed before the experiments? yes/no/some.
- Were the monkeys tested with familiar in-group members? yes/no.
    - If yes, would you say most of them were bonded? yes/no.
    - If yes, were there dominance differences between test individuals? yes/no.
- Were the monkeys housed in a social group with more than one social partner? yes/no.
    - If yes, with more than five partners? yes/no.
    - Are the monkeys housed in mixed-sex groups? yes/no.
- Was the group established at least one year prior to the experiment? yes/no.
- Was the dominance hierarchy stable when the experiment(s) was (were) conducted? yes/no.
- If there are any studies that did not follow these general practices, we would greatly appreciate if you could point them out to us.

Food Before Testing

Hypotheses: Individuals tested before feeding, i.e. when they are a bit hungry, might be more likely to engage in the task and less likely to reject a low-value offer compared to those that were tested directly after feeding, i.e. they are less hungry. For instance, children tested in the morning without breakfast showed a decline in memory and attention (Wesnes et al. 2003). Hence capuchin monkeys without their breakfast might also pay less attention to the task and are more likely to take any option. Hungrier individuals might also show the opposite response because they may perceive the high-value offer as so rewarding, that they are more likely to protest and reject the low-value offer in order to try to get access to the better food. A similar effect was observed in Norway rats (*Rattus norvegicus*). When given the choice between three food options, here normal chow, Special K and chocolate-flavoured cereals, hungry rats preferred the chocolate cereals, whereas satiated rats showed no difference between the two cereals (Barbano and Cador 2005). This result suggests that hungry rats show a stronger response for high-value food, which might be similar to the monkeys under such conditions.

Results: We found that monkeys that were either fed before the experiment, e.g. a little reward for entering the testing compartment, or still had food from a previous feeding available (e.g. *ad libitum* feeding) were more likely to show inequity (3 facilities). In contrast, monkeys that received their food after the test or had to wait in a transport box without food before the test were less likely show inequity aversion (3 facilities). Please note that there is one exception, i.e. a lab that only feeds their monkeys in the evening and tests them in the morning (#1 in Table 11.3). The studies in this lab were consistently conducted as first studies in the morning and resulted in support for inequity aversion. Hunger has not received much attention in such tasks. A study on rats that were food restricted still found evidence for inequity aversion (Oberliessen et al. 2016). Similarly, the time of the last meal did not change inequity aversion in children (Bowie 2013), nor the evaluation of unequal situations (Huppert et al. 2020). However, we are not aware of a study manipulating hunger levels to determine its effects on inequity aversion. For example, hunger might change impulsivity that can lead to the acceptance of any food rewards, even if this is a suboptimal choice (Laude et al. 2012; Mayack and Naug 2015).

Food After Testing

Hypotheses: Individuals who have constant access to food, i.e. also after testing, might be more likely to engage with the task and accept any, i.e. also the low-value, offers simply because they might prefer working for food than just receiving food for free during feeding times (Inglis et al. 1997; but see Anselme et al. 2018). In contrast, individuals who have access to food after the testing could be more likely to reject low-value food than those who do not have such access because they can afford rejecting food options in light of the constant food availability. In other words, a cucumber might be of so low value that it might be better to wait until returned to the main cage to feed on monkey chow.

Results: All research facilities that reported *ad libitum* access to food or fed goodies to their subjects after testing find consistent evidence for inequity aversion in capuchin monkeys (2 facilities). One facility (#4, Table 11.3) provides no *ad libitum* food to their monkeys but finds evidence for inequity aversion in capuchins. Here, the monkeys are fed four times a day in addition to receiving regular snacks, which is probably comparable to *ad libitum* feeding. The monkeys of lab #1 (Table 11.3) were fed every evening and tests were done early in the morning. Although their access to food is more restricted, all studies of this lab resulted in evidence for inequity aversion. It is important to note, that these monkeys are used to several tests a day and the studies, which we included here, were consistently conducted the first in the morning. The monkeys probably learned that there will be other tests, involving food, right after this test. Furthermore, they commonly received the remainder of their daily ration when returned to their home cage. Hence these monkeys had the expectation of receiving additional food right after testing. Research facilities that feed their monkeys at specific times (usually once or twice a day) and where there is usually no left over available before or after the testing, find no evidence for inequity aversion (3 facilities).

Taken together these results seem to indicate that monkeys who had access to (potentially better) food after the test are more likely to reject low-quality food compared to monkeys whose access to food is more regimented. In a sense, one could argue that inequity aversion is a trait of monkeys who can 'afford' it. Food availability can alter the perception of risk in this situation, which is known to change food-related decisions (Watson and Platt 2008). Facilities that found no evidence for inequity aversion could repeat some of their tasks and vary whether monkeys receive plenty of food for some days in contrast to their normal feedings scheme. This would help to investigate the role of *ad libitum* feeding experimentally. In addition, the risk-taking behaviour of monkeys from different facilities could be compared to study different risk perceptions of the monkeys depending on their feeding scheme.

Food Quality of the Rewards in Comparison to Daily Food

Hypotheses: If the food rewards in the test are perceived as more valuable compared to their normal diet, e.g. monkey chow, the individuals might be more likely to engage with the task and they might be more likely to accept also low-value offers. Dogs, for instance, prefer variation in their training rewards (Bremhorst et al. 2018) and hence monkeys could aim for a similar effect by accepting any offer to increase variation. Especially, if the low-value offer is comparable to the daily diet but not the high-value offer, individuals might be more likely to reject the low-value offer. Children, for instance, request and eat more high-value food, if this type of food was not in their normal diet for an extended period of time compared to when the same food was part of their diet (Fisher and Birch 1999).

Results: Almost all the rewards used in the studies were part of the monkeys' diet in some form. They were either used in training or preference phases, regularly used in other experiments or part of their normal daily diet. We found some variation regarding which reward was part of their diet, i.e. both rewards or only the low–/high-value reward. Of those facilities that commonly report positive evidence, all used rewards that are part of the monkeys' diet. More specifically, three based their experiments on rewards that were part of their diet, while one facility used rewards dependent on the roles of the participants that were either both or only one part of their diet. Although it seems that using rewards that are part of the diet increases the likelihood of finding inequity aversion, this finding stands in contrast to our predictions and should be investigated further before drawing any conclusions. Further, we only had information from two facilities that consistently find no evidence. Both facilities reported that only one reward was part of the monkeys' diet. However, this was in one situation the high and in the other the low-value reward. Hence, we conclude that familiarity with the rewards is unlikely to explain the variation in the findings. Still, future studies should record this detail to investigate whether the relative value of rewards influences the likelihood of finding a certain outcome in these studies.

Familiarity and Tolerance Between Participants

Hypotheses: The more individuals are familiar and tolerant towards each other, the less likely they might protest and reject the low-value option compared to when they are tested with a stranger because they do not begrudge their familiar partner the food. While this effect was shown in a study on chimpanzees, *Pan troglodytes* (Brosnan et al. 2005), it was not found in long-tailed macaques, *Macaca fascicularis* (Massen et al. 2012) and in another study on chimpanzees (Brosnan et al. 2015).

Results: All researchers reported that the subjects were familiar and highly tolerant with each other. Therefore, this cannot explain the variation in the available data. One study, set out to investigate inequity aversion in dependence of relationship quality, found that the effect was more pronounced between familiar partners (de Waal et al. 2008). Future studies could also compare inequity responses between strongly and less strongly bonded individuals to assess the impact of relationships.

Social Housing

Hypotheses: Individual- or pair-housed individuals might lack the same opportunities that group-housed individuals encounter to practise fairness. Hence, such monkeys might be less likely to reject low-value offers because they might not have faced such situations often enough to respond adequately. In contrast, individual- or pair-housed individuals might be more likely to reject low-value offers because the consequences of their behaviour are minimal since punishment can only be exhibited by maximally one partner instead of a whole group. This effect was suggested for chimpanzees housed in pairs or in a group (Brosnan et al. 2005).

Results: Four facilities reported that they house their monkeys in small groups, i.e. below five individuals. Two of them report consistently no evidence for inequity aversion. It should be noted, however, that the other two facilities kept their monkeys in pairs or trios and still found evidence for inequity aversion (#3 & #5 in Table 11.3). This shows that social housing is unlikely to impact the likelihood of finding inequity aversion in these monkeys.

Group Stability

Hypotheses: The longer individuals have lived together, the more likely they have formed stable social bonds and dominance hierarchies. Based on increased predictability of their partner's behaviour, the individuals might be more likely to engage in the task, shown in chimpanzees (Brosnan et al. 2015), and less likely to reject the low-value offer, also demonstrated in chimpanzees (Brosnan et al. 2005). Alternatively, one could predict that the longer individuals have lived together, the more they pay attention to fairness because they repeatedly experience fair or unfair situations together. Hence, they might be more likely to reject low-value offers in the presence of such a partner. Although plausible, this effect was not shown in chimpanzees that show inequity aversion independent of how stable their group was (Brosnan et al. 2015).

Results: All research facilities reported that they worked with stable groups, which precludes us from drawing any conclusions on whether it might influence responses to unequal situations. This demonstrates that group stability cannot explain the variation in the published studies. This factor is difficult to assess in future studies because it is questionable to interrupt the group stability for research purposes. Social structures change if some individuals are temporarily removed or die (Byrne et al. 1996), which can be used for future studies. Naturally, however, stable and unstable groups differ in many aspects and unstable groups are rare, complicating the interpretation of a finding.

Dominance

Hypotheses: Subordinates might be more likely to accept a low-value food offer in the presence of a more dominant individual than vice versa. This effect was found, for instance, in chimpanzees (Brosnan et al. 2010, but see Brosnan et al. 2005) and rats (Oberliessen et al. 2016). This effect is likely to be different in dominant partners that might be less likely to accept low-value offers. Nevertheless, a study investigating inequity aversion in apes found no evidence that inequity aversion was exhibited more strongly in dominants (Bräuer et al. 2006). In addition in long-tailed macaques, dominant and subordinate individuals did not differ in their response (Massen et al. 2012).

Results: Female and male capuchin monkeys form dominance hierarchies (e.g. Janson 1985). Accordingly, all researchers reported that there were some dominance differences between the participants. However, in retrospect it is difficult to assess how strong these differences were. Dominance can change the behaviour of these monkeys in such tasks (see above). However, to what extend is currently unclear. Hence, future studies could assess the dominance rank of individuals and incorporate this in their findings.

Conclusions

In this chapter, we reviewed the evidence for inequity aversion in capuchin monkeys. Ever since the first report in 2003, several studies have followed, conducted in nine different laboratories with several monkey populations and using various methods. While nine datasets resulted in evidence for inequity aversion, six did not support this finding and five provided results that are difficult to interpret. This variation cannot solely be explained by different task designs. While free-food distribution tasks do not provide any evidence for inequity aversion in capuchin monkeys, the token-exchange and bar/drawer pulling task result in overall positive, but also mixed evidence. An interesting pattern is, however, that different laboratories seem to provide consistently either positive or negative findings. By investigating methodological details, we aimed at exploring this variation. We found that social factors, e.g. familiarity or dominance between participants and social housing, explained no variation in the data. In contrast, feeding schemes explained some of the variation.

There was a good match between studies that resulted in positive evidence and participants having access to food after the test. Related to this, monkeys that had access to food shortly before testing, because they were either fed or had food from the last feeding still around, were also more likely to show inequity aversion compared to those that waited in a box without food or their breakfast was delayed because of the testing. Monkeys provided with plenty of food might perceive the situation as less risky or might respond less impulsive compared to monkeys provided with more temporally scattered food. These results suggest that capuchin monkeys differ in their response to unequal situations based on their feeding regimes.

Although we did not find evidence for the other proposed factors, additional factors might be interesting to explore in future studies, which might increase the variation further but are difficult to assess retrospectively. For example, the energetic state or weight of the participants might be interesting. Individuals that are in greater need of food because they are in a poorer or more energy-demanding (e.g. pregnant) conditions might be more likely to accept any food offer and are thus less likely to show inequity aversion. In addition, personality differences, e.g. in extroversion and agreeableness, have been shown to affect inequity aversion in chimpanzees (Brosnan et al. 2015) and might also affect responses in capuchins. Personality is heritable (Penke et al. 2007) and often colonies are based on a few maternal lines. Hence, capuchin populations might also differ in their personalities, leading to variation in the findings.

We would like to end with a word of caution. We acknowledge that our findings are based on only five research facilities and thus represent a small sample size, even though they represent almost all studies on this species. Broadening this approach to other species might give insights into the generalizability of our finding. Currently, we cannot establish a causal relationship between the different feeding schemes and responses to inequality. Instead, we hope that our findings will stimulate future research, using carefully controlled experimental methods to test the here generated hypotheses. Eventually, this will help us to refine the theory of inequity aversion and clarify under which conditions individuals show this behaviour that is so common among humans of all cultures.

Acknowledgements We would like to thank James Anderson and Hika Kuroshima for their kind invitation to contribute to this book and the anonymous reviewer for the valuable feedback on an earlier version of the manuscript. Further, we are very grateful to all the researchers who discussed their study protocol with us in great detail and provided us with additional highly valuable comments and suggestions. In addition, we wish to thank Kate Grounds for providing us with the picture. We would also like to acknowledge the financial support we received from the European Research Council (Synergy grant 609819 SOMICS provided to Josep Call) and the Swiss National Science Foundation (P2BEP3 175269 provided to Manon K. Schweinfurth).

References

Amici F, Call J, Aureli F (2012) Aversion to violation of expectations of food distribution: the role of social tolerance and relative dominance in seven primate species. Behaviour 149 (3–4):345–368. https://doi.org/10.1163/156853912X637833

Amici F, Visalberghi E, Call J (2014) Lack of prosociality in great apes, capuchin monkeys and spider monkeys: convergent evidence from two different food distribution tasks. Proc R Soc B Biol Sci 281(1793):20141699. https://doi.org/10.1098/rspb.2014.1699

André JB, Baumard N (2011) The evolution of fairness in a biological market. Evolution 65 (5):1447–1456. https://doi.org/10.1111/j.1558-5646.2011.01232.x

Anselme P, Dreher T, Güntürkün O (2018) Pigeons consistently prefer easy over harder access to food: no reversal after direct dopaminergic stimulation. Behav Neurosci 132(4):293–301. https://doi.org/10.1037/bne0000249

Barbano MF, Cador M (2005) Various aspects of feeding behavior can be partially dissociated in the rat by the incentive properties of food and the physiological state. Behav Neurosci 119 (5):1244–1253. https://doi.org/10.1037/0735-7044.119.5.1244

Blake PR, McAuliffe K, Corbit J, Callaghan TC, Barry O, Bowie A et al (2015) The ontogeny of fairness in seven societies. Nature 528:258–261. https://doi.org/10.1038/nature15703

Bowie A (2013) The development of inequality aversion in Ugandan children (BSc thesis, Harvard University). doi:https://doi.org/10.1017/CBO9781107415324.004

Bräuer J, Call J, Tomasello M (2006) Are apes really inequity averse? Proc R Soc B Biol Sci 273 (1605):3123–3128. https://doi.org/10.1098/rspb.2006.3693

Bräuer J, Hanus D (2012) Fairness in non-human primates? Soc Justice Res 25(3):256–276. https://doi.org/10.1007/s11211-012-0159-6

Bremhorst A, Bütler S, Würbel H, Riemer S (2018) Incentive motivation in pet dogs - preference for constant vs varied food rewards. Sci Rep 8:9756. https://doi.org/10.1038/s41598-018-28079-5

Brosnan SF (2011) A hypothesis of the co-evolution of cooperation and responses to inequity. Front Neurosci 5:43. https://doi.org/10.3389/fnins.2011.00043

Brosnan SF (2013) Justice- and fairness-related behaviors in nonhuman primates. Proc Natl Acad Sci USA 110:10416–10423. https://doi.org/10.1073/pnas.1301194110

Brosnan SF, de Waal FBM (2003) Monkeys reject unequal pay. Nature 425(6955):297–299. https://doi.org/10.1038/nature01963

Brosnan SF, de Waal FBM (2014) Evolution of responses to (un)fairness. Science 346 (6207):1251776. https://doi.org/10.1126/science.1251776

Brosnan SF, Freeman C, de Waal FBM (2006) Partner's behavior, not reward distribution, determines success in an unequal cooperative task in capuchin monkeys. Am J Primatol 68:713–724. https://doi.org/10.1002/ajp

Brosnan SF, Hopper LM, Richey S, Freeman HD, Talbot CF, Gosling SD et al (2015) Personality influences responses to inequity and contrast in chimpanzees. Anim Behav 101:75–87. https://doi.org/10.1016/j.anbehav.2014.12.019

Brosnan SF, Schiff HC, de Waal FBM (2005) Tolerance for inequity may increase with social closeness in chimpanzees. Proc R Soc B Biol Sci 272(1560):253–258. https://doi.org/10.1098/rspb.2004.2947

Brosnan SF, Talbot CF, Ahlgren M, Lambeth SP, Schapiro SJ (2010) Mechanisms underlying responses to inequitable outcomes in chimpanzees, *Pan troglodytes*. Anim Behav 79 (6):1229–1237. https://doi.org/10.1016/j.anbehav.2010.02.019

Byrne G, Abbott KM, Suomi SJ (1996) Reorganization of dominance rank among adult males in a captive group of tufted capuchins (*Cebus apella*). Lab Primate Newsl 35(1):1–4

Colquhoun D (2014) An investigation of the false discovery rate and the misinterpretation of p-values. R Soc Open Sci 1:140216. https://doi.org/10.1098/rsos.140216

Cowell JM, Sommerville JA, Decety J (2019) That's not fair: children's neural computations of fairness and their impact on resource allocation behaviors and judgments. Dev Psychol 55 (11):2299–2310. https://doi.org/10.1037/dev0000813

de Waal FBM, Leimgruber KL, Greenberg AR (2008) Giving is self-rewarding for monkeys. Proc Natl Acad Sci USA 105(36):13685–13689. https://doi.org/10.1073/pnas.0807060105

Debove S, Baumard N, André JB (2017) On the evolutionary origins of equity. PLoS One 12(3): e0173636. https://doi.org/10.1371/journal.pone.0173636

Dindo M, de Waal FBM (2007) Partner effects on food consumption in brown capuchin monkeys. Am J Primatol 69:448–456. https://doi.org/10.1002/ajp

Dubreuil D, Gentile MS, Visalberghi E (2006) Are capuchin monkeys (*Cebus apella*) inequity averse? Proc R Soc B Biol Sci 273(1591):1223–1228. https://doi.org/10.1098/rspb.2005.3433

Farrar BG, Ostojić L (2019) The illusion of science in comparative cognition. Psyarxiv

Fehr E, Bernhard H, Rockenbach B (2008) Egalitarianism in young children. Nature 454 (7208):1079–1083. https://doi.org/10.1038/nature07155

Fehr E, Schmidt KM (1999) A theory of fairness, competition, and cooperation. Q J Econ 114 (3):817–868

Fisher JO, Birch LL (1999) Restricting access to palatable foods affects children's behavioral response, food selection, and intake. Am J Clin Nutr 69(6):1264–1272. https://doi.org/10.1093/ajcn/69.6.1264

Fletcher GE (2008) Attending to the outcome of others: disadvantageous inequity aversion in male capuchin monkeys (*Cebus apella*). Am J Primatol 70(8):901–905. https://doi.org/10.1002/ajp.20576

Fontenot MB, Watson SL, Roberts KA, Miller RW (2007) Effects of food preferences on token exchange and behavioural responses to inequality in tufted capuchin monkeys, *Cebus apella*. Anim Behav 74(3):487–496. https://doi.org/10.1016/j.anbehav.2007.01.015

Henrich J (2004) Inequity aversion in capuchins? Nature 428(6979):139–139. https://doi.org/10.1038/428139a

Henrich J, Boyd R, Bowles S, Camerer C, Fehr E, Gintis H et al (2005) "Economic man" in cross-cultural perspective: behavioral experiments in 15 small-scale societies. Behav Brain Sci 28 (6):795–815. https://doi.org/10.1017/S0140525X05000142

Henrich J, McElreath R, Barr A, Ensminger J, Barrett C, Bolyanatz A et al (2006) Costly punishment across human societies. Science 312(5781):1767–1770. https://doi.org/10.1126/science.1127333

Huppert E, Shaw A, Decety J (2020) The effect of hunger on children's sharing behavior and fairness preferences. J Exp Child Psychol 192:104786. https://doi.org/10.1016/j.jecp.2019.104786

Inglis IR, Forkman B, Lazarus J (1997) Free food or earned food? A review and fuzzy model of contrafreeloading. Anim Behav 53(6):1171–1191. https://doi.org/10.1006/anbe.1996.0320

Ioannidis JPA, Munafò MR, Fusar-Poli P, Nosek BA, David SP (2014) Publication and other reporting biases in cognitive sciences: detection, prevalence and prevention. Trends Cogn Sci 18 (5):235–241. https://doi.org/10.1038/jid.2014.371

Janson CH (1985) Aggressive competition and individual food consumption in wild brown capuchin monkeys (*Cebus apella*). Behav Ecol Sociobiol 18(2):125–138

Laude JR, Pattison KF, Zentall TR (2012) Hungry pigeons make suboptimal choices, less hungry pigeons do not. Psychon Bull Rev 19(5):884–891. https://doi.org/10.3758/s13423-012-0282-2

Marshall-Pescini S, Dale R, Quervel-Chaumette M, Range F (2016) Critical issues in experimental studies of prosociality in non-human species. Anim Cogn 19(4):679–705. https://doi.org/10.1007/s10071-016-0973-6

Massen JJM, van den Berg LM, Spruijt BM, Sterck EHM (2012) Inequity aversion in relation to effort and relationship quality in long-tailed macaques (*Macaca fascicularis*). Am J Primatol 74 (2):145–156. https://doi.org/10.1002/ajp.21014

Mayack C, Naug D (2015) Starving honeybees lose self-control. Biol Lett 11:20140820. https://doi.org/10.1098/rsbl.2014.0820

McAuliffe K, Chang LW, Leimgruber KL, Spaulding R, Blake PR, Santos LR (2015) Capuchin monkeys, *Cebus apella*, show no evidence for inequity aversion in a costly choice task. Anim Behav 103:65–74. https://doi.org/10.1016/j.anbehav.2015.02.014

McGetrick J, Range F (2018) Inequity aversion in dogs: a review. Learn Behav 46:479–500. https://doi.org/10.3758/s13420-018-0338-x

Melis AP, Semmann D (2010) How is human cooperation different? Philos Trans R Soc B Biol Sci 365(1553):2663–2274. https://doi.org/10.1098/rstb.2010.0157

Noë R, Hammerstein P (1995) Biological markets. Trends Ecol Evol 10(8):336–339

Nosek BA, Errington TM (2020) What is replication? PLoS Biol 18(3):e3000691. https://doi.org/10.1371/journal.pbio.3000691

Oberliessen L, Hernandez-Lallement J, Schäble S, van Wingerden M, Seinstra M, Kalenscher T (2016) Inequity aversion in rats, *Rattus norvegicus*. Anim Behav 115:157–166. https://doi.org/10.1016/j.anbehav.2016.03.007

Oosterbeek H, Sloof R, van de Kuilen G (2004) Cultural differences in ultimatum game experiments: evidence from a meta-analysis. Exp Econ 7:171–188

Penke L, Denissen JJA, Miller GF (2007) The evolutionary genetics of personality. Eur J Pers 21:549–587. https://doi.org/10.1002/per

Price SA, Brosnan SF (2012) To each according to his need? variability in the responses to inequity in non-human primates. Soc Justice Res 25(2):140–169. https://doi.org/10.1007/s11211-012-0153-z

Roma PG, Silberberg A, Ruggiero AM, Suomi SJ (2006) Capuchin monkeys, inequity aversion, and the frustration effect. J Comp Psychol 120(1):67–73. https://doi.org/10.1037/0735-7036.120.1.67

Schmidt MFH, Svetlova M, Johe J, Tomasello M (2016) Children's developing understanding of legitimate reasons for allocating resources unequally. Cogn Dev 37:42–52. https://doi.org/10.1016/j.cogdev.2015.11.001

Sheskin M, Ashayeri K, Skerry A, Santos LR (2014) Capuchin monkeys (*Cebus apella*) fail to show inequality aversion in a no-cost situation. Evol Hum Behav 35(2):80–88. https://doi.org/10.1016/j.evolhumbehav.2013.10.004

Silberberg A, Crescimbene L, Addessi E, Anderson JR, Visalberghi E (2009) Does inequity aversion depend on a frustration effect? A test with capuchin monkeys (*Cebus apella*). Anim Cogn 12(3):505–509. https://doi.org/10.1007/s10071-009-0211-6

Stevens JR (2017) Replicability and reproducibility in comparative psychology. Front Psychol 8:862. https://doi.org/10.3389/fpsyg.2017.00862

Takimoto A, Fujita K (2011) I acknowledge your help: capuchin monkeys' sensitivity to others' labor. Anim Cogn 14(5):715–725. https://doi.org/10.1007/s10071-011-0406-5

Takimoto A, Kuroshima H, Fujita K (2010) Capuchin monkeys (*Cebus apella*) are sensitive to others' reward: an experimental analysis of food-choice for conspecifics. Anim Cogn 13 (2):249–261. https://doi.org/10.1007/s10071-009-0262-8

Talbot CF, Parrish AE, Watzek J, Essler JL, Leverett KL, Paukner A, Brosnan SF (2018) The influence of reward quality and quantity and spatial proximity on the responses to inequity and contrast in capuchin monkeys (*Cebus [Sapajus] apella*). J Comp Psychol 132(1):75–87. https://doi.org/10.1037/com0000088

van Wolkenten M, Brosnan SF, de Waal FBM (2007) Inequity responses of monkeys modified by effort. Proc Natl Acad Sci USA 104(47):18854–18859. https://doi.org/10.1073/pnas.0707182104

Watson KK, Platt ML (2008) Neuroethology of reward and decision making. Phil Trans R Soc B 363(1511):3825–3835. https://doi.org/10.1098/rstb.2008.0159

Wesnes KA, Pincock C, Richardson D, Helm G, Hails S (2003) Breakfast reduces declines in attention and memory over the morning in schoolchildren. Appetite 41(3):329–331. https://doi.org/10.1016/j.appet.2003.08.009

Wynne CDL (2004) Fair refusal by capuchin. Nature 428(2002):447–448

Chapter 12
Evolutionary Perspective on Prosocial Behaviors in Nonhuman Animals

Ayaka Takimoto-Inose

Abstract Prosocial behaviors have long been considered a hallmark of humans. However, observational and experimental studies have revealed that nonhuman animals also show a variety of prosocial behaviors, but the likely evolutionary path of prosocial behaviors has remained unclear. In this chapter, I hypothesize that some psychological factors, namely, inequity aversion, interdependence and tolerance, may have played facilitating roles in the convergent evolution of prosocial behaviors. In experimental studies, some primates who show inequity aversion, such as chimpanzees and capuchin monkeys, have also shown robust prosocial behaviors, although the partner's begging behaviors may be influential. Proactive prosocial behaviors occur in cooperative breeder primates with high levels of allomaternal care, who also show high levels of interdependence. High levels of tolerance and prosocial behaviors toward partners including strangers have been reported particularly in bonobos. I also discuss how the convergent evolution hypotheses of prosocial behaviors applies not only to primates but also non-primate species, by reviewing recent experimental studies on birds and domestic animals. Finally, I propose some possible future study directions for extending explorations of the evolutionary route of prosocial behaviors.

Keywords Prosocial behavior · Convergent evolution · Inequity aversion · Interdependence · Tolerance

Introduction

We humans often help others when we see them in need, even if we do not know them. For example, in a train or on a bus we might give up our seat to an elderly person, or a pregnant woman. We also sometimes feel sympathy to disaster victims

---

A. Takimoto-Inose (✉)
Department of Behavioral Science, Graduate School of Humanities and Human Sciences, Hokkaido University, Sapporo, Japan
e-mail: atakimoto@let.hokudai.ac.jp

J. R. Anderson, H. Kuroshima (eds.), *Comparative Cognition*,
https://doi.org/10.1007/978-981-16-2028-7_12

even if they live far away, and donate money or resources to them anonymously. Such behaviors that result in a benefit for others are referred to as prosocial behaviors (Cronin 2012). To be considered prosocial, individuals need not incur any costs in giving benefit to others; thus, the concept is distinct from altruistic behavior, which requires that the actor incurs some cost when providing a benefit to another. Prosocial behaviors can thus include altruistic behaviors.

The human prosocial proclivity for helping others is present even in 18-month-old infants (e.g., Warneken and Tomasello 2006) and the human tendency toward egalitarian reward sharing in anonymous situations is seen in most 7- to 8-year-old children (Fehr et al. 2008). For students of animal behavior, the following questions arise: are any prosocial behaviors seen in nonhuman species, and if so, how have such behaviors evolved?

Prosocial Behaviors in Nonhuman Primates

Over the past decades many examples of nonhuman primate prosocial behaviors have been reported in observational and experimental studies in the wild and captivity (e.g., de Waal and Suchak 2010; Jaeggi and van Schaik 2011; Marshall-Pescini et al. 2016). Observational data are essential for constructing a fuller picture of a species' natural behaviors. For example, observational studies have reported numerous examples of spontaneous helping behaviors, ranging from traveling slower to accommodate wounded companions, to taking special care of severely disabled individuals, as in chimpanzees (*Pan troglodytes*), Japanese macaques (*Macaca fuscata*), capuchin monkeys (*Sapajus apella*), and so on (e.g., Boesch 1991; de Waal 1996; Matsumoto et al. 2016). Chimpanzees and capuchin monkeys also share meat after a group hunt (e.g., Boesch and Boesch 1989; Perry and Rose 1994) and chimpanzees occasionally share food for grooming (de Waal 1997) or sex (Gomes and Boesch 2009). Bonobos (*Pan paniscus*) share food for social bonding (e.g., Yamamoto 2015). Moreover, great apes show consolation behaviors (e.g., Cordoni et al. 2006; de Waal and van Roosmalen 1979; Palagi et al. 2004). However, observational studies usually have little or no control over the context and conditions in which the behaviors of interest occur, making quantitative and qualitative evaluations of participants' benefits and costs difficult. By contrast experimental studies allow us to investigate details of prosocial behaviors, for example by setting situations in which an actor receives no immediate or direct benefit as a result of a prosocial act. Below, I consider experimental studies that illustrate the latter approach to studying nonhuman primate prosocial behaviors.

Experimental Tasks for Prosocial Behaviors in Nonhuman Primates

Three main kinds of experimental tasks have been used to test nonhuman primates: the prosocial choice task, the food sharing task, and the targeted-helping task (Fig. 12.1). In the prosocial choice task, an actor is given an opportunity to provide benefits to a recipient at little or no cost to itself (Fig. 12.1a). The actor can choose one of two options: a prosocial option in which the actor can give valuable rewards

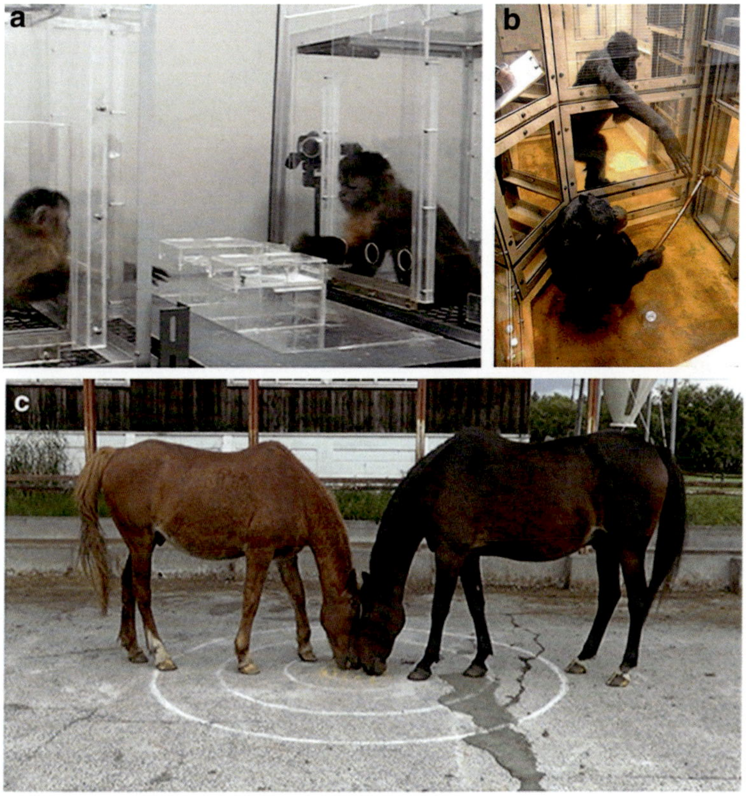

Figs. 12.1 Test tasks for testing animals' prosocial behaviors. (**a**) Prosocial choice task: Animals are presented with two options: one is just self-rewarding (selfish), and the other benefits both an actor and a recipient (prosocial). In this picture, the actor (right) is choosing prosocial option (back) in which both the actor and the recipient (left) received a preferred food (photograph by Ayaka Takimoto-Inose). (**b**) Food sharing task: Animals are tested whether they share food with a conspecific partner. In this picture, an actor (right) is permitting a partner (left) to approach him and share food with him (photograph by Mao Inoue). (**c**) Targeted helping task: Animals are tested whether they help others. In this picture, an actor (front) is giving a tool based on begging behavior from the partner (back) who can subsequently use the tool to get her out-of-reach reward. (photograph by Shinya Yamamoto; cited from Yamamoto et al. (2009). Chimpanzees help each other upon request. *PLoS ONE,* 4, e7416. Figure 2)

to both self and the recipient, or a selfish option in which only the actor can obtain valuable rewards. Whatever the choice, the actor obtains the same amount and quality of reward; therefore, choosing the prosocial option is not costly (other than the minimum labor required to pull the bar connected to the food tray, for example). The important point is to compare the actor's choices when another individual (recipient) is present versus a recipient-absent condition. Without this comparison, it is possible that the actor might just choose the prosocial option because it includes more or higher-value food rewards than the selfish option. However, if the actor chooses the prosocial option significantly more often when the recipient-present than the recipient-absent condition, we can more confidently conclude that the actor's prosocial preference is "intentional." This prosocial choice task has become a popular experimental procedure, and it has revealed some species differences. A prosocial tendency, i.e., choosing the prosocial option more frequently when the recipient is present than absent, has been observed in various species including New World monkeys (e.g., capuchin monkeys: de Waal et al. 2008; Lakshminarayanan and Santos 2008; Takimoto and Fujita 2011; Takimoto et al. 2010, common marmosets (*Callithrix jacchus*): Burkart et al. 2007, cotton-top tamarins (*Saguinus oedipus*): Cronin et al. 2010, but see also Cronin et al. 2009; Stevens 2010), and Old World monkeys (rhesus monkeys (*Macaca mulatta*): Chang et al. 2011; Masserman et al. 1964; long-tailed macaques (*Macaca fascicularis*): Massen et al. 2010). Surprisingly, however, chimpanzees have shown indifference to the recipient's outcomes in most studies using this task: they did not discriminate between the prosocial and selfish options (Brosnan et al. 2009; Jensen et al. 2006; Silk et al. 2005; Vonk et al. 2008; Yamamoto and Tanaka 2010). Claidière et al. (2015) suggested that such negative results may reflect lack of understanding the tasks, and they reported positive prosocial choice results in chimpanzees who were shown to understand the apparatus and the consequences of the proposer's actions. Horner et al. (2011) also obtained positive results in chimpanzees when the actor was prevented from developing a side bias and by ensuring the actor's attention to the difference between the two options (this was done by noisily unwrapping the rewards—banana slices—while giving them to the recipient). Additionally, Schmelz et al. (2017) demonstrated that chimpanzees are willing to behave prosocially only to a partner that previously assisted them, and particularly when the assistance was risky, in a task similar to the prosocial choice task. This latter study suggests that chimpanzees do not care to be prosocial unless there is some reason for them to do so.

The food sharing task is also widely used, for assessing an actor's tendency for prosocially sharing food with a partner (Fig. 12.1b). Food sharing has been defined as the unresisted transfer of food from one food-motivated individual to another (Feistner and McGrew 1986), but the definition of the phenomenon has been the subject of debate: some researchers exclude any incidents of theft of food (Feistner and McGrew 1986), whereas others accept "tolerated theft" and joint foraging as examples of food sharing (Stevens and Gilby 2004). Food sharing is universal in human societies, and it has received considerable attention for its potential importance in the evolutionary emergence of prosocial behaviors, including altruistic acts (e.g., Jaeggi et al. 2010a). Like humans, nonhuman primates show various kinds of food sharing, though most instances are passive food transfers (>95% in most species, reviewed in Jaeggi and van Schaik, 2011). The following have been reported: (a) spontaneous giving, in which a proactive food transfer (hand to hand; mouth to mouth; hand to mouth or vice versa) is initiated by the owner without any kind of begging behaviors by the recipient; (b) reactive sharing, which is an active food transfer by the owner in response to begging behaviors by the recipient; (c) tolerated theft or scrounging, which is a passive transfer in that the owner allows the recipient to take food directly from his or her possession; (d) co-feeding, another passive transfer, in which the owner allows the recipient to take food that is near enough to be monopolizable by the owner. For example, bonobos, one of our closest living evolutionary relatives, have shown such prosocial tendencies in the form of food-sharing (Hare and Kwetuenda 2010), but not in the standard prosocial choice task (Tan et al. 2015). In a study of capuchin monkeys, owners' tolerance of food transfers appeared to reflect an understanding of the recipient's motivational state (having seen her eating food or not) and judging whether the recipient's desire for food matched the owner's (Hattori et al. 2012).

The targeted helping task examines whether an actor helps a partner when the latter is faced with a problem which cannot be solved by one individual alone (Fig. 12.1c). For example, the experimenter trains an animal (the partner) to solve a problem by using a particular object, but then presents the object out of the partner's reach. However, the object is within the actor's reach, and the actor can pick it up and pass it to the partner. In a control condition, the actor has access to the same object, but the partner has no need of it. This experimental setup would allow us to conclude that an actor helps the partner "intentionally" if the actor picks up and passes the object to the partner when it is needed significantly more frequently than in the control condition. Results to date from experiments using this task have appeared inconsistent with those using the other tasks (Marshall-Pescini et al. 2016). For example, chimpanzees are reportedly the most prosocial nonhuman primates in the targeted helping task (Melis et al. 2010; Warneken and Tomasello 2006; Warneken et al. 2007; Yamamoto et al. 2009, 2012). By contrast, capuchin monkeys are robustly prosocial in the prosocial choice task, but much are much less so in the targeted helping task (Barnes et al. 2008; Skerry et al. 2011). Additionally, targeted helping has been observed in only a few species (e.g., de Waal 2008) compared to the other two tasks, that is, the prosocial choice and food sharing tasks.

Why are there the differences between the prosocial choice and food sharing task, and the targeted-helping task? In the two former tasks, if the actor is "other-regarding," he or she needs only to discriminate between the food reward distribution of the two options and then decide whether to share food with the recipient or not. Therefore, these two tasks are useful for evaluating whether animals have other-regarding preferences and whether they can understand the experimental situation, especially, what food is their own and what can be made available for the recipient. In addition, the desires of the actor and the recipient are nearly always the same in these tasks, therefore it is relatively easy for the actor to guess the recipient's desire for food. However, to complete a targeted helping task, the actor has to understand desires (of the partner) that may be different from the actor's own. For example, the actor must infer what help the partner desires to obtain food, even if the actor needs to do very little to get its own food. Therefore, the targeted helping task requires more elaborate cognitive abilities, i.e., perspective-taking (Yamamoto and Takimoto 2012). In fact, capuchin monkeys' results seem to fit with this view. These monkeys show prosociality in the simpler, prosocial choice test (de Waal et al. 2008; Lakshminarayanan and Santos 2008; Takimoto and Fujita 2011; Takimoto et al. 2010), but rather less in the targeted helping test (Barnes et al. 2008; Skerry et al. 2011), and they appear capable of limited perspective-taking compared to chimpanzees (capuchin monkeys: Hare et al. 2003; chimpanzees: Hare et al. 2000; Hare et al. 2001; Krupenye et al. 2016). Chimpanzees, by contrast, have shown robust prosociality in targeted helping tasks (Melis et al. 2010; Warneken and Tomasello 2006; Warneken et al. 2007; Yamamoto et al. 2009, 2012), but much less in the prosocial choice tasks (Brosnan et al. 2009; Jensen et al. 2006; Silk et al. 2005; Vonk et al. 2008; Yamamoto and Tanaka 2010; but see also Claidière et al. 2015; Horner et al. 2011). One highly plausible explanation is that chimpanzees recognize another's desire for a tool, but not food (e.g., Yamamoto et al. 2009; Yamamoto et al. 2012); as a chimpanzee actor is currently strongly food motivated, they chimpanzees might simply overlook or pay little attention to the partner's food reward. Yamamoto et al. (2012) also demonstrated that chimpanzees can observe and understand the partner's desires, and adjust their targeted helping interventions accordingly. Conceivably, chimpanzees may not always behave prosocially in the prosocial choice task using food because they perform naturally more skillfully in competitive, rather than in cooperative, cognitive tasks (e.g., Hare and Tomasello 2004).

Characteristics of Prosocial Behaviors in Nonhuman Primates

Cronin (2012) and de Waal and Suchak (2010) nicely reviewed experimental studies of proximate factors that influence prosocial behaviors in nonhuman primates. Below, I focus on three such factors: begging behaviors, reciprocity and social closeness, with attention to the difference between human and nonhuman primates' prosocial behaviors.

There are two kinds of begging behaviors. One shows the recipient's interest in the reward, for example, reaching toward it with an arm or hand. The other involves begging directly from the actor for the reward (Cronin 2012). Both kinds have been reported to elicit prosocial behaviors. For example, Takimoto et al. (2010) reported that capuchin monkeys' prosocial choices are facilitated by the former kind of begging behavior. In this study, capuchin monkeys behaved prosocially to a subordinate recipient but not to a dominant one. However, when the actor's view of the recipient's begging behaviors was blocked by an opaque barrier, prosocial choices to the subordinate recipient also ceased. Moreover, chimpanzees help a partner more significantly when the partner shows both kinds of begging behaviors such as reaching or touching the actor directly, or vocalizing (Yamamoto et al. 2009). And Yamamoto et al. (2012) reported that chimpanzees do not help a partner who does not beg, even if they have seen what the partner desires. The importance of the partners' begging behaviors has also been reported in other experimental studies (Jaeggi et al. 2010a; Melis et al. 2010; Warneken and Tomasello 2006; Warneken et al. 2007). However, begging behaviors do not influence (Vonk et al. 2008) or might even decrease (Horner et al. 2011) chimpanzee prosociality in some kinds of prosocial choice tasks, which suggests that begging works sometimes, and sometimes functions as a kind of harassment which might in fact hinder prosociality by the partner. There has been no experimental demonstration that any kind of begging facilitates prosocial behaviors in Old World monkeys (e.g., Chang et al. 2011; Massen et al. 2010, 2011), but begging was reported to facilitate the exchange of social grooming roles, a kind of prosocial behavior, in Japanese macaques (Ueno et al. 2014). Interestingly, however, Burkart et al. (2014) reported that cooperatively breeding primate species with extensive allomaternal care, such as cotton-top tamarins, lion tamarins (*Leontopithecus chrysomelas*), common marmosets, sakis (*Pithecia pithecia*), and siamangs (*Hylobates syndactylus*) show frequent proactive prosociality in the absence of begging. Those authors tested 15 primate species using the same group service task as follows: The actor could choose whether to pull the handle of a platform to allow only the recipient to obtain a food reward which was otherwise out of reach. Actors of cooperative breeder species frequently pulled the handle even though they themselves received no food reward. Control tests ruled out the possibilities that the handle pulling was simply a kind of exploratory behavior and that the actors did not understand how the test worked.

Reciprocity means that A helps B or A gives some benefit to B, and B helps A or returns some benefit to A. Reciprocity has also been shown to influence nonhuman primates' prosocial behaviors. Reciprocity has been tested in 11 nonhuman primate species in more than 30 experimental studies, about 80% of which has reported positive evidence of reciprocity (Schweinfurth and Call 2019). In capuchin monkeys, reciprocal food sharing has been confirmed repeatedly (reviewed by Schweinfurth and Call 2019). We also found a kind of reciprocity in capuchin monkeys' food sharing in the prosocial choice task (Takimoto and Fujita 2011): Actors chose the prosocial container for both relatively higher- and lower-ranked recipients significantly more often when the recipient helped the actor to operate the containers than when the actor operated alone, without the recipient's help. Actors also chose the prosocial container significantly more often when a recipient was present than not, but only when the recipient helped the actor to operate the containers. These results suggest that capuchin monkeys reciprocate recipients' help. Jaeggi and van Schaik (2011) used phylogenetic analyses to explore the occurrence of food sharing in 68 primate species. They found that food was traded for mating and coalitionary support, that is, food sharing co-occurred with partner choice. Thus, reciprocal exchange seems to accompany the presence of food sharing among unrelated adults across many primates. For example, Jaeggi et al. (2010b) found that chimpanzees shared food more frequently and more tolerantly than bonobos, and they shared especially with recipients from whom food had been exchanged reciprocally.

Some research has also been done on the influence of social closeness on nonhuman primates' prosocial behaviors. Using the prosocial choice task, de Waal et al. (2008) reported that capuchin monkeys behaved prosocially to familiar group members even if they were not relatives, but they did not behave prosocially toward strangers. Among Old World monkeys, rhesus macaques avoided delivering an aversive stimulus to a familiar individual (Masserman et al. 1964). Perhaps surprisingly, however, Tan and Hare (2013) reported that bonobos behave prosocially not only toward their own group members but also toward strangers. In that study, bonobos also shared food tolerantly with strangers and behaved prosocially even if a recipient showed no begging behaviors. One limit on bonobos' proactive prosociality toward strangers, however, is that they share food only if they are also able to interact socially with the recipient.

To summarize this section: three main points to emerge from studies of prosocial behaviors in nonhuman primates are as follows: (1) selective prosocial behaviors based on the partner's begging behaviors have been reported notably in chimpanzees and capuchin monkeys, (2) proactive prosocial behaviors occur in cooperatively breeding primate species with high levels of allomaternal care, (3) tolerance and prosocial behaviors toward a range of partners including strangers are reported in bonobos. These three kinds of prosocial behaviors may have evolved with species-typical psychological characteristics, which have developed in different socioecological environments. Below, I consider three potential factors which seem likely to have produced cases of convergent evolution of prosocial behaviors in nonhuman primates.

Psychological Factors in Convergent Evolution of Prosocial Behaviors in Nonhuman Primates

This chapter opened with a statement about humans' proclivity for spontaneous prosocial behaviors. One likely factor in this is our species' theory-of-mind abilities: ready understanding others' desires, perspective-taking, and shared intentionality (Yamamoto and Takimoto 2012). However, unsolicited prosocial behaviors can sometimes become "meddling," which the recipient neither needs nor appreciates. Prosocial behaviors do not increase the actor's benefit, so meddling is not helpful to anyone; it is wasted labor. Moreover, when humans behave prosocially toward strangers or anonymous others, they often experience only the cost of the prosocial behavior (similar to meddling); they are unlikely to obtain any direct benefit from the recipients. Why has such a risky prosocial tendency developed in humans? Conceivably, potential benefits in terms of reputation and indirect reciprocity have been influential (Nowak and Sigmund 1998; Ohtsuki and Iwasa 2006). Prosocial actors can improve their reputation among third-party individuals who directly observe their behaviors, or through reputation, as third-party individuals who directly observe their behaviors inform others and spread word among the community. This might apply even if the prosocial actors are responded negatively by recipients. Therefore, at society level, third-party evaluations and their diffusion may build a kind of social norm, which in turn could be potentially beneficial for prosocial actors and lead to negative evaluations and even punishment of selfish individuals or free-riders.

Humans can communicate reputational information through language; indirect reciprocity is therefore much more prevalent than in nonhuman societies. This probably explains why both proactiveness and broad prosociality are often thought to have developed as human socio-cognitive specialities. However, as we have just seen, some proactive and tolerant prosocial behaviors toward strangers are also seen in some nonhuman primates. Below, I look in more detail at three psychological factors that might have supported the convergent evolution of prosocial behaviors in nonhuman primates.

Inequity Aversion

Prosocial acts directed specifically toward beggars may be more important for maintaining prosocial behaviors in species in which language-based indirect reciprocity does not occur. We propose that *disadvantageous inequity aversion* has played an important role alongside such selectivity (Yamamoto and Takimoto 2012). Inequity aversion (IA) refers to negative reactions to unfair outcomes (Fehr and Schmidt 1999). Humans are averse not only to receiving less than someone else for the same effort (disadvantageous inequity aversion), but also often to earning more than another for the same effort (advantageous inequity aversion). Thus, humans often tend to prefer equality with others, even if this means decreased personal benefit. Advantageous IA directly facilitates prosocial behavior because it helps others who are relatively worse off (Yamamoto and Takimoto 2012). Nonhuman primates, however, have mostly demonstrated only disadvantageous IA in experimental studies (e.g., Brosnan and de Waal 2003; Brosnan et al. 2005, 2010; Massen et al. 2012; van Wolkenten et al. 2007); they show little or no strong evidence of advantageous IA (e.g., Brosnan et al. 2010). Thus, IA in nonhuman primates does not directly promote prosocial behavior. What about disadvantageous IA? For prosocial behaviors to have evolved, prosocial individuals must receive beneficial returns at some point in the future; if not, they would accumulate too many losses. Therefore, individuals need to refrain from being prosocial toward free-riders; instead, they should choose fair partners with whom reciprocal relationships can be established. We propose that disadvantageous IA has played an important role in achieving this (e.g., Yamamoto and Takimoto 2012; see also Brosnan 2011, 2013, Brosnan and de Waal 2014).

Many experimental studies have suggested that disadvantageous IA and prosocial behaviors are related (e.g., Yamamoto and Takimoto 2012), as primates who show one often show the other. For example, Brosnan and de Waal (2003) first reported capuchin monkeys' disadvantageous IA when a partner received a higher-value food reward than the subject monkey, when both completed the same task. The experimenter gave a token to each monkey, and then gave a food reward if the monkey returned the token to her. The subject monkey always received only a low-value food reward (a piece of cucumber), whereas the partner monkey received different-value rewards depending on the experimental conditions. In the fair condition, both monkeys received the low-valued food reward, but in the unfair conditions, the partner received the higher-value food reward (a piece of fresh grape) or the partner received the higher-value reward without returning the token to the experimenter. In each condition, the partner received one trial, followed by the subject. In this study monkeys refused to return the token to the experimenter, or rejected the low-value food reward significantly more frequently in the unfair conditions than in the fair condition. Similar instances of disadvantageous IA have been reported in chimpanzees (Brosnan et al. 2005; Brosnan et al. 2010), long-tailed macaques (Massen et al. 2012), and rhesus macaques (Hopper et al. 2013), and appropriate control conditions generally rule out explanations such as simple expectancy violation.

In a study using new methods, Anderson et al. (2013) showed capuchin monkeys are also sensitive to third-party unfairness. When given a choice after witnessing either a fair or an unfair exchange of objects between two people, capuchins more readily accepted food from fair reciprocators compared to unfair non-reciprocators. If capuchin monkeys can perform third-party social evaluations such as these, their likelihood of behaving prosocially may also be based on observed third-party unfairness. As already mentioned, evidence suggests that species who show disadvantageous IA also show prosocial behaviors in experimental studies (e.g., chimpanzees: Yamamoto et al. 2009, 2012; long-tailed macaques: Massen et al. 2010, 2011; rhesus macaques: Chang et al. 2011; capuchin monkeys: de Waal et al. 2008; Takimoto and Fujita 2011; Takimoto et al. 2010). Unfortunately, there are still no experimental studies on the relationship between disadvantageous IA and partner-specific prosocial behaviors. However, if it emerges that individuals that are highly sensitive to inequity preferentially choose fair individuals as targets of prosocial behaviors, the hypothesis of co-evolution of disadvantageous IA and prosocial behaviors would be supported.

Chimpanzees, long-tailed and rhesus macaques, and capuchin monkeys all show high levels of maternal care and relatively low levels of allomaternal care (e.g., Burkart et al. 2014; Isler and van Schaik 2012; Ross and MacLarnon 2000). In species with low levels of interdependence among group members during infant rearing, it is less important for individuals to build long-term prosocial relationships, and the cost of switching partners is small. Furthermore, the frequency of switching prosocial partners and choosing a fair partner through disadvantageous IA increases. Therefore, disadvantageous IA becomes more important in low-interdependent than high-interdependent species.

Interdependence

What kinds of prosocial behaviors have evolved and what factors have supported their evolution primate species in which disadvantageous inequity aversion has not developed? Proactive prosocial behaviors have been reported in cooperative breeder primates that apparently lack disadvantageous IA. Bi-parental care involves cooperation by the mother and the father, and cooperative breeding involves cooperation by group members other than the biological parents. Among primates, species such as siamangs and sakis show the former pattern, while humans, marmosets, and tamarins show the latter. Species with highly developed allomaternal care also show high interdependence among individuals (Hrdy 2009): cooperation in infant rearing and individual survival is essential, and the cost of switching social partners is heavier than in low-interdependent species. Therefore, in high-interdependent species, the emergence of disadvantageous IA is less beneficial than in low-interdependent species. This can explain why disadvantageous IA appears rare in common marmosets (Freeman et al. 2013; but see Mustoe et al. 2016) and cotton-top tamarins (Neiworth et al. 2009).

However, high interdependence and low probability of severing the partner relationship permit "investment" (Tomasello et al. 2012), as there will be plenty of opportunities for reciprocating prosocial acts, even unsolicited ones. Thus, the cost of proactive prosocial behaviors is much smaller in high —than low-interdependent species, promoting the emergence of such behaviors. This explanation receives support from experimental work. As previously described, Burkart et al. (2014) found that in a sample of 15 primate species there was a significant positive correlation relationship between the extent of allomaternal care and the ratio of proactive prosocial choices, the latter characterizing notably humans (children), siamangs, tamarins, marmosets, and sakis. Thus, the cooperative breeding hypothesis appears to present a valid explanation for the convergent evolution of proactive prosociality in primates.

Tolerance

There is one non-cooperatively breeding nonhuman species in which prosocial behaviors have been reported in experimental situations (e.g., the food sharing task: Hare and Kwetuenda 2010), but in which disadvantageous IA has not been seen yet (Bräuer et al. 2009). The species is the bonobo. As only one study has reported disadvantageous IA in bonobos, it may be too early to conclude that they are not sensitive to disadvantageous inequity; further studies are desirable. However, bonobos' reaction to disadvantageous inequity was not statistically significant, although they did increase their refusal rates in the disadvantageous inequity condition relative to the equality condition (Bräuer et al. 2009). By contrast, bonobos may behave prosocially toward a wide range of partners, including strangers (Tan and

Hare 2013). Such non-selectivity may be supported by high levels of tolerance (Hare et al. 2007). The "emotional-reactivity hypothesis" predicts that because of their higher levels of social tolerance bonobos should cooperate more than chimpanzees. This prediction derives from studies of domesticated animals, which suggests that selection for emotional reactivity can influence social cognitive abilities (Hare et al. 2005; Hare and Tomasello 2005). Hare et al. (2005) tested social cognitive skills of silver foxes (*Vulpes vulpes*) from a population selectively bred over 45 years to approach humans fearlessly and non-aggressively and a control population not bred for tameness. The experimentally domesticated foxes were found to be not only as skillful as dog puppies at using human social gestures such as pointing and gaze cues, but also more skilled than foxes from the untamed population. These results suggest that social cognitive skills have developed as by-product of fearless, nonaggressive, and tolerant traits toward humans in the course of domestication.

Although the bonobo is not a human-domesticated species, bonobos share several traits with domesticated animals; for example, their canine teeth are smaller and they are overall less aggressive than chimpanzees (e.g., Hare et al. 2012). These phenomena are considered to have occurred during human evolution, i.e., in a process of "self-domestication," in which our ancestors gradually tamed themselves in ways that allowed them to live successfully, with typical changes in morphology, physiology, psychology, and behavior. The emotional reactivity hypothesis predicts that, in self-domesticated species such as humans and bonobos, social tolerance has increased markedly and social cognitive abilities including prosociality have also more than in closely related but non self-domesticated species (Hare et al. 2007).

In fact, bonobos continue to show social tolerance whereas chimpanzees tend to become less tolerant with maturity (Wobber et al. 2010), although the extent of social tolerance in the two species may be overestimated and underestimated, respectively. Indeed, Jaeggi et al. (2010b) reported opposite social tolerance tendencies in bonobos and chimpanzees. Tolerance here means allowing potential recipients to approach possessors. In bonobos, such tolerance is seen in foraging contexts that typically lead to high levels of competition in most other primates, and bonobos may forage in close proximity to others. In addition, bonobos were shown to cooperate more successfully to obtain food and co-fed with their more than chimpanzees (Hare et al. 2007). Bonobos behave prosocially even toward strangers in the food sharing task (Tan and Hare 2013). This prosocial tendency may be a by-product of the species' inter-group tolerance: whereas most other primates are generally xenophobic in the wild (Crofoot and Wrangham 2010), bonobos may intermingle with neighboring groups, move around with them for a couple days, forage with them, and show affiliation including socio-sexual behaviors (Furuichi 2011). Therefore, both experimental and observational studies suggest that tolerance facilitates development of stranger-directed prosociality in bonobos.

Prosocial Behaviors in Non-Primate Animals

As reviewed above, most experimental studies on prosocial behaviors in nonhumans have focused on primates, with recent efforts to compare across species with different social or ecological environments (e.g. Burkart et al. 2014). Clearly, there is a need to broaden the range of non-primate studied. In fact, some experimental work has started on birds and domestic animals using all three of the tasks described earlier. Here I review these studies (Table 12.1), although the number of species studied is still too small to conclude whether the convergent evolution hypothesis of prosocial behaviors can also be applied to non-primates.

Prosocial choice tasks have yielded evidence of prosocial behaviors in species as diverse as jackdaws (*Corvus monedula*: Schwab et al. 2012), azure-winged magpies (*Cyanopica cyanus*: Horn et al. 2016), and dogs (*Canis lupus familiaris*: Quervel-Chaumette et al. 2015, Dale et al. 2016). Food sharing tasks have also led to positive results in non-primates, albeit with the focus has been mainly on corvids (jackdaws: de Kort et al. 2006; von Bayern et al. 2007, rooks (*Corvus frugilegus*): Scheid et al. 2008, Eurasian jays (*Garrulus glandarius*): Ostojić et al. 2013, 2014), dogs (Dale et al. 2017), and horses (*Equus caballus*; Takimoto, Inoue and Kawai, in prep). In

Table 12.1 Experimental studies of prosocial behaviors in non-primate animals

Species	Task	Effects/characteristics	References
Non-cooperative breeder species			
Jackdaws	Prosocial choice	Begging behavior/gender	Schwab et al. (2012)
	Food sharing	Begging behavior/ reciprocity	de Kort et al. (2006)
	Food sharing	Social closeness	von Bayern et al. (2007)
Dogs	Helping	Begging behavior	Bräuer et al. (2013)
	Prosocial choice	Social closeness	Quervel-Chaumette et al. (2015)
	Prosocial choice	Social closeness	Dale et al. (2016)
	Food sharing	Social closeness	Dale et al. (2017)
Eurasian jays	Food sharing	Desired state	Ostojić et al. (2013)
	Food sharing	Desired state	Ostojić et al. (2014)
Rats	Helping	Desired state	Ben-Ami Bartel et al. (2011)
	Helping	Desired state	Sato et al. (2015)
Rooks	Food sharing	Reciprocity/dominance/ gender	Scheid et al. (2008)
Horses	Food sharing	Social closeness	Takimoto et al. (in preparation)
Cooperative breeder species			
Azure-winged magpies	Prosocial choice	Proactive	Horn et al. (2016)

studies using targeted helping tasks, prosocial behaviors have been reported only in dogs (Bräuer et al. 2013) and rats (*Rattus norvegicus*: Ben-Ami Bartal et al. 2011; Sato et al. 2015; Silberberg et al. 2013).

Selective Prosocial Behaviors in Non-cooperative Breeder Species

In non-cooperative breeder species, selective prosocial behaviors that are influenced by begging behaviors, reciprocity, or social closeness have been reported in numerous experimental studies (Table 11.1). First, I introduce studies of solicited prosocial behaviors. Studying birds, Schwab et al. (2012) used a prosocial choice task and found that jackdaws provided significantly more food to a conspecific opposite-sex than a same-sex recipient, but only in trials where the recipient approached the food, that is, showed begging behaviors. Therefore, prosocial choices in jackdaws may be facilitated by the partners' begging behaviors, but there may also be an effect of local or stimulus enhancement, directing the actor's attention towards the apparatus, rather than prosociality. de Kort et al. (2006) reported that juvenile jackdaws shared food beak-to-beak with others regardless of sex, dominance relationship or kinship, but most food transfers were solicited, as in Schwab et al. (2012) (but see von Bayern et al. 2007). In the only study of solicited prosocial behaviors in dogs, Bräuer et al. (2013) tested willingness to help a human experimenter spontaneously, without special training. The experimenter (owner or stranger) attempted to enter a compartment in a room (the "target room") to get a key. The dog was able to open the door to the target room by pushing a button. The experimenter expressed desire to enter the target room by reaching for and pushing the door, or communicating with the dog. The dogs helped when the experimenter pointed at the button and when she communicated naturally with the dogs. In these situations, dogs continued to open the door for no reward, suggesting that dogs are motivated to help, and that natural human begging behaviors facilitate their recognition of human goals.

Two studies of Eurasian jays addressed the motivation underlying food sharing. Ostojić et al. (2013) tested whether male Eurasian jays would take a female recipient's desire for one food type over another into account when choosing to share food with her. They reported that when the males observed the female recipient reaching satiation on one food type (e.g. mealworms), they shared a different food with her (e.g. wax moth larvae), suggesting that they took her desire state into account (Ostojić et al. 2013). A subsequent study obtained similar results, even though the male jays' own desire conflicted with the females' and the task was difficult for the males (Ostojić et al. 2014). Relevant research has also been done on the motivation behind prosocial behaviors in rats. Ben-Ami Bartal et al. (2011) studied helping behavior. In their task, a partner rat (the actor's cagemate) was confined in a small tube inside an enclosure while the actor could move around freely in the enclosure. After learning how to open the door of the tube to release the partner, the actor rats opened the door significantly more frequently when the tube held the partner than when it was empty. They also released the trapped partner even when social contact between the two was prevented. When actor rats were subsequently presented with a cage containing the trapped partner and another cage containing a preferred food (chocolate), they opened both doors, even though this meant that they had to share with the partner. These results suggest that rats are both empathic and prosocial. Furthermore, Sato et al. (2015) reported that actor rats with experience of getting soaked were faster to learn how to help a soaked partner (cagemate) by opening a door to allow to escape from water than actors that had never been soaked. The actor rats did not open the door if the partner was not distressed. Additionally, when actors could choose between opening the door to help the distressed partner and opening a different door to obtain food, they helped the partner first. Together, these results suggest that rats may be motivated by empathy-like feelings toward a distressed partner. Therefore, they may behave prosocially based on the recipient's desire even in the absence of explicit begging behaviors.

Second, I describe experimental studies about reciprocity of prosocial behaviors. de Kort et al. (2006) also found that juvenile jackdaws transferred more food to partners from whom they had most frequently received food and affiliative contacts (co-feeding, bill twining, and joint object manipulation). These results suggest that reciprocity applies to food sharing in this species. Moreover, in rooks, co-feeding is explained by reciprocity, although their co-feeding occurs with little or no influence of begging by recipients (Scheid et al. 2008).

Third, I review experimental studies on the influence of social closeness on prosocial behaviors. Food sharing in birds may facilitate formation and maintenance of social bonds between individuals (cf. Emery 2004, similar to wild bonobos (Yamamoto 2015)). von Bayern et al. (2007) reported that captive juvenile jackdaws transferred food to peers irrespective of sex and kinship. The direction of active transfer was initially unfocused but became increasingly selective, until each donor predominantly gave food to one particular recipient and affiliative relationships became established. Thus, in jackdaws, food-sharing may be an integral aspect of ontogeny, allowing exploration of social possibilities and promoting the formation of affiliative or pair bonds. Similarly, de Kort et al. (2006) reported that juvenile jackdaws actively gave a preferred food more than a less preferred food even though the recipients' begging rates for the two food types did not differ. Again, jackdaws reduced the number of individuals with whom they shared over time as affiliative relationships stabilized. In another corvid study, rooks co-fed with their nestmates more often than expected by chance, although active food offering (beak-to-beak) was not associated with the pair bond but was influenced by dominance relationships and gender (Scheid et al. 2008). In domestic animals, some research has shown a familiarity effect on prosocial behaviors. Quervel-Chaumette et al. (2015) presented pet dogs with a bar-pulling task and found that they continued delivering food to a familiar recipient dog for longer than in a social facilitation control condition in which the partner was present but unable to access the reward, or an empty enclosure control condition. Furthermore, the dogs delivered more food to a familiar than an unfamiliar recipient. However, the authors reported that attention-getting behaviors by the recipient had no influence on actor dogs, unlike in the study of Bräuer et al. (2013) with human recipients. Dale et al. (2016) studied dogs using a procedure based on the token exchange task, frequently used with primates. In this task, subjects touched a token with their nose to deliver food into an adjacent enclosure which contained a familiar recipient dog, a stranger or no dog. Subjects delivered more food when the room contained a familiar recipient than when empty, but less food when it contained a stranger. Dale et al. (2017) also reported a familiarity effect on co-feeding. Peaceful co-feeding was mediated by the social closeness between dogs: it occurred on 97% of trials, and the higher the affiliation score, the more time dyads spent peacefully sharing. Recently, we found a similar co-feeding effect in horses: co-feeding with affiliative partners occurred in 100% of trials compared to 75% of trials with non-affiliative partners (Takimoto et al. in preparation; Fig. 12.1c); co-feeding was also significantly longer in the presence of the affiliative partner. The literature clearly indicates that dogs and horses, both non-cooperative breeder species with a history of domestication, show relatively selective prosocial behaviors.

As above, in non-cooperative breeder birds and domestic animals, selective prosocial behaviors that are influenced by begging behaviors, reciprocity, or social closeness have been frequently reported. Interestingly, dogs and horses also show disadvantageous IA (dogs: Range et al. 2009, 2012; horses: Takimoto et al. 2015), a phenomenon which may have supported development of selective prosocial behaviors. These findings appear consistent with the hypothesis of co-evolution of disadvantageous IA and prosocial behaviors, though we consider the effects of artificial selection carefully when discussing the evolutionary process of prosocial behaviors in domestic animals. Nevertheless, again, relevant data on a wider range of species are required for further verification.

Proactive Prosocial Behaviors in Cooperative Breeder Species

Horn et al. (2016) reported that azure-winged magpies, a cooperatively breeding corvid, spontaneously delivered food to their group members even without receiving solicitation, that is, they showed proactive prosocial choices. These choices were more frequent in the presence of a recipient that could not access food than in the absence of recipient or when the latter was present but with access to the food blocked. In addition, in two control conditions the magpies eventually stopped offering food. These findings are consistent with the cooperative breeding hypothesis, but more corvid species should be tested in the same conditions, as Burkart et al. (2014) did with primates. At least, this study supports the hypothesis of co-evolution of interdependence and proactive prosocial behaviors, although again more comparative studies of cooperative and non-cooperative breeder species are necessary to further test this convergent evolution hypothesis of proactive prosocial behaviors in non-primates.

Summary and Future Directions

In this chapter, I reviewed experimental studies of prosocial behaviors in nonhuman animals with reference to several important factors central to hypotheses concerning the evolution, particularly convergent evolution of prosocial behaviors: begging, reciprocity, and social closeness. Primate prosocial behaviors were divided into three main categories as follows: (1) selective prosocial responses based on the partner's begging behaviors, mainly seen in chimpanzees and capuchin monkeys, (2) proactive prosocial behaviors seen in cooperative breeder primates with high levels of allomaternal care, (3) tolerant prosocial behaviors toward a range of partners including strangers in bonobos. These three kinds of prosocial behaviors may have co-evolved with species-typical psychological factors including inequity aversion, interdependence, and tolerance. I also pointed to the lack of experimental studies on non-primate prosocial behaviors and potential psychological factors which may have influenced these. Further studies need to be conducted especially on cooperative breeder species and phylogenetically close but non-cooperative breeder species, and domestic animals and phylogenetically close but non-domesticated animals.

A fuller understanding of the evolution of prosocial behavior will require looking at multiple pathways. The kinds of prosocial behaviors may have been different and be shown for different recipients depending on the above three pathways (with inequity aversion, interdependence, tolerance): long-term reciprocal prosocial behaviors toward familiar individuals may be supported by inequity aversion or interdependence, whereas non-reciprocal or immediate reciprocal behaviors for strangers can be supported by tolerance. Moreover, the three psychological factors, that is, inequity aversion, interdependence, and tolerance, may interact with each other to at least some extent. For example, some cooperative breeder species might have high levels of interdependence and inequity aversion. Some non-cooperative breeder species might show tolerance toward strangers. If we compare these species' prosocial behaviors and the above three psychological factors with those of genetically close species with a focus on the recipients' begging behaviors, reciprocity, and social closeness of prosocial behaviors, we can hope to identify interaction effects in the evolution of prosocial behaviors. In addition, this approach can be useful for comparisons with humans. Humans breed cooperatively, have high level of interdependence, and show inequity aversion and tolerance even toward strangers beginning in childhood (e.g., Blake and McAuliffe 2011; Hamann et al. 2011; Warneken et al. 2007; Warneken and Tomasello 2006). These interactions may make humans a highly prosocial species in which members sometimes offer anonymous donations or help strangers even at risk to our lives.

Quite recently there has been greater recognition of the benefits of a wider comparative approach and of testing multiple species with the same task (e.g., Burkart et al. 2014); the field would benefit from such an approach being applied to non-primate species (Marshall-Pescini et al. 2016). In fact, disadvantageous IA, which possibly underlies selective prosocial behaviors, has been reported in non-primate species including dogs (Range et al. 2009; Range et al. 2012), ravens

(*Corvus corax*), crows (*Corvus corone*) (Wascher and Bugnyar 2013), horses (Takimoto et al. 2015), and rats (Oberliessen et al. 2016). What is required are new experimental paradigms for comparative analyses of prosocial behaviors and underlying psychological factors, using simple tasks adjusted to particular subject species.

Additionally, more attention should be paid to subjects' and partners' motivational states in studies of prosocial behaviors. Yamamoto and Takimoto (2012) focused on empathy and a sense of fairness as psychological factors in prosociality. They proposed that, starting from a preference for similarity, empathy promotes prosocial behaviors directly by understanding others' states or desires based on simple emotional contagion without any need for explicit perspective-taking. However, a sense of fairness also starts from a preference for similarity with others; dissimilarities are experienced as negative. Such disadvantageous IA functions to stabilize prosocial behaviors through social evaluations of reciprocity based on direct or indirect interactions and choosing partners accordingly. In fact, recent reports of sensitivity to third-party reciprocity in various species (Anderson et al. 2013; Kawai et al. 2014; Chijiiwa et al. 2015) have led to the hypothesis that negativity bias toward a nonreciprocating partner might reflect fear, disliking or distrust, among other possible negative states (Anderson et al. 2017). Moreover, using a well-known self-control task ("the marshmallow test"), Michaelson and Munakata (2016) reported that 4- to 5-year-old children were less likely to wait for the entire delay period for an extra reward proposed by an "untrustworthy" adult whom the children had previously observed behaving anti-socially toward a third party, than by a trustworthy adult. It is conceivable that nonhumans might also avoid or punish a third-party individual who violates norms, to avoid exploitation of prosocial individuals. To investigate this, the same individuals should be tested in within-subject experimental designs, as these can help to reveal correlations between psychological factors and clarify individual differences. For example, if individuals showing negative emotional contagion behave more prosocially toward others, then a positive correlation between a form of empathy and prosocial behavior becomes clear. Furthermore, by conducting longitudinal studies and combining behavioral data with genetic, neural, and physiological measures we can even hope to determine causal relationships among potential psychological factors and prosocial behaviors.

Acknowledgements Preparation of this chapter was funded by a Grant-in-Aid for Young Scientists (18 K18339 to Ayaka Takimoto). I gratefully acknowledge James R. Anderson for correcting my English. I wish to thank Kazuo Fujita, Hika Kuroshima, James Anderson (Kyoto University), Toshikazu Hasegawa, Atsuko Saito, Leanne Proops (The University of Tokyo), Keita Nishiyama (Senshu University), Kazuhiro Yoshizawa (Japan Racing Association), Masahito Kawai (Hokkaido University) and the members of Prof. Fujita's Laboratory at Kyoto University, Baji Koen of Japan Racing Association, the horse riding clubs of The University of Tokyo and Hokkaido University, and Shizunai Livestock Farm of Field Science Center for Northern Biosphere of Hokkaido University, for their cooperation and help in conducting our studies. Our studies were supported by a Research Fellowships of the Japan Society for the Promotion of Science (JSPS) for Young Scientists (No. 21264 to Ayaka Takimoto), a Grant-in-Aid for Scientific Research (Nos. 17300085 and 20220004 to Kazuo Fujita) from JSPS, the twenty-first Century COE Program, D-10 to Kyoto University, from Japan Ministry of Education, Culture, Sport, Science, and Technology (MEXT) and by the MEXT Global COE Program, D-07, to Kyoto University, a Grant-in-Aid for Young Scientists (B) (15 K20946 to Ayaka Takimoto) and Grants-in-Aid for Scientific Research on Innovative Areas (No. 26118004 to Ayaka Takimoto, and No. 30172894 to Toshikazu Hasegawa).

References

Anderson JR, Bucher B, Chijiiwa H, Kuroshima H, Takimoto A, Fujita K (2017) Third-party social evaluations of humans by monkeys and dogs. Neurosci Biobehav Rev 82:95–109

Anderson JR, Takimoto A, Kuroshima H, Fujita K (2013) Capuchin monkeys judge reciprocity between humans. Cognition 127:140–146

Barnes JL, Martinez M, Langer M, Hill T, Santos LR (2008) Helping behavior and regard for others in capuchin monkeys (*Cebus apella*): an evolutionary perspective on altruism. Biol Lett 4:638–640

Ben-Ami Bartal I, Decety J, Mason P (2011) Empathy and pro-social behavior in rats. Science 334:1427

Blake PR, McAuliffe K (2011) "I had so much it didn't seem fair": eight-year-olds reject two forms of inequity. Cognition 120:215–224

Boesch C (1991) The effects of leopard predation on grouping patterns in forest chimpanzees. Behavior 117:220–242

Boesch C, Boesch H (1989) Hunting behavior of wild chimpanzees in the Tai National Park. Am J Phys Anthropol 78:547–573

Bräuer J, Call J, Tomasello M (2009) Are apes inequity averse? New data on the token-exchange paradigm. Am J Primatol 71:175–181

Bräuer J, Schönefeld K, Call J (2013) When do dogs help humans? Appl Anim Behav Sci 148:138–149

Brosnan SF (2011) A hypothesis of the coevolution of cooperation and responses to inequity. Front Neurosci 5:43

Brosnan SF (2013) Justice- and fairness related behaviors in nonhuman primates. Proc Natl Acad Sci USA 112:10416–10423

Brosnan SF, de Waal FBM (2003) Monkeys reject unequal pay. Nature 425:297–299

Brosnan SF, de Waal FBM (2014) Evolution of responses to (un)fairness. Science 346:1251776

Brosnan SF, Schiff HC, de Waal FBM (2005) Tolerance for inequity may increase with social closeness in chimpanzees. Proc R Soc B Biol Sci 272:253–258

Brosnan SF, Silk JB, Henrich J, Mareno MC, Lambeth SP, Schapiro SJ (2009) Chimpanzees (*Pan troglodytes*) do not develop contingent reciprocity in an experimental task. Anim Cogn 12:587–597

Brosnan SF, Talbot C, Ahlgren M, Lambeth SP, Schapiro SJ (2010) Mechanisms underlying responses to inequitable outcomes in chimpanzees, *Pan troglodytes*. Anim Behav 79:1229–1237

Burkart JM, Allon O, Amici F, Fichtel C, Finkenwirth C, Heschl A, Huber J, Isler K, Kosonen ZK, Martins E, Meulman EJ, Richiger R, Rueth K, Spillmann B, Wiesendager S, Van Schaik CP (2014) The evolutionary origin of human hyper-cooperation. Nat Commun 5:4747

Burkart JM, Fehr E, Efferson C, Van Schaik CP (2007) Other-regarding preferences in a non-human primate: common marmosets provision food altruistically. Proc Natl Acad Sci USA 104:19762–19766

Chang SWC, Winecoff AA, Platt ML (2011) Vicarious reinforcement in rhesus macaques (*Macaca mulatta*). Front Neurosci 5:27

Chijiiwa H, Kuroshima H, Hori Y, Anderson JR, Fujita K (2015) Dogs avoid people who behave negatively to their owner: third-party affective evaluation. Anim Behav 106:123–127

Claidière N, Whiten A, Mareno MC, Messer EJE, Brosnan SF, Hopper LM, Lambeth SP, Schapiro SJ, McGuigan N (2015) Selective and contagious prosocial resource donation in capuchin monkeys, chimpanzees and humans. Sci Rep 5:7631

Cordoni G, Palagi E, Borgognini Tarli S (2006) Reconciliation and consolation in captive western gorillas. Int J Primatol 27:1365–1382

Crofoot M, Wrangham R (2010) Intergroup aggression in primates and humans: the case for a unified theory. In: Kappeler P, Silk J (eds) Mind the gap - tracing the origins of human universals. Springer, Berlin, pp 171–195

Cronin KA (2012) Prosocial behaviour in animals: the influence of social relationships, communication and rewards. Anim Behav 84:1085–1093

Cronin KA, Schroeder KKE, Rothwell ES, Silk JB, Snowdon C (2009) Cooperatively breeding cottontop tamarins (*Saguinus oedipus*) do not donate rewards to their long-term mates. J Comp Psychol 123:231–241

Cronin KA, Schroeder KKE, Snowdon CT (2010) Prosocial behaviour emerges independent of reciprocity in cottontop tamarins. Proc R Soc B Biol Sci 277:3845–3851

Dale R, Quervel-Chaumette M, Huber L, Range F, Marshall-Pescini S (2016) Task differences and prosociality; investigating pet dogs' prosocial preferences in a token choice paradigm. PLoS One 11:e0167750

Dale R, Range F, Stott L, Kotrschal K, Marshall-Pescini S (2017) The influence of social relationship on food tolerance in wolves and dogs. Behav Ecol Sociobiol 71:107

de Kort SR, Emery NJ, Clayton NS (2006) Food sharing in jackdaws, *Corvus monedula*: what, why and with whom? Anim Behav 72:297–304

de Waal FBM (1996) Good natured: the origins of right and wrong in humans and other animals. Harvard University Press, Cambridge, MA

de Waal FBM (1997) The chimpanzee's service economy: food for grooming. Evol Hum Behav 18:375–386

de Waal FBM (2008) Putting the altruism back into altruism: the evolution of empathy. Annu Rev Psychol 59:279–300

de Waal FBM, Leimgruber K, Greenberg AR (2008) Giving is self-rewarding for monkeys. Proc Natl Acad Sci USA 105:13685–13689

de Waal FBM, Suchak M (2010) Prosocial primates: selfish and unselfish motivations. Phil Trans R Soc B 365:2711–2722

de Waal FBM, van Roosmalen A (1979) Reconciliation and consolation among chimpanzees. Behav Ecol Sociobiol 5:55–66

Emery NJ (2004) Are corvids 'feathered apes'? Cognitive evolution in crows, jays, rooks, and jackdaws. In: Watanabe S (ed) Comparative analysis of minds. Keio University Press, Tokyo, pp 181–213

Fehr E, Bernhard H, Rockenbach B (2008) Egalitarianism in young children. Nature 454:1079–1083

Fehr E, Schmidt KM (1999) A theory of fairness, competition and cooperation. Q J Econ 114:817–868

Feistner ATC, McGrew WC (1986) Food-sharing in primates: a critical review. In: Seth PK, Seth S (eds) Perspectives in primate biology, vol 3. Today and Tomorrow's Printers and Publishers, New Delhi, pp 21–36

Freeman HD, Sullivan J, Hopper LM, Talbot CF, Holmes AN, Schultz-Darken N, Williams LE, Brosnan SF (2013) Different responses to reward comparisons by three primate species. PLoS One 8:e76297

Furuichi T (2011) Female contributions to the peaceful nature of bonobo society. Evol Anthropol 20:131–142

Gomes CM, Boesch C (2009) Wild chimpanzees exchange meat for sex on a long-term basis. PLoS One 4:e5116

Hamann K, Warneken F, Greenberg JR, Tomasello M (2011) Collaboration encourages equal sharing in children but not in chimpanzees. Nature 476:328–331

Hare B, Adessi E, Call J, Tomasello M, Visalbergi E (2003) Do capuchin monkeys, *Cebus apella*, know what conspecifics do and do not see? Anim Behav 65:131–142
Hare B, Call J, Agnetta B, Tomasello M (2000) Chimpanzees know what conspecifics do and do not see. Anim Behav 59:771–785
Hare B, Call J, Tomasello M (2001) Do chimpanzees know what conspecifics know? Anim Behav 61:139–151
Hare B, Kwetuenda S (2010) Bonobos voluntarily share their own food with others. Curr Biol 20: R230–R231
Hare B, Melis A, Woods V, Hastings S, Wrangham R (2007) Tolerance allows bonobos to outperform chimpanzees on a cooperative task. Curr Biol 17:619–623
Hare B, Plyusnina I, Ignacio N, Schepina O, Stepika A, Wrangham R, Trut L (2005) Social cognitive evolution in captive foxes is a correlated by-product of experimental domestication. Curr Biol 15:226–230
Hare B, Tomasello M (2004) Chimpanzees are more skillful in competitive than in cooperative cognitive tasks. Anim Behav 68:571–581
Hare B, Tomasello M (2005) Human-like social skills in dogs? Trends Cogn Sci 9:439–444
Hare B, Wobber V, Wrangham R (2012) The self-domestication hypothesis: evolution of bonobo psychology is due to selection against aggression. Anim Behav 83:573–585
Hattori Y, Leimgruber K, Fujita K, de Waal FBM (2012) Food-related tolerance in capuchin monkeys (*Cebus apella*) varies with knowledge of the partner's previous food-consumption. Behaviour 149:171–185
Hopper LM, Lambeth SP, Schapiro SJ, Bernacky BJ, Brosnan SF (2013) The ontogeny of social comparisons in rhesus macaques (*Macaca mulatta*). J Primatol 2:1000109
Horn L, Scheer C, Bugnyar T, Massen JJM (2016) Proactive prosociality in a cooperatively breeding corvid, the azure-winged magpie (*Cyanopica cyana*). Biol Lett 12:20160649
Horner V, Carter JD, Suchak M, de Waal FBM (2011) Spontaneous prosocial choice by chimpanzees. Proc Natl Acad Sci USA 108:13847–13851
Hrdy SB (2009) Mothers and others: the evolutionary origins of mutual understanding. Harvard University Press, Cambridge, UK
Isler K, van Schaik CP (2012) Allomaternal care, life history and brain size evolution in mammals. J Hum Evol 63:52–63
Jaeggi AV, Burkart JM, van Schaik CP (2010a) On the psychology of cooperation in humans and other primates: combining the natural history and experimental evidence of prosociality. Phil Trans R Soc B 365:2723–2735
Jaeggi AV, Stevens JMG, van Schaik CP (2010b) Tolerant food sharing and reciprocity is precluded by despotism among bonobos but not chimpanzees. Am J Phys Anthropol 143:41–51
Jaeggi AV, van Schaik CP (2011) The evolution of food sharing in primates. Behav Ecol Sociobiol 65:2125
Jensen K, Hare B, Call J, Tomasello M (2006) What's in it for me? Self-regard precludes altruism and spite in chimpanzees. Proc R Soc Lond B 273:1013–1021
Kawai N, Yasue M, Banno T, Ichanohe N (2014) Marmoset monkeys evaluate third-party reciprocity. Biol Lett 10:20140058
Krupenye C, Kano F, Hirata S, Call J, Tomasello M (2016) Great apes anticipate that other individuals will act according to false beliefs. Science 354:110–114
Lakshminarayanan VR, Santos LR (2008) Capuchin monkeys are sensitive to others' welfare. Curr Biol 18:999–1000
Marshall-Pescini S, Dale R, Quervel-Chaumette M, Range F (2016) Critical issues in experimental studies of prosociality in non-human species. Anim Cogn 19:679–705
Massen JJM, Luyten IJAF, Spruijt BM, Sterck EHM (2011) Benefiting friends or dominants: prosocial choices mainly depend on rank position in longtailed macaques (*Macaca fascicularis*). Primates 52:237–247
Massen JJM, van den Berg LM, Spruijt BM, Sterck EHM (2010) Generous leaders and selfish underdogs: pro-sociality in despotic macaques. PLoS One 5:e9734

Massen JJM, van den Berg LM, Spruijt BM, Sterck EHM (2012) Inequity aversion in relation to effort and relationship quality in long-tailed macaques (*Macaca fascicularis*). Am J Primatol 74:145–156
Masserman JH, Wechkin S, Terris W (1964) 'Altruistic' behavior in rhesus monkeys. Am J Psychiatr 121:584–585
Matsumoto T, Itoh N, Inoue S, Nakamura M (2016) An observation of a severely disabled infant chimpanzee in the wild and her interactions with her mother. Primates 57:3–7
Melis AP, Warneken F, Jensen K, Schneider AC, Call J, Tomasello M (2010) Chimpanzees help conspecifics obtain food and non-food items. Proc R Soc London B Biol Sci 278:1405–1413
Michaelson LE, Munakata Y (2016) Trust matters: seeing how an adult treats another person influences preschoolers' willingness to delay gratification. Dev Sci 19(6):1011–1019
Mustoe AC, Harnisch AM, Hochfelder B, Cavanaugh J, French JA (2016) Inequity aversion strategies between marmosets are influenced by partner familiarity and sex but not by oxytocin. Anim Behav 114:69–79
Neiworth JJ, Johnson ET, Whillock K, Greenberg J, Brown V (2009) Is a sense of inequity an ancestral primate trait? Testing social inequity in cotton top tamarins (*Saguinus oedipus*). J Comp Psychol 123:10–17
Nowak MA, Sigmund K (1998) Evolution of indirect reciprocity by image scoring. Nature 393:573–577
Oberliessen L, Hernandez-Lallement J, Schäble S, van Wingerden M, Seinstra M, Kalenscher T (2016) Inequity aversion in rats, *Rattus norvegicus*. Anim Behav 115:157–166
Ohtsuki H, Iwasa Y (2006) The leading eight: social norms that can maintain cooperation by indirect reciprocity. J Theor Biol 239:435–444
Ostojić L, Legg EW, Shaw RC, Cheke LG, Mendl M, Clayton NS (2014) Can male Eurasian jays disengage from their own current desire to feed the female what she wants? Biol Lett 10:20140042
Ostojić L, Shaw RC, Cheke LG, Clayton NS (2013) Evidence suggesting that desire-state attribution may govern food sharing in Eurasian jays. Proc Natl Acad Sci USA 110:4123–4128
Palagi E, Paoli T, Borgognini Tarli S (2004) Reconciliation and consolation in captive bonobos (*Pan paniscus*). Am J Primatol 62:15–30
Perry S, Rose L (1994) Begging and transfer of coati meat by white-faced capuchin monkeys, *Cebus capucinus*. Primates 35:409–415
Quervel-Chaumette M, Dale R, Marshall-Pescini S, Range F (2015) Familiarity affects other-regarding preferences in pet dogs. Sci Rep 5:18102
Range F, Horn L, Viranyi Z, Huber L (2009) The absence of reward induces inequity aversion in dogs. Proc Natl Acad Sci USA 106:340–345
Range F, Leitner K, Virányi Z (2012) The influence of the relationship and motivation on inequity aversion in dogs. Soc Justice Res 25:170–194
Ross C, Mac Larnon A (2000) The evolution of non-maternal care in anthropoid primates: a test of the hypotheses. Folia Primatol 71:93–113
Sato N, Tan L, Tate K, Okada M (2015) Rats demonstrate helping behavior toward a soaked conspecific. Anim Cogn 18:1039–1047
Scheid C, Schmidt J, Noe R (2008) Distinct patterns of food offering and co-feeding in rooks. Anim Behav 76:1701–1707
Schmelz M, Grueneisen S, Kabalak A, Jost J, Tomasello M (2017) Chimpanzees return favors at a personal cost. Proc Natl Acad Sci USA 114:7462–7467
Schwab C, Swoboda R, Kotrschal K, Bugnyar T (2012) Recipients affect prosocial and altruistic choices in jackdaws (*Corvus monedula*). PLoS One 7:e34922
Schweinfurth M, Call J (2019) Revisiting the possibility of reciprocal help in non-human primates. Neurosci Biobehav Rev 104:73–86
Silberberg A, Allouch C, Sandfort S, Kearns D, Karpel H, Slotnick B (2013) Desire for social contact, not empathy, may explain "rescue" behavior in rats. Anim Cogn 17:609–618

Silk JB, Brosnan SF, Vonk J, Henrich J, Povinelli DJ, Richardson AS, Lambeth SP, Mascaro J, Schapiro SJ (2005) Chimpanzees are indifferent to the welfare of unrelated group members. Nature 437:1357–1359

Skerry AE, Sheskin M, Santos LR (2011) Capuchin monkeys are not prosocial in an instrumental helping task. Anim Cogn 14:647–654

Stevens JR (2010) Donor payoffs and other-regarding preferences in cotton-top tamarins (*Saguinus oedipus*). Anim Cogn 13:663–670

Stevens JR, Gilby IC (2004) A conceptual framework for nonkin food sharing: timing and currency of benefits. Anim Behav 67:603–614

Takimoto A, Fujita K (2011) I acknowledge your help: capuchin monkeys' sensitivity to others' labor. Anim Cogn 14:715–725

Takimoto, A., Inoue, M. and Kawai, M. in preparation Domestic horses (*Equus caballus*) tend to share more food with familiar group members

Takimoto A, Kuroshima H, Fujita K (2010) Capuchin monkeys (*Cebus apella*) are sensitive to others' reward: an experimental analysis of food-choice for conspecifics. Anim Cogn 13:249–261

Takimoto A, Proops L, Saito A Hasegawa T (2015) Inequity aversion in horses (*Equus caballus*). Behaviour 2015 (a joint meeting of the international ethological conference (IEC), Australasian Society for the Study of animal behaviour (ASSAB), Australasian evolution society (AES), and Australasia, New Zealand and Africa region of applied ethology), Poster-897, Cairns, Australia, 9–14, August, 2015

Tan J, Hare B (2013) Bonobos share with strangers. PLoS One 8:e51922

Tan J, Kwetuenda S, Hare B (2015) Preference or paradigm? Bonobos show no evidence of other-regard in the standard prosocial choice task. Behaviour 152:521–544

Tomasello M, Melis AP, Tennie C, Wyman E, Herrmann E (2012) Two key steps in the evolution of human cooperation: the interdependence hypothesis. Curr Anthropol 53:673–692

Ueno M, Yamada K, Nakamichi M (2014) The effect of solicitations on grooming exchanges among Japanese macaques at Katsuyama. Primates 55:81–87

van Wolkenten M, Brosnan SF, de Waal FBM (2007) Inequity responses of monkeys modified by effort. Proc Natl Acad Sci USA 104:18854–18859

von Bayern AMP, de Kort SR, Clayton NS, Emery NJ (2007) The role of food-and object-sharing in the development of social bonds in juvenile jackdaws (*Corvus monedula*). Behaviour 144:711–733

Vonk J, Brosnan SF, Silk JB, Henrich J, Richardson AS, Lambeth SP, Schapiro SJ, Povinelli DJ (2008) Chimpanzees do not take advantage of very low cost opportunities to deliver food to unrelated group members. Anim Behav 75:1757–1770

Warneken F, Hare B, Melis AP, Hanus D, Tomasello M (2007) Spontaneous altruism by chimpanzees and young children. PLoS Biol 5:e184

Warneken F, Tomasello M (2006) Altruistic helping in human infants and young chimpanzees. Science 311:1301–1303

Wascher C, Bugnyar T (2013) Behavioral responses to inequity in reward distribution and working effort in crows and ravens. PLoS One 8:e56885

Wobber W, Wrangham R, Hare B (2010) Bonobos exhibit delayed development of social behavior and cognition relative to chimpanzees. Curr Biol 20:226–230

Yamamoto S (2015) Non-reciprocal but peaceful fruit sharing in the wild bonobos of Wamba. Behaviour 152:335–357

Yamamoto S, Humle T, Tanaka M (2009) Chimpanzees help each other upon begging. PLoS One 4:e7416

Yamamoto S, Humle T, Tanaka M (2012) Chimpanzees' flexible targeted helping based on an understanding of conspecifics' goals. Proc Natl Acad Sci USA 109:3588–3592

Yamamoto S, Takimoto A (2012) Empathy and fairness: psychological mechanisms for eliciting and maintaining prosociality and cooperation in primates. Soc Justice Res 25:233–255

Yamamoto S, Tanaka M (2010) The influence of kin relationship and a reciprocal context on chimpanzees' (*Pan troglodytes*) sensitivity to a partner's payoff. Anim Behav 79:595–602

Chapter 13
Social Evaluation in Non-human Animals

Hitomi Chijiiwa

Abstract Humans evaluate others based on not only direct interactions, but also third-party interactions that are often of no direct relevance to us. For example, we might feel happy to see someone acting kindly, and praise that person, whereas we might feel upset when we see someone bullying another person, and reprimand the bully. Such sensitivity to the content of interactions between third parties appears to develop in early infancy. Comparative psychologists have become increasingly interested in third party-based social evaluations in other species. Here, I introduce experimental studies by Kazuo Fujita's research group, in which monkeys and companion animals observe humans behave helpfully or unhelpfully, and fairly or unfairly, in third-party contexts. These studies revealed avoidance of non-helpful and unfair humans by capuchin monkeys (*Sapajus apella*) and dogs (*Canis familiaris*). I discuss the correspondence between this asymmetric result, which is referred to as a "negativity bias," and findings in human children. I suggest that third-party evaluation—a likely first step toward large-scale cooperative society in humans—may be shared with non-human animals including other primate species, and that comparative studies of this phenomenon provides an important and interesting perspective for understanding the evolution of both small-scale and large-scale cooperation in human societies. Finally, I propose future directions for comparative research on third party-based social evaluations.

Keywords Social evaluation · Third-party evaluation · Eavesdropping · Helping · Reciprocity · Negativity bias

H. Chijiiwa (✉)
Department of Psychology, Kyoto University Graduate School of Letters, Kyoto, Japan

J. R. Anderson, H. Kuroshima (eds.), *Comparative Cognition*,
https://doi.org/10.1007/978-981-16-2028-7_13

Introduction

We humans build long-term relationships with others and repeatedly interact with those individuals; we also have less frequent interactions with many other people. If we can usefully evaluate characteristics such as personality and behavioral tendencies from the words and actions of others, we can adjust our interactions and relevant behaviors more flexibly. In particular, humans evaluate others based on a "moral sense," which is to positively evaluate "prosocial" others, and disapprove of "antisocial" others (Hamlin 2013). For example, we may want to have a long-lasting relationship with a person who is kind to us and who helps us with our problems, whereas we are more likely to eschew interacting with someone who harasses or defames us.

We are sensitive not only to words or actions that are directed to us, but also to interactions between third parties. Rumors, word of mouth, and other forms of reputation can be used as information to guide our interactions with strangers, either in person or, for example, via the Internet. In addition, even in the absence of actions that are of direct relevance to ourselves, we tend to form an impression of "kind" or "good" for a person who reaches out to others in need, and an impression of "mean" or "bad" for a person who bullies others. The ability to evaluate others from a third-party perspective is one of the factors that enable humans to maintain largely cooperative societies (Nowak 2006; Wu et al. 2016). Sensitivity to the content of third-party interactions appears to develop early in human development. When 3-year-old children are asked to give a ball that will be needed later in a game to either a person who takes and destroys another's property or to a neutral person, they avoid the former and give to the latter (Vaish et al. 2010). Even younger infants (6–10 months) prefer geometric shapes and puppets that appear behave in a prosocial manner toward others (e.g., helping them to climb a hill or to open a box), and they avoid agents that hinder others (Hamlin et al. 2007, 2010; Hamlin and Wynn 2011).

Comparative psychologists have been studying third party-based social evaluations in non-human animals. Individuals that frequently interact with others distinguish between prosocial and antisocial others and behave differently toward them (for review, Abdai and Miklósi 2016). For example, in a cleaning symbiosis situation, client fish may first observe a cleaner fish (*Labroides dimidiatus*) interacting with another client, and then avoid individual cleaners that betray (eat the client's mucus) and approach those that cooperate (remove the client's parasites only) (Bshary 2002). However, in early studies of third-party social evaluation in non-human animals, one problem was that most studies use experimental procedures in which subjects observed exchanges of food between humans (chimpanzees, *Pan troglodytes*: Herrmann et al. 2013; Russell et al. 2008; Subiaul et al. 2008, orangutans, *Pongo pygmaeus*: Herrmann et al. 2013, dogs, *Canis familiaris*: Kundey et al. 2011; Marshall-Pescini et al. 2011) and then chose between two humans. In these experiments, subjects generally preferred a person who had behaved positively by giving food to a begging person (or sometimes a conspecific of the subject), over another person who did not give food. However, with these kinds of situations we cannot rule out the possibility that the subjects simply chose the person who appeared most likely to give them some food.

To address this issue, some studies of third-party social evaluations by animals have adapted methods used with pre-verbal human infants (e.g., Hamlin et al. 2007), for example, showing animations of simple shapes (bottlenose dolphins, *Tursiops* spp.: Johnson et al. 2018; bonobos, *Pan paniscus*: Krupenye and Hare 2018; dogs: McAuliffe et al. 2019). These studies have produced variable results: based on their looking patterns, dolphins appeared to predict that an animated neutral shape would interact with a "helper" shape (Johnson et al. 2018), whereas bonobos preferred a "hinderer" that obstructed another agent's goal (Krupenye and Hare 2018); dogs showed no clear preference for either "helper" or "hinderer" (McAuliffe et al. 2019). Although the use of simple figure animations eliminates the potential problem of associative learning between experimenter and food, the relevance of the social context in the movements of the abstract figures might be compromised by species' perceptual and cognitive abilities.

As mentioned above, humans use a moral sense to judge the rightness or wrongness of social behaviors, such as helping/hindering others' actions, giving/ depriving, and fairness/unfairness (Hamlin 2013), and at least a precursor of this morality appears to be present in young infants. As members of Kazuo Fujita's research group, we have conducted several experiments in which several species of animals could observe these kinds of behaviors in third-party contexts. In the following section, I introduce our studies of third-party social evaluation in primates and companion animals.

Third-Party Evaluation of "Helpful vs Non-helpful" and "Fair vs Unfair" People by Capuchin Monkeys

Capuchin monkeys (*Sapajus apella*) are a species of the new world monkeys whose ancestors diverged from old world primates about 35–40 million years ago. In various settings they show well-documented prosocial tendencies and a high degree of social tolerance, including cooperating on a range of tasks (de Waal et al. 2008; Sabbatini et al. 2012; Takimoto et al. 2010). Anderson et al. (2013a) tested whether capuchin monkeys socially evaluated humans after witnessing third-party interactions in a situation in which a helpful act could take place or be withheld. The monkeys were shown the following interaction between two experimenters. One experimenter (attempter) tried to open the lid of the container to get an object that was inside it, but failed. The attempter then asked the other person (responder) for help. In the Helper condition, the responder held the container stable so that the attempter was able to remove the lid and get the object. After briefly manipulating the object, the attempter returned it to the container, replaced the lid, and placed the container on the table. In the non-helper condition, when the attempter requested help the responder turned away, refusing to help. The attempter continued trying but

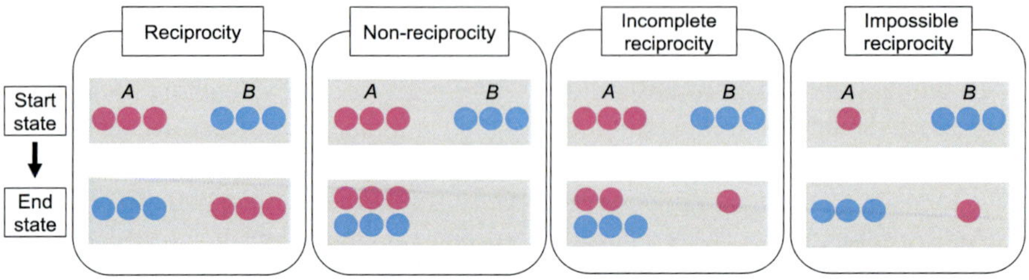

Fig. 13.1 Schematic illustration of the start and end states of balls in reciprocity, non-reciprocity, incomplete reciprocity, and impossible reciprocity conditions (Anderson et al. 2013b)

failed to open the lid, and placed the unopened container on the table. After the interaction, the attempter and responder presented the same amount of food to the monkeys, and the monkeys were allowed to take food from only one of them.

In these situations, the monkeys were less willing to accept food from a person who refused to help, although they showed no tendency to prefer the person who responded positively to the request for help. The monkeys' judgments were not based on any association with food, because food was not used during the observed interaction. In addition, several control conditions showed that avoidance of the person who refused to help was not caused by the act of turning away or simply whether or not the attempter was able to retrieve the object. Thus, it was concluded that the monkeys evaluated the "non-helpful" actor negatively.

Anderson et al. (2013b) then examined whether capuchin monkeys would make judgements about reciprocity from third-party interactions. Individually tested monkeys watched the following interaction. Two experimenters each started out in possession of three small plastic balls. One experimenter (*A*) nonverbally asked the other (*B*) for her balls (by holding out an empty container in a begging gesture), and *B* gave all three of her balls to *A*. Next, *B* requested *A* for her balls, and *A* responded differently depending on which of the following four conditions was tested (Fig. 13.1): (1) In the Reciprocity condition, *A* gave *B* all three balls that she started out with, so that both ended up with each other's original three balls. (2) In the non-reciprocity condition, *A* refused to give the balls to *B*, meaning that *A* ended up with six balls, name the three she originally had, and the three she received from *B*. (3) In the incomplete reciprocity condition, *A* received all three balls from *B* but stopped transferring after giving only one ball in response to *B*'s request, meaning that *A* ended up with five balls, and *B* with only one. (4) In the impossible reciprocity conditions, *A* started out with only one ball, while *B* had 3. *A* received all three balls from *B*, *A* then gave the only ball she started out with. In all conditions, after the interaction with the balls, both actors offered an identical piece of food to the monkeys, who were allowed to accept one offer.

The results showed no preference for either actor in reciprocity sessions. By contrast, the monkeys avoided taking food from *A* when she behaved in a non-reciprocal manner, i.e., refused to return the balls or when she gave only one of her three balls. Such avoidance was reduced when *A* transferred the only one ball she had to begin with. Overall, these results suggested the capuchin monkeys avoided actors who behaved unfairly.

The research described above showed that capuchin monkeys pay attention to interactions between third parties, and evaluate humans based on these interactions. It expanded upon findings from earlier studies in which chimpanzees were more likely to approach, or they learned to beg food from, a human who shared food with another than one who did not (Russell et al. 2008; Subiaul et al. 2008), because the monkeys (a) watched an interaction over non-food objects that were of no relevance to them, and (b) were then offered identical food by the two people; there was no foraging advantage from preferentially taking food from either actor. These results suggest that the cognitive or emotional (affective) bases for evaluating helpfulness and reciprocity in relatively "neutral" third-party interactions exist in at least one other primate species.

Sensitivity to Reciprocity Violations in Other Primates

In a study of another New World monkey, the cooperatively breeding common marmoset (*Callithrix jacchus*) (capuchin monkeys are not cooperative breeders), subjects were again reported to accept food less frequently from a non-reciprocator than a reciprocator, in a variant of the exchange situation used with capuchins (Kawai et al. 2014). Kawai et al. (2014) suggested that the sensitivity to reciprocal vs. non-reciprocal third-party exchanges emerge from species' natural cooperative and prosocial tendencies, and therefore can be found in some primate species other than capuchins.

To test whether social evaluation of third-party exchange behavior is indeed related to a species' inherent cooperativeness, Anderson et al. (2016) tested squirrel monkeys (*Saimiri sciureus*), a species not renowned for cooperative behaviors. Like capuchins (Anderson et al. 2013b) and marmosets (Kawai et al. 2014), the squirrel monkeys were found to accept food less frequently from non-reciprocators than reciprocators. Importantly, however, the squirrel monkeys not only avoided non-reciprocators, but also positively preferred taking food from reciprocator. This pattern suggests that the squirrel monkeys had a simple strategy of choosing the last actor to transfer items to the other actor. Therefore, although the mechanisms underlying social evaluation processes in the squirrel monkeys might differ from those in capuchins, we can now rule out the possibility that a given species' cooperative tendencies are critical in monkeys' responses to unfair exchanges.

More recently, a study using an Old World monkey species, the Japanese macaque (*Macaca fuscata*), categorized as having "despotic" social relationships, showed no preference in either reciprocal or non-reciprocal exchange conditions (Kawai et al. 2019). To summarize the current state of knowledge: sensitivity to unfair third-party exchanges, leading to differential willingness to engage with the exchangers, appears to exist among several species of monkeys (capuchins, marmosets, squirrel monkeys) tested so far, albeit possibly based on different mechanisms, but not among primates in general. Further comparative research is needed to test the general validity of this perspective on responses to unfair third-party exchanges in primate species.

Third-Party Evaluation of Helpful Versus Non-helpful People by Companion Animals

Although the studies discussed so far indicate that at least some non-human primates engage in social evaluations of humans, their ecological relevance appears restricted because primates do not usually have the opportunity to choose cooperative humans as partners. In this light, domestic dogs are arguably an ideal species for studying social evaluations of humans. Dogs' social cognitive abilities are of interest from both phylogenetic and individual developmental perspectives (Freidin et al. 2013; McAuliffe et al. 2019). Dogs have evolved outstanding abilities to read human communicative signals, over at least 15,000 years of domestication (Hare and Tomasello 2005). They also have countless opportunities to learn human about behavior through direct and indirect interactions with humans in their daily life.

Research on dogs' social evaluations in indirect contexts has produced mixed results. For example, dogs preferred a winner of a tug-of-war game with another dog to a loser (Rooney and Bradshaw 2006). Dogs also more frequently approached a person who generously gave food to a begging person compared to another who withheld it (Kundey et al. 2011; Marshall-Pescini et al. 2011). When two people were equally generous towards the beggar, dogs showed a preference for the donor who received positive reactions from the beggar (Freidin et al. 2013). By contrast, Nitzschner et al. (2012) claimed that dogs' evaluations of humans took into account only their direct experiences: dogs preferred an actor who behaved friendly to them over an actor who ignored them, whereas they did not show any preference after seeing the actor behave in those ways toward another dog. Nitzschner et al. (2014) raised the possibility that dogs simply choose the location where a beggar received food, rather than a person, in tests involving food transfer.

Chijiiwa et al. (2015) tested dogs' evaluation of people in interactions depicting either helpfulness or refusal to help, similar to the study with capuchin monkeys (Anderson et al. 2013a), again avoiding any interaction between third parties involving food. We showed the dogs scenes in which three people were present, one of which was the dog's owner (Fig. 13.2). An "actor" and a "neutral person" sat on either side of the owner. First, the owner tried to open the lid of the container to take out an object that was inside, but without success. Next, the owner showed the container to the actor and asked for help. In the Helper condition, the actor responded positively to the request and held the bottom of the container, allowing the owner to open the lid, remove the object and show it to the dog. In the non-helper condition, the actor refused to help by briefly turning away from the requesting owner, leaving the latter to continue his or her unsuccessful attempts to open the container

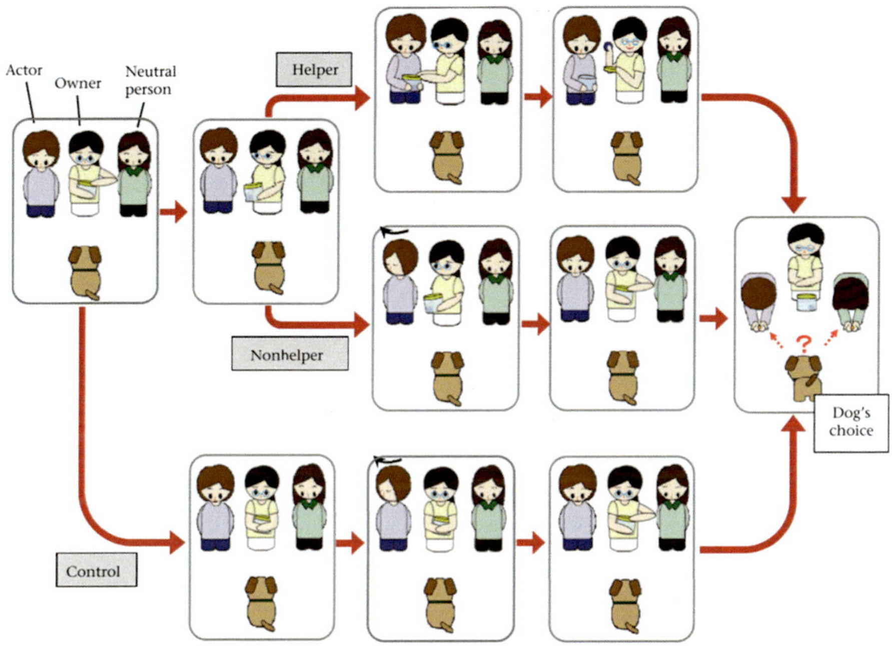

Fig. 13.2 Illustration of the procedures used in Chijiiwa et al. (2015)

Fig. 13.3 A dog in Chijiiwa et al. (2015) watches as an unhelpful actor declines to help the dog owner's help in opening a box

(Fig. 13.3). In a control condition, the actor turned away while the owner briefly stopped manipulating and simply looked at the container. The owner then again tried to open the container, in vain. In all conditions, the neutral person remained passive

Fig. 13.4 A cat watches as a helper assists the cat's owner in opening a container

during the interaction. After the interaction, the actor and the neutral person simultaneously offered a piece of food to the dog, which was allowed to take the food from one of the two.

The results showed that dogs chose the actor less frequently in the non-helper condition, but chose randomly in the Helper and the Control conditions. This bias against the non-helpful actor is consistent with the finding with capuchin monkeys in Anderson et al. (2013a). The dogs' avoidance of people who behaved negatively to their owner suggests that third-party social evaluations of humans can also be seen in at least one non-primate species.

For comparative psychologists, the interest of studying dogs' tendencies to socially evaluate humans appears obvious. Cats (*Felis catus*) are also domesticated animals that live in close association with humans as companions, with exposure to various forms of human social behavior. In the last few years, evidence has been increasing that cats also have advanced social cognitive abilities when processing human signals (for a review, Shreve and Udell 2015), and aspects of their behavior may be influenced by their owner's interactions with third parties (Bucher et al. 2020). Given these features, cats also present a potentially useful species for examining social evaluations of humans. To our knowledge, to date only one study has addressed this issue, and the finding was that cats did not discriminate between "friendly" and "aggressive" experimenters, either from their own direct experience or from a third-party perspective (Leete et al. 2020).

Chijiiwa et al. (2021) tested whether cats evaluate people in third-party contexts, using the same Helper vs. Non-helper procedure used in our dog experiments, to enable direct comparisons between the two species (Fig. 13.4). However, we found no evidence that cats are like dogs when it comes to evaluating people in third-party interactions: the felines neither avoided the non-helper nor preferred the helper. In interpreting this difference, it is important to note that "cooperation" is not as crucial to cats as it is to dogs. Unlike dogs, cats have not been selected to cooperate with humans in a range of contexts. Furthermore, feline species generally tend to be asocial (but see Caro et al. 1989; Stander 1992). Future work on cats' social evaluation abilities needs to give greater consideration to ecological validity, and to explore different contexts.

Our studies on third party-based social evaluation in companion animals yielded positive results with dogs, despite the species being phylogenetically much more distant from humans, compared to primates. This suggests that the range of species capable of third-party evaluation probably extends beyond primates.

Negativity Bias

Our studies with monkeys and dogs revealed no preference for helpful or fair actors, but clear avoidance of non-helpful or unfair actors. This asymmetric result reflects what is referred to as a negativity bias, and is consistent with findings in human children (Hamlin et al. 2010; Vaish et al. 2010). Negative information is prioritized over positive information in a variety of psychological situations and tasks (for a review, Vaish et al. 2008). A negativity bias toward antisocial others might be based on any of several possible affective states, such as fear, disliking, or distrust (Anderson et al. 2017).

In humans, a negativity bias is detectable soon after birth (Hamlin et al. 2010). However, as they develop, young humans give increasingly positive evaluations to prosocial others; a shift that can be seen at around 6 months of age (Hamlin et al. 2007). As responding appropriately to negative stimuli is undoubtedly important for an individual's survival, negativity bias may be an evolutionarily old and widespread trait. Abdai and Miklósi (2016) proposed that the term "social evaluation" should be restricted to cases when both negativity and positivity bias can be detected, that is, the individual shows clear preferences in the following three situations: (1) prosocial vs antisocial, (2) prosocial vs neutral, and (3) neutral vs antisocial. If future studies show a clear positivity bias in monkeys and dogs, it will be stronger evidence that those species' third-party-based evaluations are similar to those of humans.

Future Directions

In the series of studies that I have described, the actors involved in the interactions were not conspecifics, but humans. It seems likely that monkeys (at least some species) are capable of generalizing social evaluative processes across a range of agents and situations, somewhat similar to human infants, who even evaluate inanimate objects such as geometric shapes and puppets from a third-party viewpoint (Hamlin et al. 2007, 2010; Hamlin and Wynn 2011). Nonetheless, it is important to confirm whether these social evaluations also occur in situations involving conspecifics. In the case of dogs, it would help to clarify whether their social evaluation abilities reflect an object-specific specialization (focused on humans), acquired by their history of coexistence with humans, or whether it is a more general social ability.

Moreover, for social evaluation ability to reach the level of reputation formation and cooperative sociality, and for it to function it in real world, it is important to take the presence of others into account and to adjust one's own behavior. For example, one strategy to gain a good reputation at minimal cost is to behave cooperatively only when being observed by others (for a review, Izuma 2012). Human children attempt to manage their reputations: they behave more prosocially when they are being watched by others than when they are alone (Engelmann et al. 2012). However, if other individuals notice someone adopting such a deliberate or calculating strategy, any accumulated positive reputation might be negated. Kishimoto et al. (2020) reported that not only human adults but even preschool children show a reduced preference for people who provide help primarily to gain personal reputational benefits. Whether non-human animals engage in similar impression management strategies awaits new studies.

Humans often display cooperative acts, including altruistic behaviors for non-relatives, even for no apparent benefits to themselves. Indirect reciprocity—"Individual *A* has helped *B*, then *C* helps *A* based on reputation"—has been proposed as an important factor supporting this phenomenon (Nowak and Sigmund 2005). A first step in establishing this mechanism is the ability to evaluate others based on interactions between third parties. The demonstration of such social evaluative processes in various species provides an interesting and important perspective for understanding the evolution of humans' highly developed, cooperative societies. Studies of more species from diverse phylogenies and ecological backgrounds will contribute new information to this exciting field of research. Other highly social species, such as corvids and horses, are possible subjects for future studies. Carrion crows (*Corvus corone corone*) appear to monitor relationships between group members and use third-party affiliation as post-conflict behavior (Sima et al. 2018). Horses (*Equus caballus*) reacted differently to the positive and negative interactions between a horse and a human seen on video and responded in ways that could be interpreted as attempted appeasement toward a negative experimenter (Trösch et al. 2020). It would be interesting to test third party-based social evaluations in helping and unfair exchange situations with those and other species.

References

Abdai J, Miklósi Á (2016) The origin of social evaluation, social eavesdropping, reputation formation, image scoring or what you will. Front Psychol 7:1772

Anderson JR, Kuroshima H, Takimoto A, Fujita K (2013a) Third-party social evaluation of humans by monkeys. Nat Commun 4:1561

Anderson JR, Takimoto A, Kuroshima H, Fujita K (2013b) Capuchin monkeys judge third-party reciprocity. Cognition 127:140–146

Anderson JR, Bucher B, Kuroshima H, Fujita K (2016) Evaluation of third-party reciprocity by squirrel monkeys (Saimiri sciureus) and the question of mechanisms. Anim Cogn 19:813–818

Anderson JR, Bucher B, Chijiiwa H, Kuroshima H, Takimoto A, Fujita K (2017) Third-party social evaluations of humans by monkeys and dogs. Neurosci Biobehav Rev 82:95–109
Bshary R (2002) Biting cleaner fish use altruism to deceive image–scoring client reef fish. Proc R Soc Lond Ser B 269:2087–2093
Bucher B, Arahori M, Chijiiwa H, Takagi S, Fujita K (2020) Domestic cats' reactions to their owner and an unknown individual petting a potential rival. Pet Behav Sci 9:16–33
Caro TM, Fitzgibbon CD, Holt ME (1989) Physiological costs of behavioural strategies for male cheetahs. Anim Behav 38:309–317
Chijiiwa H, Kuroshima H, Hori Y, Anderson JR, Fujita K (2015) Dogs avoid people who behave negatively to their owner: third-party affective evaluation. Anim Behav 106:123–127
Chijiiwa H, Takagi S, Arahori M, Anderson JR, Fujita K, Kuroshima H (2021) Cats (Felis catus) show no avoidance of people who behave negatively to their owner. Anim Behav Cogn 8:23–35
de Waal FBM, Leimgruber K, Greenberg AR (2008) Giving is self-rewarding for monkeys. Proc Natl Acad Sci 105:13685–13689
Engelmann JM, Herrmann E, Tomasello M (2012) Five-year olds, but not chimpanzees, attempt to manage their reputations. PLoS ONE 7:e48433
Freidin E, Putrino N, D'Orazio M, Bentosela M (2013) Dogs' eavesdropping from people's reactions in third party interactions. PLoS ONE 8:e79198
Hamlin JK (2013) Moral judgment and action in preverbal infants and toddlers: evidence for an innate moral core. Curr Dir Psychol Sci 22:186–193
Hamlin JK, Wynn K (2011) Young infants prefer prosocial to antisocial others. Cogn Dev 26:30–39
Hamlin JK, Wynn K, Bloom P (2007) Social evaluation by preverbal infants. Nature 450:557–559
Hamlin JK, Wynn K, Bloom P (2010) Three-month-olds show a negativity bias in their social evaluations. Dev Sci 13:923–929
Hare B, Tomasello M (2005) Human-like social skills in dogs? Trends Cogn Sci 9:439–444
Herrmann E, Keupp S, Hare B, Vaish A, Tomasello M (2013) Direct and indirect reputation formation in nonhuman great apes (Pan paniscus, Pan troglodytes, Gorilla gorilla, Pongo pygmaeus) and human children (Homo sapiens). J Comp Psychol 127:63–75
Izuma K (2012) The social neuroscience of reputation. Neurosci Res 72:283–288
Johnson CM, Sullivan J, Jensen J, Buck C, Trexel J, St. Leger J (2018) Prosocial predictions by bottlenose dolphins (Tursiops spp.) based on motion patterns in visual stimuli. Psychol Sci 29:1405–1413
Kawai N, Yasue M, Banno T, Ichinohe N (2014) Marmoset monkeys evaluate third-party reciprocity. Biol Lett 10:20140058
Kawai N, Nakagami A, Yasue M, Koda H, Ichinohe N (2019) Common marmosets (Callithrix jacchus) evaluate third-party social interactions of human actors but Japanese monkeys (Macaca fuscata) do not. J Comp Psychol 133:488–495
Kishimoto R, Itakura S, Fujita K, Hashiya K (2020) Evaluation of "calculating" helpers based on third-party observation in adults and children. Psychologia 61:185–199
Krupenye C, Hare B (2018) Bonobos prefer individuals that hinder others over those that help. Curr Biol 28:280–286
Kundey SMA, De Los Reyes A, Royer E, Molina S, Monnier B, German R, Coshun A (2011) Reputation-like inference in domestic dogs (Canis familiaris). Anim Cogn 14:291–302
Leete J, Vonk J, Oriani S, Eaton T, Lieb J (2020) Do domestic cats (Felis silvestris catus) infer reputation in humans after direct and indirect experience? Hum Anim Interact Bull 8:35–53
Marshall-Pescini S, Passalacqua C, Ferrario A, Valsecchi P, Prato-Previde E (2011) Social eavesdropping in the domestic dog. Anim Behav 81:1177–1183
McAuliffe K, Bogese M, Chang LW, Andrews CE, Mayer T, Faranda A, Hamlin JK, Santos LR (2019) Do dogs prefer helpers in an infant-based social evaluation task? Front Psychol 10:591
Nitzschner M, Melis AP, Kaminski J, Tomasello M (2012) Dogs (Canis familiaris) evaluate humans on the basis of direct experiences only. PLoS ONE 7:e46880
Nitzschner M, Kaminski J, Melis A, Tomasello M (2014) Side matters: potential mechanisms under-lying dogs' performance in a social eaves dropping paradigm. Anim Behav 90:263–271

Nowak MA (2006) Five rules for the evolution of cooperation. Science 314:1560–1563
Nowak MA, Sigmund K (2005) Evolution of indirect reciprocity. Nature 437:1291–1298
Rooney NJ, Bradshaw JW (2006) Social cognition in the domestic dog: behaviour of spectators towards participants in interspecific games. Anim Behav 72:343–352
Russell YI, Call J, Dunbar RI (2008) Image scoring in great apes. Behav Process 78:108–111
Sabbatini G, de Bortoli Vizioli A, Visalberghi E, Schino G (2012) Food transfers in capuchin monkeys: an experiment on partner choice. Biol Lett 8:757–759
Shreve KRV, Udell MA (2015) What's inside your cat's head? A review of cat (Felis silvestris catus) cognition research past, present and future. Anim Cogn 18:1195–1206
Sima MJ, Matzinger T, Bugnyar T, Pika S (2018) Reconciliation and third-party affiliation in carrion crows. Ethology 124:33–44
Stander PE (1992) Cooperative hunting in lions: the role of the individual. Behav Ecol Sociobiol 29:445–454
Subiaul F, Vonk J, Okamoto-Barth S, Barth J (2008) Do chimpanzees learn reputation by observation? Evidence from direct and indirect experience with generous and selfish strangers. Anim Cogn 11:611–623
Takimoto A, Kuroshima H, Fujita K (2010) Capuchin monkeys (Cebus apella) are sensitive to others' reward: an experimental analysis of food-choice for conspecifics. Anim Cogn 13:249–261
Trösch M, Pellon S, Cuzol F, Parias C, Nowak R, Calandreau L, Lansade L (2020) Horses feel emotions when they watch positive and negative horse–human interactions in a video and transpose what they saw to real life. Anim Cogn 23:643–653
Vaish A, Grossmann T, Woodward A (2008) Not all emotions are created equal: the negativity bias in social-emotional development. Psychol Bull 134:383–403
Vaish A, Carpenter M, Tomasello M (2010) Young children selectively avoid helping people with harmful intentions. Child Dev 81:1661–1669
Wu J, Balliet D, Van Lange PA (2016) Reputation, gossip, and human cooperation. Soc Personal Psychol Compass 10:350–364

Part V
Metacognition

Chapter 14
Planning Abilities in Nonhuman Animals: In Search of the Evolutionary Origins of "Thought"

Hiromitsu Miyata

Abstract Historically, studies in comparative psychology have attempted to elucidate abilities of "thought" in nonhuman animals by assessing how they perform on various behavioral tasks. One extensively studied topic in modern comparative cognition research is planning, which refers to an internal process of establishing systematic and organized courses of action about one's future behavior. Planning is deemed to be a form of sophisticated cognition in humans, but planning should also be advantageous for nonhuman species when they are confronted with various situations in daily life. Empirical studies in recent decades have uncovered planning abilities in nonhuman species including apes and monkeys, as well as avian species. These findings range from great apes' abilities to select and store tools that they might use hours later, to pigeons' determining directions of future steps when navigating on computerized maze and "traveling salesperson" problems. The data from these studies support the notion that planning capabilities are shared among species from a wider taxonomic range than previously presumed. Both neuroanatomical and ecological perspectives should help to shed further light on the evolutionary origins of planning. Future studies may involve various perspectives including metacognition of planning, to better elucidate the evolutionary origins of thought and consciousness.

Keywords Planning · Problem solving · Maze · Pigeon · Kea · Thought

Introduction

The ability to "think" forms a crucial part of human (*Homo sapiens*) intelligence. Higher thought processes such as combining multiple physical and/or social elements and manipulating complex concepts or ideas help us to reduce redundant

---

H. Miyata (✉)
Faculty of Letters, Arts and Sciences, Waseda University, Tokyo, Japan
e-mail: miyata@waseda.jp

J. R. Anderson, H. Kuroshima (eds.), *Comparative Cognition*,
https://doi.org/10.1007/978-981-16-2028-7_14

behavior and avoid inefficient and risky actions in our daily life. What are the evolutionary origins of such "thought"? In other words, to what extent do nonhuman animals "think"? The present chapter intends to address these fundamental questions by reviewing empirical studies on animals, in particular by focusing on other species' abilities to plan their future behaviors.

Both humans and nonhuman species should benefit from having the capability to think, as this can help to solve complex physical and social problems. This may lead to the assumption that humans and nonhuman animals from multiple taxonomic groups may potentially possess at least partially similar or even equivalent thinking abilities. However, brain structures and ecological environments vary widely across taxa. These factors may also influence the nature of thinking abilities of animals (Miyata 2014). To examine these issues, it would be advantageous to compare thinking abilities of diverse living species, including mammals and birds.

Historical and Recent Studies on Thought in Animals

Both traditional experimental and comparative psychology, and more recent comparative cognitive research have addressed thinking abilities of animals in various settings. One historically well-known set of studies is that by Edward Thorndike, who examined behavior of cats (*Felis catus*) as they tried to escape from inside a puzzle box with one or more locking devices. Time to exit the boxes decreased across successive trials, referred to as problem-solving by trial-and-error (Thorndike 1898). In another well-known study, Wolfgang Köhler observed captive chimpanzees (*Pan troglodytes*) combining multiple boxes (to make a climbing frame) and utilizing sticks to obtain food that was otherwise out of reach. Köhler proposed that such sudden and accurate problem-solving was achieved by "insight" (Köhler 1925). From today's perspectives, trial-and-error learning and insight are not regarded as mutually exclusive; both strategies might operate depending on the problem-solving situations animals are faced with.

More recent empirical studies on animal cognition have uncovered various aspects of logical behaviors, understanding of causality, etc. For example, transitive inference refers to an internal process of inferring relations between two elements that had not been trained, e.g., the ability to infer $B > D$ following training on relationships $A > B > C > D > E$. Multiple species including squirrel monkeys (*Saimiri sciureus*), rats (*Rattus norvegicus*), and pigeons (*Columba livia*) have been shown to demonstrate transitive responses in laboratory settings (e.g., Davis 1992; McGonigle and Chalmers 1977; von Fersen et al. 1991). In social contexts, the ability to use transitive inference should help individuals to know their own place in the dominance hierarchy, thereby allowing them to avoid unnecessary fights. Paz-y-Miño et al. (2004) demonstrated that pinyon jays (*Gymnorhinus cyanocephalus*), a highly social corvid species, are capable of correctly inferring their own dominance status against an unfamiliar individual after observing an interaction between that unfamiliar individual and a familiar one. Comparable evidence for social inference in a cichlid fish (*Julidochromis transcriptus*) has also been reported (Hotta et al. 2015).

Another frequently studied behavior to explore thought in animals is tool use. Beyond classical studies such as Köhler's (1925), empirical studies over the past three decades have addressed whether and how animals may understand causality between a tool, a food reward, and the environment in which animals use the tool. Visalberghi and Limongelli (1994) examined behavior of capuchin monkeys (*Sapajus apella*) to use a stick as a tool to push a food reward out of a horizontal Plexiglas tube that had a trap (container). Three out of the four monkeys performed at a chance level; the remaining, successful monkey was shown to have used a simple associative rule to avoid pushing the food toward a trap, with no real understanding of the causal relationships between the tool, reward, and the trap. Fujita et al. (2003) contributed to this field by presenting capuchin monkeys with the task of selecting one of the two hook-like tools to obtain a reward. After learning a basic choice task, the monkeys generalized their learning to different tools of different colors and shapes, but they were initially unsuccessful when an obstacle was introduced on the path along which the tool had to be dragged. Fujita et al. (2011) went on to demonstrate that the monkeys not only quickly learned to avoid a hindrance (either a trap or an obstacle), but also generalized their performance to novel types of hindrances, consistent with the notion that the monkeys can understand complex relationships between a tool, a reward, and a hindrance. Cummins-Sebree and Fragaszy (2005) had earlier demonstrated that capuchin monkeys could pull in a hook-shaped tool to retrieve food, and that they also sometimes repositioned the tools to enable them to obtain food.

Defining Animal Thought and Planning

Given the aforementioned complex problem-solving performances and logical-like behavior reported in various species, it is important to ask what kind of internal processes may underlie those behaviors. In other words, can we identify more clearly the nonverbal forms of thought that are potentially shared across nonhuman species? Based on Fujita (2004), in this chapter I propose a definition of thought as an internal process of operating on mental representations (see also Miyata 2014). First, an animal takes in information from the external world and stores it in the brain, which is called a first-order representation. Next, the animal internally transforms various properties of that information to create representations that differ from the initial one; this is higher-order representation. This latter representation enables problem-solving behavior as motor output. Such processes of generating higher-order representations in the brain can be operationally defined as "thought" (Fujita 2004). This definition allows us to design appropriate behavioral experiments to study nonverbal forms of thought in nonhuman species.

One important form of thought as defined above is planning, which refers to an internal process of formulating a systematic and organized method about one's behavior in advance (e.g., Friedman et al. 1987). Planning is deemed to involve higher mental processes such as keeping in mind the goal of the actions and monitoring of the outcomes relating to one's own behavior; planning is thus an important part of the internal processes that guide our daily behavior. Although planning has long been studied in the field of artificial intelligence, in the past few decades psychological and neurophysiological studies have also addressed neurocognitive underpinnings of human planning. Shallice (1982) reported that patients with cerebral lesions showed diminished abilities on planning tasks such as the Tower of London (TOL) task. In this task an individual is required to rearrange disks or beads sequentially to match a model in a minimum number of stepwise movements. Using functional magnetic resonance imaging (fMRI), van den Heuvel et al. (2003) reported that dorsolateral prefrontal cortex and caudate nucleus of the basal ganglia both showed enhanced activation in adult humans performing TOL tasks, suggesting involvement of fronto-striatal networks in human planning. Planning abilities may well have important adaptive significance not only for human adults and children (Friedman et al. 1987; Friedman and Scholnick 1997) but also for other species, because planning should help promote efficient decision-making in novel and challenging situations in various ecological niches.

Empirical Studies of Planning in Animals

Whether and to what extent are nonhuman animals endowed with abilities for planning? Despite several anecdotal reports during the twentieth century suggesting future-oriented behavior in apes, the predominant view among researchers until the last few decades was that planning ability was unique to humans (Roberts 2002). For example, the Bischof–Köhler hypothesis proposed that nonhuman animals are primarily governed by current motivational states, with no role for anticipation of future motivational states (Suddendorf and Corballis 1997). However, many animals encounter situations that require sequential and/or future-oriented actions by integrating spatio-temporal information, such as selecting efficient routes to feeding sites or storing food for future consumption. In these situations, planning should enhance appropriate and efficient decision-making.

In fact, evidence from numerous well-controlled experiments has uncovered planning abilities in nonhuman species, notably primates, and birds (for reviews see Miyata 2014; Miyata and Fujita 2011a). These taxa are considered to have diverged from a common ancestor approximately 300 million years ago. They differ not only in phylogenetic status but also in neuroanatomical characteristics: neocortex with increased volume in primates is absent in avian species (Striedter 2005). If bird and primate species nevertheless show similarities in planning capabilities, that would suggest common selective pressures in their physical or social environments that led to convergence in higher cognitive abilities. Accumulating evidence of planning in these taxa should thus help to unravel the evolutionary history of planning and pertinent interactions between phylogeny and ecology.

When studying planning in animals, it is important to differentiate planning as an advanced form of cognition from simpler operant learning, that is, situation-specific response selection based on prior experience of response-outcome contingencies. If a given operant response has reliably led to a desirable outcome in a specific situation, the animal need not use planning when making a subsequent selection among responses in that same situation. To interpret a behavior as that based on planning instead of simple operant responses, at least one of the following two conditions must be met: (1) the behavior is a novel (not familiar) one which involves more than one sequential components, and (2) the behavior is based on future, as opposed to present, motivational states (Miyata and Fujita 2011a). The following sections review empirical data that meet these criteria and thus suggest planning in animals.

Planning to Meet Current Motivational States

Although planning can occur on multiple different levels (e.g., physical, social), when considering animals it seems particularly important to distinguish between planning based on present vs. future motivational states. Many species in the wild may solve complex sequential problems in order to meet current needs, such as satisfying hunger. Thus, planning to meet current motivational states is expected to be present in diverse taxa. In laboratory settings, such planning has been examined through the use of behavioral paradigms such as navigation and maze tasks, and puzzle boxes, as discussed below.

Navigation and Maze Problems

Navigation behavior has been frequently studied to examine potential mechanisms that animals use to determine goal locations and the paths to reach goals when moving through actual or virtual space (e.g., Tolman 1948). Fragaszy et al. (2003) examined strategies used by chimpanzees and capuchin monkeys to navigate by manipulating joysticks in two-dimensional computerized mazes with up to five choice points. There were both individual and species differences in the types of errors made, with chimpanzees making fewer errors at the choice points than capuchins. Nevertheless, both species showed flexible strategies such as selecting a "correct" detour that led away from the goal before providing access to it, as well as correcting the paths after selecting an erroneous route. These findings supported the view that the subjects planned their choices at these points. Fragaszy et al. (2009) tested their subjects on a wider variety of mazes either in a random or a fixed presentation order. Regardless of presentation order, chimpanzees made fewer errors at the choice points and more frequent self-correction of errors than capuchin monkeys. Fragaszy et al. (2009) suggested that differences in attention and motor processes may underlie the observed performance differences between the species, with chimpanzees more likely than capuchins to use higher levels of planning by integrating information about continuity of the path and directedness to the goal. Pan et al. (2011) further trained capuchin monkeys on a large number of computerized mazes and showed that they used vector or topological information, with some individuals making correct choices at a rate comparable to chimpanzees.

In a behavioral neurophysiological study, Mushiake et al. (2006) trained two Japanese macaques (*Macaca fuscata*) to solve virtual maze tasks presented on a computer screen by directing a cursor to a goal location. The monkeys were trained to move the cursor in a step-by step manner by operating two manipulanda by moving their wrists. A population of prefrontal neurons exhibited increased activity associated with the first, second, and third of the three cursor movements, respectively. These neurons also showed increased activity during a preparatory period immediately before the monkeys started to move the cursor. By contrast, neuronal activity in the primary motor cortex was associated with the arm movements but not the preparatory period. These data suggest that the monkeys planned the stepwise movements of the cursor before starting to physically solve the mazes, at least at the neurophysiological level (see also Shima et al. 2007).

Puzzle Box

Dunbar et al. (2005) examined abilities of chimpanzees, orangutans (*Pongo pygmaeus*), and 3- to 7-year-old human children to plan solutions to puzzle boxes before starting to physically solve them. Several different versions of puzzle boxes were used, requiring one or more sequential steps to open them to obtain a reward inside. Latencies to solve these tasks were compared between a "prior view" condition in which the box was presented to the subjects before they started to manipulate it and a "no prior view" condition in which the box was not presented in advance. For chimpanzees and orangutans, the box was located adjacent to their cages for 24–48 h so that they could visually examine the box. For human children, in the "prior view" condition they were instructed to draw pictures of the box for 20 min. Results showed that the human children, but not the great apes, solved the problem more quickly in the "prior view" than in the "no prior view" condition, suggesting that the former planned their strategies for solving the tasks before physically starting to solve them. Although Dunbar et al. (2005) used the term "mental rehearsal" to explain the internal processes involved in these tasks, this could be considered as equivalent to planning before initiating a set of sequential actions (see also Dunbar 2000).

Planning to Meet Future Motivational States

Planning to meet future, as opposed to present, motivational states is an advanced mental process that involves "mental time travel" in which the individual generates internal representations about its past and the future; it also requires sophisticated self-control abilities. This type of planning can enable flexible control of behavior over relatively large temporal scales, i.e., hours, days, or longer (Osvath and Osvath 2008). Evidence of planning to meet future motivational states in nonhuman species would suggest that higher mental systems that characterize human intelligence are shared across species (Miyata and Fujita 2011a). Recent empirical studies have also uncovered abilities of planning to fulfill the future needs independently of the current ones in several species of primates and birds, by using paradigms such as tool use and food caching.

Tool Use

Some tool-using behaviors of both primates and birds have been suggested to involve planning to meet future motivational states, not only present ones. Mulcahy and Call (2006), for example, showed that both bonobos (*Pan paniscus*) and orangutans appropriately selected, transported, and saved tools that they would need later. After basic training to obtain food rewards using several different types to tools (e.g., tubes, hooks), the apes first selected an appropriate one out of the multiple options at the test room, and brought the tool into a waiting room adjacent to the test room. One hour later, they brought the tool back to the test room and used it to successfully obtain reward. Successful tool-selections occurred even after the waiting period was extended to 14 h. These data support the idea that the apes planned the future use of tools independently of their current needs. Because Mulcahy and Call (2006) used an experimental situation with a minimal amount of pretraining, simple trial-and-error learning is unlikely to explain the apes' behavior.

In a similar tool-selection context, Osvath and Osvath (2008) also found that chimpanzees and orangutans (*Pongo abelli*) selected and stored appropriate tools (e.g., hoses, pipes) that they would need 70 min later. The apes selected appropriate tools even when a favorite fruit was included among the possible choices in the selection phase. As Osvath and Osvath (2008) suggest, these findings support the view that the apes' ability to plan for the future needs involved sophisticated cognitive processes including internally pre-experiencing a future event and self-control to overcome current drives.

Among avian taxa, Kabadayi and Osvath (2017) reported that ravens (*Corvus corax*) successfully selected, saved, and used a tool (and an exchangeable token) after a delay of 15 min. The birds maintained high levels of performance when the delay was 17 h, and also showed self-control abilities, selecting the tool (token) instead of a favorite food reward that could be immediately obtained. These results are similar to those obtained in studies with apes (Mulcahy and Call 2006; Osvath and Osvath 2008), and suggest that the primate ability to plan for the future motivational states as opposed to present ones may be shared by at least one corvid species.

Food Caching

Some avian species including corvids are known to cache food instead of immediately consuming it. Does such behavior involve planning for the future? Emery and Clayton (2001) examined planning in scrub-jays (*Aphelocoma coerulescens*) in a study of food-caching behavior. After initially caching food, birds were found to re-cache food in novel sites if a conspecific had observed the initial caching behavior. Notably, re-caching was observed in jays with prior experience of pilfering another bird's caches, but not in those without pilfering experience. These results suggest that scrub-jays can flexibly adjust their food-caching strategies so as to minimize the risk of potential theft by other birds. The data also appear consistent with the idea that the birds may use planning to secure future access to food.

Raby et al. (2007) showed that western scrub-jays (*Aphelocoma californica*) adjusted their food caching for a future need. First, each morning an experimenter provided food on one of two caching trays, each located in a different compartment. The jays experienced these two compartments on alternate days. After this training, the birds were tested one evening to assess on which tray they would cache food for the next morning. The jays cached food on the tray in which they would not have food next morning, more frequently than the other tray. In a second experiment, two different foods (kibble and peanuts) were given, one in each compartment. The birds were tested one evening to determine which of the two foods they would cache. They showed a significant tendency to cache the food that they would not have the next morning. Based on these data, Raby et al. (2007) argued that the jays were capable of spontaneously "planning for breakfast." Correia et al. (2007) further suggested that western scrub-jays anticipate recovery of the cached food by making a distinction between their current and future motivational states (see also Cheke and Clayton 2012 for comparable data on Eurasian jays *Garrulus glandarius*).

Exploring Pigeon Planning Using Computerized Navigation Tasks

As outlined above, not only apes and monkeys but also some corvid species have been suggested to exhibit planning abilities to meet both present and future needs. Regarding birds, however, most studies have focused on behavior that might reflect a domain-specific adaptation, such as food caching. There are a few exceptions such as the study by Kabadayi and Osvath (2017), who used a tool and a token although ravens are not habitual users of tools or tokens in the wild. Given that the scrub-jays are habitual food cachers, their behavioral flexibility and cognitive prowess in that specific domain might not generalize to other contexts or species, and it remains possible that planning abilities are restricted to only a few avian species. It is therefore desirable to examine planning abilities in birds by using more general learning tasks that can be employed to directly compare a wide range of species. Furthermore, most evidence of planning in birds is limited to corvids. If similar evidence of planning can be obtained in more phylogenetically and ecologically distant species, such data would suggest common selective pressures for convergent planning abilities, and allow us to better understand the evolutionary history of planning (Miyata 2014; Miyata and Fujita 2011a).

Navigation and Maze Planning

Against the background outlined above, Miyata et al. (2006) examined whether pigeons plan before physically starting to solve maze problems presented on an LCD screen. First, we developed an original computerized navigation task that involved moving a red square (the target) to a blue square (the goal) by a chain of pecking responses. Four individually tested pigeons pecked at one of four small dots ("guides") located around the target, to determine the direction of each movement, until the target arrived at the goal. After this navigation training, the birds were tested on various maze problems that had a straight or L-shaped line as a barrier. No guide dots appeared beyond the line, so that subjects had to make a detour around the barrier (see Fig. 14.1a for an illustration of the experiment). As the barrier became longer and bent to form an L-shape, the birds came to select direct routes more frequently than indirect ones. They were also presented with the mazes in pale colors for a few seconds before being allowed to solve the tasks. The question here was whether this preview phase would facilitate performance on the task. All pigeons performed well on various versions of the navigation and maze tasks. However, there was no evidence that the preview phase facilitated performance on novel mazes. On well-practiced mazes having an L-shaped barrier, performance was worse when the maze changed from the preview to the solution phase than when the change did not occur, suggesting that birds at least identified each maze among familiar sets of problems during the preview phase.

Miyata and Fujita (2008) introduced novel mazes to examine whether pigeons planned one or more steps ahead either (1) while solving the maze or (2) before starting to solve the maze. Experiment 1 used a plus-shaped maze (Fig. 14.1b) and revealed that the pigeons frequently moved the target toward the previous goal immediately after the goal jumped to another corner when the target arrived at the center point. When the birds correctly responded on the new goal, response time was longer than in other trials. These data suggest that pigeons plan one step beyond the

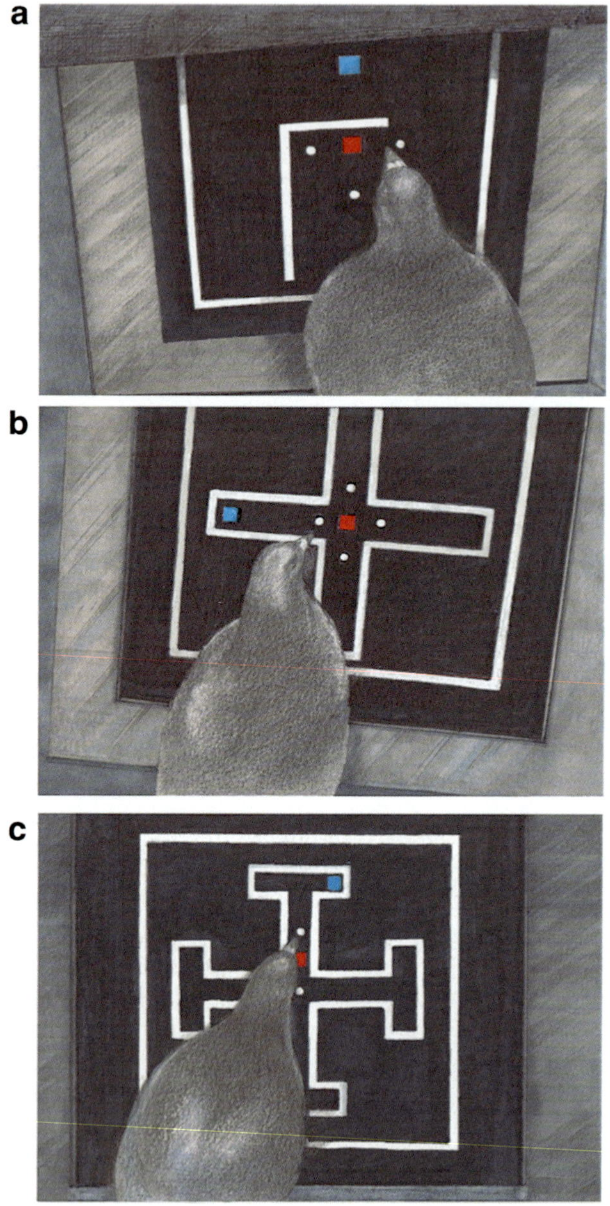

Fig. 14.1 Illustrations of experiments with detour and maze problems to test planning in pigeons. (**a**) A pigeon (Kanta) solving a detour task with an L-shaped line as a barrier by pecking on an LCD screen in an operant chamber (Miyata et al. 2006). (**b**) A pigeon (Caesar) solving a plus-shaped maze task (Miyata and Fujita 2008). (**c**) The same pigeon as in (**b**) solving an 8-arm variation of the plus-shaped maze task (Miyata and Fujita 2008)

center of the maze, and that the birds sometimes correctly adjusted previously planned behavior. When the goal-shift occurred one to three steps before the target was at the center point, relatively few errors occurred, suggesting that the birds do not plan multiple steps ahead. Experiment 2 introduced an eight-arm variation of the previous maze (Fig. 14.1c). The pigeons frequently started by erroneously going towards the old goal when the goal jumped to another arm-end immediately after the preview period. Response time was significantly longer when the birds correctly started toward a new goal. The results of these experiments suggest that pigeons may plan at least one step forward, not only during solution of the mazes but also before physically starting to solve them (see also Miyata and Fujita 2012).

Route Selection Strategies in "Traveling Salesperson" Tasks

To find out whether pigeons also use a relatively longer-term strategy of estimating which route to take, Miyata and Fujita (2010) examined how they solved multi-goal navigation tasks presented on a computer monitor. The birds were required to move the target to two or three goals successively by making a chain of pecking responses on the LCD screen, a task that represented a traveling salesperson problem (TSP) in a city-block metric (Yamamoto and Kubo 1997). The ability to find efficient routes in situations analogous to TSPs should be important for nonhuman species in daily situations such as determining efficient routes to multiple feeding sites (Gallistel and Cramer 1996; Gibson et al. 2012). In Experiment 1, the pigeons solved two-goal TSPs by frequently selecting routes to visit the nearer goal first and the farther one next (Fig. 14.2a). Experiment 2 used three-goal TSPs in which the goals and the starting location occupied each corner of a square (Fig. 14.2b). In this situation pigeons frequently visited the nearest goal first; however, they also showed tendencies to travel either counterclockwise or clockwise. In doing so, the birds did not necessarily use the "nearest-neighbor" algorithm, a strategy to consistently visit the next nearest goal. The pigeons also selected routes with relatively short traveling distances. In Experiment 3, three goals were located along a straight line (Fig. 14.2c), and in this test the pigeons frequently selected the nearest goal as the first to visit. In Experiment 4, in which two of the three goals were located adjacent to each other in a group (Fig. 14.2d), the birds frequently started by visiting one of the grouped goals that was nearer to the starting location.

Because the pigeons consistently started by visiting the nearest goal in Miyata and Fujita (2010) and Miyata and Fujita (2011b) further exposed pigeons to a variation of two-goal "traveling salesperson" navigation tasks that either did or did not involve an L-shaped line as a barrier between the target starting location and the nearer goal. Whereas the pigeons often started by visiting the nearer goal in tasks having no barrier, in tasks having a barrier the birds frequently visited the farther goal first. In other words, their strategy of visiting the nearest goal was not necessarily maintained in the presence of a barrier that forced a detour.

Results from these sets of studies showed that pigeons efficiently solved a variety of navigation, maze/detour, and TSP tasks presented on an LCD screen, and thereby validated use of an innovative methodology for examining cognitive processing in avian species. Despite their efficiency on the tasks, the pigeons' data collectively suggest that they primarily used local or proximate strategies to solve these problems. Nevertheless, pigeons may also use more advanced processes, including

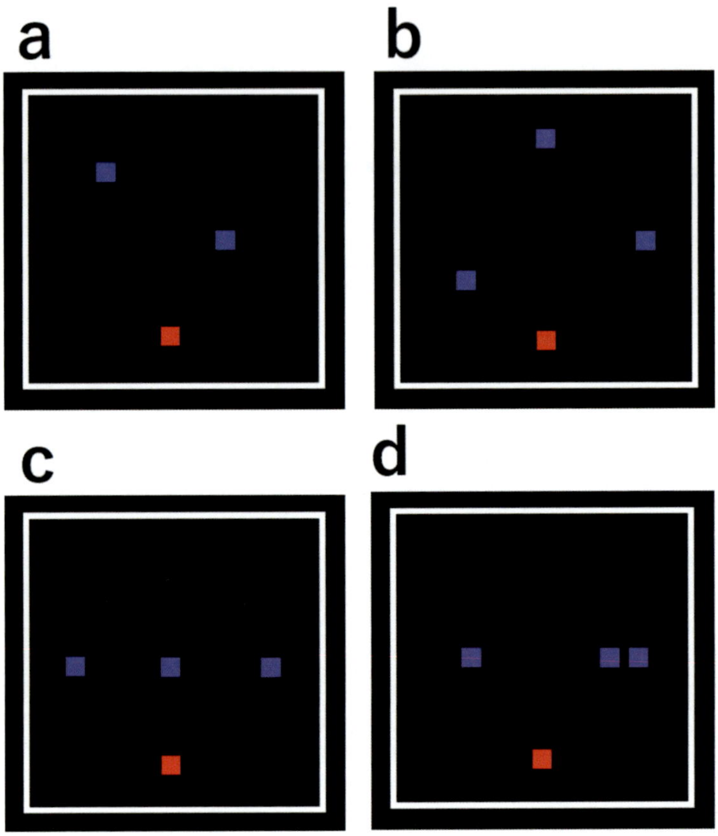

Fig. 14.2 Variations of TSPs that pigeons solved (Miyata and Fujita 2010). Each picture (**a–d**) shows an example of the TSP used in experiments 1–4, respectively. The red square represents the target starting location and the blue squares represent the goals

planning one step ahead (Miyata and Fujita 2008), selecting "round" routes to efficiently visit multiple goals (Miyata and Fujita 2010), and avoiding visiting the nearest goal in the presence of an obstacle (Miyata and Fujita 2011b).

Comparisons with Young Human Children

A strong point of these computerized navigation tasks is that they can be applied to species from multiple taxa for direct comparisons of solution strategies. Miyata et al. (2009) examined 3- to 4-year-old human children's performances on touchscreen maze tasks comparable to those previously used for pigeons (Miyata and Fujita 2008; Miyata et al. 2006). In Experiment 1, children around the age of 3 successfully solved a navigation task involving moving a target (a picture of a dog) to a goal (a picture of a bone) by making serial responses on the touch monitor with their fingers. Similar to pigeons (Miyata et al. 2006), children made a detour to reach the goal while avoiding an L-shaped barrier by taking direct routes more frequently than chance. In Experiment 2, which used a plus-shaped maze as in Miyata and Fujita (2008), 3- and 4-year-olds showed responses and reaction times comparable to those of pigeons. These parallel results from different taxa appear to indicate similar planning and problem-solving processes across species. Furthermore, whereas 3-year-olds had difficulty immediately adjusting their previously planned strategies after a change of goal locations, 4-year-olds did so more readily. These data suggest developmental change in planning abilities that involves response inhibition and reengagement following an unexpected event. The overall behavioral similarity between these phylogenetically and developmentally distant species on these tasks suggests common selective pressures for planning, which should be an adaptive mental system for diverse species living in various physical and social environments (see also Miyata et al. 2006, 2014).

Life History and Planning: Lock-Opening Behavior of Keas

Results from the experiments described in the previous section support the view that pigeons have at least some forms of planning ability, which may be shared across avian species or even more taxa. Might the nature of species' planning ability vary across diverse ecological niches? Keas (*Nestor notabilis*), New Zealand parrots, are an interesting species for these comparative purposes because of their distinctive ecology and behaviors, including wide seasonal variability in diet ("opportunistic foraging"), persistent and destructive object manipulation with a lack of neophobia, and highly playful tendencies especially among juveniles (Diamond and Bond 1999; Huber and Gajon 2006). Such behavioral flexibility and manipulative skills led us to hypothesize that keas might be endowed with sophisticated planning capabilities.

Miyata et al. (2011a, b) examined whether keas solved "artificial-fruit" lock-opening problems by planning before starting to physically solve them. The keas were first trained to remove a metal rod from a Plexiglas lid to obtain a reward from inside a container, then they were exposed to several problems having multiple locks (Fig. 14.3a). A preview period was also introduced, during which the keas were or were not allowed to observe the task for 10–30 s before starting to solve it. In a series of tests with two or three locks, one of which required a manipulation, the keas either selected the appropriate lock right at the start of the solution phase or showed a side preference—consistently selecting the lock on a particular side; there was no strong evidence that the preview period facilitated performance. However, in a more complex version of the task that required manipulating rods in a fixed sequence (i.e., Rod B → Rod A in Fig. 14.3b), the keas always manipulated the incorrect rod (Rod A) first, but corrected these inappropriate responses faster if they had first been allowed to observe the problems than if they had not. This may suggest that keas engaged in "implicit" or "covert" planning, i.e., planning the correct sequence (Rod

Fig. 14.3 Experiments with lock-opening problems in keas (Miyata et al. 2011a, b). (**a**) A kea (Zappel) solving a lock problem on the experimental device. The birds observed the task in a waiting compartment on the right-hand side before starting to solve it. (**b**) A variation of the locks that required two sequential manipulations (Rod B → Rod A). Over the locks is a transparent (or opaque) board which allowed (or prevented) prior observation of the lock. The bird removed the preview board and Rod A from over the transparent lid, thus opening the lid and gaining access to the reward in the food container (upper-left corner of the picture). (Photos by Hiromitsu Miyata)

B and then Rod A) in advance, but always checking Rod A first to see whether a single manipulation would suffice.

In summary, the keas primarily exhibited explorative, trial-and-error tendencies when solving the lock problems, but they may also have used some planning for the more cognitively demanding versions. These results may be understood from the perspective of keas' specific adaptations to their ecological environment (Diamond and Bond 1999), which include explorative and persistent foraging strategies in a strongly seasonal habitat with limited food resources and low predation risk. Nevertheless, like other avian or primate species, keas may have been exposed to common selective pressure for some planning ability in their physical or social environment. Our data are compatible with some planning abilities being shared among avian species but some abilities also be significantly influenced by species-specific ecological factors.

Evolutionary Origins of "Thought" and Consciousness

The abovementioned studies with pigeons, keas, and human children provide positive evidence that some planning abilities are shared across a wider range of taxa than previously presumed, including birds and primates. The research outcomes also suggest that planning abilities may be influenced by species-specific ecological factors. Both neuroanatomical and ecological perspectives can shed light on the evolution of thought, with promising future perspectives for addressing evolutionary origins of consciousness including metacognition of planning.

Evolution of the Brain and Avian Planning

Birds and mammals including primates have developed largely different anatomical structures of the brain since diverging from their common ancestors (Watanabe and Kojima 2007; see also Striedter 2005). Unlike mammals, birds do not have a neocortex but instead have a large structure named the dorsal ventricular ridge (DVR). Within the DVR, nidopallium is relatively voluminous in corvids compared with other groups of birds. Despite these differences, behavioral experiments suggest that planning abilities may be shared across multiple species of birds and primates. Although neural correlates of avian planning remain to be uncovered, one hypothesis would be that part of the avian DVR such as the nidopallium has functions similar to mammalian prefrontal cortex in achieving planning. In support, recent reports have indicated that cells in the pigeon nidopallium caudolaterale serve advanced cognitive functions, such as processing reward value (Dykes et al. 2018).

What are the potential factors that led to such structural divergence and cognitive convergence? Shimizu (2000) notes that mammals and reptiles (modern forms of ancestors of birds), differ in both metabolic rate and reproductive strategies. With a relatively low metabolic rate, reptiles rely on quick and successive movements that involve midbrain and DVR systems. By contrast, because of their higher metabolic rate mammals consistently require relatively large amount of energy from their diet. In addition, whereas parental care is relatively rare among reptiles, many mammals devote large amounts of energy on rearing offspring (maternal investment). As numbers of offspring per individual decrease, sexual selection can become more critical and courtship displays more complex. Thus, for mammals deliberate and accurate decision-making—involving the prefrontal cortex—could be an adaptive strategy.

Birds hold a unique standpoint in this general framework (Miyata 2014; Watanabe and Kojima 2007). Like mammals, birds are homothermic, with a high metabolic rate. Birds also expend large amounts of energy on offspring rearing. Thus, deliberate and flexible decision-making strategies would again be adaptive. However, birds have evolved from reptiles, and therefore might be expected to have possessed a developed DVR since early in their evolution. Alternatively, birds may have undergone further DVR development to meet the requirements of homothermic animals, rather than developing a neocortex. Also, because birds fly, mammal-like brain enlargement would be costly. Therefore, brain evolution in birds conceivably maintained the quick and accurate systems of reptiles, with additional improvements—within size restraints—to enable some more advanced cognitive processes such as planning by operating on mental representations. These views are consistent with the evidence from behavioral experiments suggesting planning in pigeons and keas, as well as corvids.

Avian Ecology and Planning

The results of studies of birds' planning in navigation tasks can largely be explained in relation to their navigation behavior in the wild. For example, a typical foraging strategy of pigeons is to seek pieces of grain distributed on the ground. During such small-scale navigation, a "nearest-neighbor" strategy consisting of taking a food item that is close to the forager and then looking for the next closest one is likely to be an efficient, if not optimal, strategy. In these situations, using a proximate strategy while planning at most one step ahead could enable the birds to maximize their energy gain by minimizing cognitive and time costs. Trial-and-error strategies might also be effective, for example, when a goal location is invisible from the pigeon's current location. These views are consistent with pigeons' planning in maze and TSP tasks (Miyata and Fujita 2008, 2010, 2011b). In other words, for pigeons *planning one step ahead* is much more advantageous than never planning, and probably also better than planning too much, even if the latter seems counterintuitive.

In addition to foraging on the ground, some avian species exhibit outstanding large-scale navigation abilities, including homing, to fly over long distances. Because such navigation is far more demanding than ground foraging, inefficient strategies such as selecting routes with relatively long traveling distances are too costly time and energy. In addition, pigeons and other—migrating—birds are known to use more than one of several available cues, including airborne olfactory cues, the sun, magnetic fields, and star patterns (Gould 2009). Although the mental processes involved in these behaviors remain largely unknown, it would be of interest to conduct comparisons between experimental situations that represent ground foraging and longer-distance navigation. For example, Gibson et al. (2012) demonstrated that pigeons tested in an experimental room selected routes primarily based on local proximity along with some possible planning when solving TSPs. Such behavior may give way to optimal or more efficient route selection strategies if analogous but larger-scale spatial tasks are presented to pigeons (e.g., with nodes more than several kilometers apart from each other).

Towards a Better Understanding of Thought and Consciousness

What approaches might help us to better understand the evolutionary origins of planning and thought? First, there should be greater integration of different levels of planning (Miyata and Fujita 2011a). As mentioned above, most studies of planning in nonhuman animals have involved either serial, complex manipulations in a short time scale to meet current needs, or a simple manipulation to meet future needs. There should be cases in which animals need to plan complex, sequential behavior for future motivational states. This could be examined by integrating procedures used in several of the preceding studies. For example, sequential navigation (Miyata and Fujita 2008, 2010) and serial learning tasks (Johnston et al. 2018; Scarf and Colombo 2010) could be used in naturalistic settings before animals are allowed potential access to high-value rewards several hours later.

Second, greater consideration should be given to the metacognition of planning. Adapting to changing environments can be facilitated by being able to compare multiple options for future behavior before deciding which option to select, and here, metacognition of planning may be advantageous not only for humans but also other species (Miyata and Fujita 2011a). Metacognition forms a core part of consciousness because it involves conscious and active access to one's own internal states (Fujita 2010). Thus, metacognition of planning is important because it is deemed as one of the highly cognitive dimensions of consciousness. For example, before starting to solve a given problem, metacognitive insight into whether or not the individual knows the solution would influence the decision to initiate planning or not. Such metacognition might be revealed if hint-seeking behaviors leading to correct responses are observed more frequently in novel or relatively demanding tasks than in familiar or less demanding ones (Iwasaki et al. 2018; Kornell et al. 2007). Also, after planning has occurred, metacognition might be involved in evaluating whether it was efficient or useful in terms of obtaining reward. This process could be examined by, for example, requiring subjects to discriminate between responses that were correct/efficient or incorrect/inefficient. Following an incorrect or inefficient behavior, animals might be relatively uncertain about their planning, in which case the incidence of "escape" responses that result in low-value reward instead of the discrimination responses might increase (Smith et al. 1997). From a comparative perspective, it would be interesting to compare planning among species that differ only in specific domains of ecology or phylogeny. For example, among avian taxa, ravens and keas have similar body and brain sizes, but they have quite different ecologies. Whereas keas are typically characterized by explorative, trial-and-error, and playful behavior, ravens live in a relatively competitive social environment that includes, for example, risk of pilfering (Bugnyar 2013). These differences may lead to more highly developed capabilities for metacognition of planning in ravens than in keas. Such comparative research would be valuable for enriching our understanding of the evolutionary origins of thought and consciousness.

Acknowledgments The author would like to acknowledge distinguished contributions by Kazuo Fujita in comparative studies of thought. In this chapter the focus is on planning, but Fujita has also conducted and supervised projects in several other relevant domains including tool use, transitive inference, serial learning, number recognition, etc. My studies on pigeons, keas, and human children reviewed in this chapter were financially supported by a Research Fellowship of the Japan Society for the Promotion of Science (JSPS) for Young Scientists, Grants-in-Aid for Scientific Research Nos. 14651020, 17300085, and 20220004 from JSPS to Kazuo Fujita and No. 16500161 to Shoji Itakura, Austrian Science Fund (FWF) Grant P19087-B17 (to Ludwig Huber), Nissan Science Foundation grant to Shoji Itakura, and by the 21st Century Center of Excellence Program, D-10 and the Global COE Program, D-07, to Kyoto University from the Ministry of Education, Culture, Sports, Science, and Technology, Japan. The author also thanks James R. Anderson, Kyoto University, for carefully editing and providing useful advice on an earlier version of this chapter. Finally, the author is grateful to Chizuko Ariizumi for drawing Fig. 14.1.

References

Bugnyar T (2013) Social cognition in ravens. Comp Cogn Behav Rev 8:1–12

Cheke LG, Clayton NS (2012) Eurasian jays (*Garrulus glandarius*) overcome their current desires to anticipate two distinct future needs and plan for them appropriately. Biol Lett 8:171–175

Correia SPC, Dickinson A, Clayton NS (2007) Western scrub-jays (*Aphelocoma californica*) anticipate future needs independently of their current motivational state. Curr Biol 17:856–861

Cummins-Sebree SE, Fragaszy DM (2005) Choosing and using tools: Capuchins (*Cebus apella*) use a different metric than tamarins (*Saguinus oedipus*). J Comp Psychol 119:210–219

Davis H (1992) Transitive inference in rats (*Rattus norvegicus*). J Comp Psychol 106:342–349

Diamond J, Bond AB (1999) Kea, bird of paradox: the evolution and behavior of a New Zealand parrot. University of California Press, Berkeley

Dunbar RIM (2000) Causal reasoning, mental rehearsal, and primate cognition. In: Heyes C, Huber L (eds) The evolution of cognition. MIT Press, Cambridge, pp 205–219

Dunbar RIM, McAdam MR, O'Connell S (2005) Mental rehearsal in great apes (*Pan troglodytes* and *Pongo pygmaeus*) and children. Behav Process 69:323–330

Dykes M, Klarer A, Porter B, Rose J, Colombo M (2018) Neurons in the pigeon nidopallium caudolaterale display value-related activity. Sci Rep 8:5377

Emery N, Clayton N (2001) Effects of experience and social context on prospective caching strategies in scrub jays. Nature 414:443–446

Fragaszy D, Johnson-Pynn E, Hirsh E, Brakke K (2003) Strategic navigation of two-dimensional alley mazes: comparing capuchin monkeys and chimpanzees. Anim Cogn 6:149–160

Fragaszy D, Kennedy E, Mumane A, Menzel C, Brewer G, Johnson-Pynn J et al (2009) Navigating two-dimensional mazes: Chimpanzees (*Pan troglodytes*) and capuchins (*Cebus apella sp.*) profit from experience differently. Anim Cogn 12:491–504

Friedman SL, Scholnick EK (1997) The developmental psychology of planning: why, how, and when do we plan? Lawrence Erlbaum Associates, Mahwah

Friedman SL, Scholnick EK, Cocking RR (1987) Blueprints for thinking: the role of planning in cognitive development. Cambridge University Press, Cambridge

Fujita K (2004) Hikaku ninchi kagaku (comparative cognitive science). In: Otsu Y, Hatano G (eds) Ninchi-kagaku he no shotai: Kokoro no kenkyu no omoshirosa ni semaru (an introduction to cognitive science: an approach to interesting studies of mind). Kenkyusha, Tokyo, pp 122–140

Fujita K (2010) Comparative studies of metacognition. Jpn Psychol Rev 53:270–294

Fujita K, Kuroshima H, Asai S (2003) How do tufted capuchin monkeys (*Cebus apella*) understand causality involved in tool use? J Exp Psychol Anim Behav Process 29:233–242

Fujita K, Sato Y, Kuroshima H (2011) Learning and generalization of tool use by tufted capuchin monkeys (*Cebus apella*) in tasks involving three factors: reward, tool, and hindrance. J Exp Psychol Anim Behav Process 37:10–19

Gallistel CR, Cramer AE (1996) Computations on metric maps in mammals: getting oriented and choosing a multi-destination route. J Exp Biol 199:211–217

Gibson B, Wilkinson M, Kelly D (2012) Let the pigeons drive the bus: pigeons can plan future routes in a room. Anim Cogn 15:379–391

Gould JL (2009) Animal navigation: a wake-up call for homing. Curr Biol 19:R338–R339

Hotta T, Takeyama T, Heg D, Awata S, Jordan LA, Kohda M (2015) The use of multiple sources of social information in contest behavior: testing the social cognitive abilities of a cichlid fish. Front Ecol Evol 3:85

Huber L, Gajon GK (2006) Technical intelligence in animals: the kea model. Anim Cogn 9:295–305

Iwasaki S, Watanabe S, Fujita K (2018) Pigeons (*Columba livia*) know when they will need hints: prospective metacognition for reference memory? Anim Cogn 21:207–217

Johnston M, Clarkson AN, Gowing EK, Scarf D, Colombo M (2018) Effects of nidopallium caudolaterale inactivation on serial-order behavior in pigeons (*Columba livia*). J Neurophysiol 120:1143–1152

Kabadayi C, Osvath M (2017) Ravens parallel great apes in flexible planning for tool-use and bartering. Science 357(6347):202–204

Köhler W (1925) The mentality of apes. Routledge & Kegan Paul, London

Kornell N, Son LK, Terrace HS (2007) Transfer of metacognitive skills and hint seeking in monkeys. Psychol Sci 18:64–71

McGonigle BO, Chalmers M (1977) Are monkeys logical? Nature 267:694–696

Miyata H (2014) Doubutsu no keikaku nouryoku: "Shikou" no sinka wo saguru (planning abilities in animals: searching for the evolutionary origins of "thought"). Kyoto University Press, Kyoto

Miyata H, Fujita K (2008) Pigeons (*Columba livia*) plan future moves on computerized maze tasks. Anim Cogn 11:505–516

Miyata H, Fujita K (2010) Route selection by pigeons (*Columba livia*) in "traveling salesperson" navigation tasks presented on an LCD screen. J Comp Psychol 124:433–446

Miyata H, Fujita K (2011a) Planning abilities in non-human animals: New findings in primates and birds. Jpn J Anim Psychol 61:69–82

Miyata H, Fujita K (2011b) Flexible route selection by pigeons (*Columba livia*) on a computerized multi-goal navigation task with and without an "obstacle". J Comp Psychol 125:431–435

Miyata H, Fujita K (2012) Further tests of pigeons' (*Columba livia*) planning behavior using a computerized plus-shaped maze task. Percept Mot Skills 115:27–42

Miyata H, Ushitani T, Adachi I, Fujita K (2006) Performance of pigeons (*Columba livia*) on maze problems presented on the LCD screen: In search for preplanning ability in an avian species. J Comp Psychol 120:358–366

Miyata H, Itakura S, Fujita K (2009) Planning in human children (*Homo sapiens*) assessed by maze problems on the touch screen. J Comp Psychol 123:69–78

Miyata H, Watanabe S, Minagawa-Kawai Y (2011a) Two successive neurocognitive processes captured by near-infrared spectroscopy: prefrontal activation during a computerized plus-shaped maze task. Brain Res 1374:90–99

Miyata H, Gajdon GK, Huber L, Fujita K (2011b) How do keas (*Nestor notabilis*) solve artificial-fruit problems with multiple locks? Anim Cogn 14:45–58

Miyata H, Watanabe S, Minagawa Y (2014) Performance of young children on "traveling salesperson" navigation tasks presented on a touch screen. PLoS ONE 9(12):e115292

Mulcahy NJ, Call J (2006) Apes save tools for future use. Science 312(5776):1038–1040

Mushiake H, Saito N, Sakamoto K, Itoyama Y, Tanji J (2006) Activity in the lateral prefrontal cortex reflects multiple steps of future events in action plans. Neuron 50:631–641

Osvath M, Osvath H (2008) Chimpanzee (*Pan troglodytes*) and orangutan (*Pongo abelli*) forethought: self-control and pre-experience in the face of future tool use. Anim Cogn 11:661–674

Pan J, Kennedy EH, Pickering T, Menzel CR, Stone BW, Fragaszy DM (2011) Development of maze navigation in tufted capuchins (*Cebus apella*). Behav Process 86:206–215

Paz-y-Miño GC, Bond AB, Kamil AC, Balda RP (2004) Pinyon jays use transitive inference to predict social dominance. Nature 430:778–781

Raby CR, Alexis DM, Dickinson A, Clayton NS (2007) Planning for the future by western scrub-jays. Nature 445:919–921

Roberts WA (2002) Are animals stuck in time? Psychol Bull 128:473–489

Scarf D, Colombo M (2010) The formation and execution of sequential plans in pigeons (*Columba livia*). Behav Process 83:179–182

Shallice T (1982) Specific impairments of planning. Phil Trans R Soc Lond B 298:199–209

Shima K, Isoda M, Mushiake H, Tanji J (2007) Categorization of behavioural sequences in the prefrontal cortex. Nature 445:315–318

Shimizu T (2000) Kokoro no shinka to nou no shinka (The evolution of mind and the evolution of the brain). In: Watanabe S (ed) Kokoro no hikaku-ninchi-kagaku (comparative cognitive science of the mind). Minervashobo, Kyoto, pp 27–81

Smith JD, Shields WE, Schull J, Washburn DA (1997) The uncertain response in humans and animals. Cognition 62:75–97

Striedter GF (2005) Principles of brain evolution. Sinauer Associates, Sunderland

Suddendorf T, Corballis MC (1997) Mental time travel and the evolution of the human mind. Genet Soc Gen Psychol Monogr 123:133–167

Thorndike EL (1898) Animal intelligence: an experimental study of the associative processes in animals. Psychol Monogr Gen Appl 2:1–109

Tolman EC (1948) Cognitive map in rats and men. Psychol Rev 55:189–208

van den Heuvel OA, Groenewegen HJ, Barkhof F, Lazeron RHC, van Dyck R, Veltman DJ (2003) Frontostriatal system in planning complexity: a parametric functional magnetic resonance version of tower of London task. NeuroImage 18:367–374

Visalberghi E, Limongelli L (1994) Lack of comprehension of cause-effect relations in tool-using capuchin monkeys (*Cebus apella*). J Comp Psychol 108:15–22

von Fersen L, Wynne CD, Delius JD, Staddon JE (1991) Transitive inference formation in pigeons. J Exp Psychol Anim Behav Process 17:334–341

Watanabe S, Kojima S (2007) No-kagaku to kokoro no shinka (Brain science and the evolution of mind). Iwanamishoten, Tokyo

Yamamoto Y, Kubo M (1997) Junkai seirusuman mondai he no shotai (an introduction to the traveling salesperson problem). Asakurashoten, Tokyo

Chapter 15
Studies of Prospective Information-Seeking in Capuchin Monkeys, Pigeons, and Human Children

Sumie Iwasaki and Reiki Kishimoto

Abstract Metacognition allows us to monitor our own cognitive states and thereby better control our behavior. Comparative psychologists have developed several non-verbal procedures to examine metacognition in non-human species. Among these, information-seeking paradigms appear particularly naturalistic, and they can be adapted to various experimental settings (e.g., face-to-face contexts, or operant experiments). In this chapter, we describe three of our three experiments of prospective information-seeking. First, we introduce a study that uses a reference memory task in pigeons. Although avian species have shown weak evidence of metacognition in a delayed matching to sample test, some of our pigeons prospectively sought information during the reference memory task, which required fewer working memory resources. This finding indicated the importance of using cognitive tasks that are suitable for the subject species. Second, we describe studies of human children that used non-verbal procedures, as in studies of metacognition in animals. These kinds of developmental studies may also contribute to understanding the evolution of metacognition. Finally, we introduce a non-verbal information-seeking procedure to examine preparatory behavior for potential forgetting. Our studies of capuchin monkeys revealed no compelling evidence of preparatory information-seeking; however, the procedure appears promising for studying other species too. As preparatory information-seeking is related to future-oriented cognition, our procedure might help to shed new light on metacognition across a range of non-human animals.

Keywords Metacognition · Information-seeking · Pigeons · Human children · Capuchin monkeys

S. Iwasaki (✉)
Institute of Human and Social Sciences, Kanazawa University, Kanazawa, Japan

R. Kishimoto
Graduate School of Arts and Sciences, University of Tokyo, Tokyo, Japan

J. R. Anderson, H. Kuroshima (eds.), *Comparative Cognition*,
https://doi.org/10.1007/978-981-16-2028-7_15

Prospective Information-Seeking as a Non-Verbal Metacognitive Procedure

Metacognition is a cognitive function that allows us to monitor our own cognitive states and thereby better control our behavior (Nelson 1996). It was long thought that verbal reports were necessary for metacognition studies, and therefore it was assumed to be difficult to demonstrate this cognitive function in non-human animals (Smith et al. 1997, 1998). However, comparative psychologists have developed several non-verbal procedures to examine metacognition in non-human species, and research has revealed that metacognitive abilities might not be unique to humans. These procedures include escape response paradigms, risk–safe reward paradigms, and information-seeking paradigms, which we review below.

In the typical escape response paradigm, subjects are given an option to escape a test before or during a perceptual discrimination task or memory task. Subjects receive a high-value reward for a correct response (therefore no escape response), whereas an incorrect response leads to a timeout period and no reward. Choosing the escape option always results in a medium-value reward or avoiding the timeout. In these situations, monkeys, rats, and dolphins have been shown to escape difficult trials, but they accept easy trials (e.g., Foote and Crystal 2007; Hampton 2001; Smith et al. 1995).

In the risk–safe reward paradigm, subjects have to choose between "risk" and "safe" options after responding to a task. The "risk" option leads to a high-value reward after a correct response, but no reward after an incorrect response. By contrast, the "safe" option delivers a small reward regardless of whether the response was correct or incorrect. Monkeys and birds choose the risk option more frequently after easy discrimination tasks than difficult ones (Goto and Watanabe 2012; Kornell et al. 2007; Nakamura et al. 2011).

Compared to the escape response and risk–safe reward paradigms, information-seeking paradigms appear more naturalistic because they require simpler set-ups and less preliminary training of subjects (Call 2012). In a pioneering information-seeking paradigm study by Call and Carpenter (2000), an experimenter hid a sticker in one of the three tubes in front of individually tested subjects and allowed them to choose a tube. Chimpanzees looked into the tubes more often when they had not seen the hiding event than when they had seen it; in other words, they sought information when they did not know the correct answer. Information-seeking paradigms can be applied not only in face-to-face contexts but also in operant experiments, and results of such studies have suggested metacognition in various species, albeit with individual differences (chimpanzees: Call and Carpenter 2000; orangutans: Suda-King 2008; macaques: Marsh 2014; rhesus monkeys: Beran and Smith 2011; Brown et al. 2019; Hampton et al. 2004; Kornell et al. 2007; capuchin monkeys: Basile et al. 2009; Vining and Marsh 2015; dogs: Belger and Bräuer 2018; rats: Foote and Crystal 2012, ravens: Lambert and Osvath 2020; scrub-jays: Watanabe and Clayton 2016; Watanabe et al. 2014; pigeons: Castro and Wasserman 2012; Iwasaki et al. 2013).

Although non-verbal metacognitive studies have provided evidence of metacognition in non-human animals, the studies have come in for criticism, including the claim that metacognitive-like behavior could emerge from lower-level processes (e.g., associative learning) or external cues, rather than metacognition (e.g., Jozefowiez et al. 2009; Le Pelley 2012; Smith et al. 2008). Hampton (2009, 2019) proposed that metacognition depends on "private" information which only a subject can use, and that this information is different from observable, "public" information. Public information includes stimulus characteristics such as similarity and ambiguity and a long response latency. Another possible reason for metacognitive-like behaviors is response competition, when the escape option was given simultaneously with a test. In difficult trials in which subjects take longer to respond, this prolonged time might increase the likelihood of an escape response. To reduce the possibility of using publicly available information, Hampton (2009) advocated using prospective tasks in which metacognitive responses are available before the test materials are presented. Based on Hampton's proposal, our laboratory investigated prospective (or preparatory) information-seeking abilities in diverse species, namely pigeons, human children, and capuchin monkeys.

Pigeons Show Prospective Information-Seeking During a Reference Memory Task

Monkeys and rats have shown prospective metacognition (Fujita 2009; Hampton 2001; Templer et al. 2017; Yuki and Okanoya 2017); they prospectively avoided taking a memory matching test when their memory traces were weak, but took the test when their memory traces were strong. In contrast to those mammalian species, birds (pigeons and crows) have shown weaker evidence of prospective metacognition in the delayed matching-to-sample (DMTS) task (Goto and Watanabe 2012; Inman and Shettleworth 1999; Sutton and Shettleworth 2008). As discussed by Goto and Watanabe (2012), two cues could be simultaneously available immediately before the prospective metacognitive judgment, and a delay interval cue might have been more salient and effective in controlling crows' behavior than an introspective cue. However, there is another possibility, which is that the DMTS task might not be suitable for birds due to its large working memory requirements. During this task, samples change on every trial, so that the subjects are required to remember a sample while avoiding proactive interference from previous trials; subjects must use some working memory resources to solve the task. Requiring too much cognitive load can have a considerably negative impact on metacognitive judgment (Hampton 2019; Smith et al. 2013).

However, some studies have demonstrated metacognitive-like responses in pigeons during tasks requiring less working memory capacity. Nakamura et al. (2011) reported that pigeons could make appropriate retrospective confidence judgments about their decisions in perceptual tasks. Castro and Wasserman (2012) used a

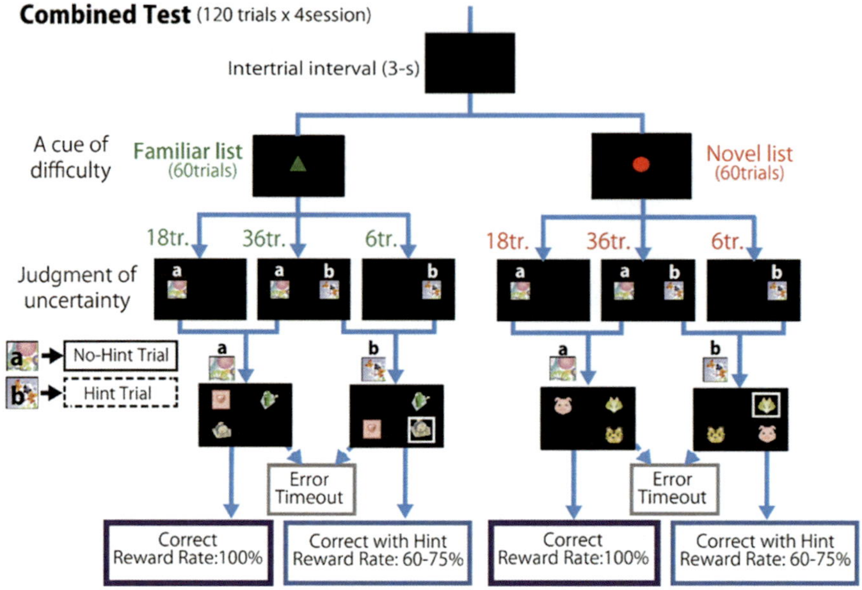

Fig. 15.1 A schematic representation of a combined test in Iwasaki et al. (2018)

concurrent metacognition task overlain on a homogeneity discrimination task and showed that pigeons often opted to change difficult trials into easy ones. Iwasaki et al. (2013) used an information-seeking paradigm in a reference memory task (a serial learning task). In this task, subjects pecked three arbitrary photographs in a predetermined correct order and were sometimes given a hint icon. A peck on the hint icon produced a white frame on an item to be pecked next. This memory task requires fewer working memory resources than a DMTS task, because the same stimuli are presented in all trials. The results showed that some pigeons sought hints more often when their accuracy on the serial learning task was low than when it was high. However, those metacognitive-like responses leave open the possibility that the birds used public information (e.g., response competition). To further explore metacognition in pigeons, in a subsequent study we investigated prospective information-seeking during a serial learning task.

In Iwasaki et al. (2018), pigeons were first trained to discriminate "familiar" and "novel" lists signaled by different colored self-start icons. After the discrimination training, the subjects were familiarized with hint-available and hint-unavailable trials. When pigeons responded to a "hint-unavailable trial" icon after pecking a neutral self-start icon, they had to solve normal three-item serial learning tasks. However, the hint-available trial icon led to a hint in the form of a shape that visibly framed the correct item in the proper sequence. After training, subjects took a test that combined a hint option with "familiar" and "novel" lists (Fig. 15.1). At the start of the test, the self-start icon signaling either "familiar" or "novel" lists was presented. After pigeons pecked the self-start icon, an icon (hint-available or hint-unavailable icon) was presented. The subjects had to peck the icon, which then led to a serial learning task. All correct serial responses without hint were reinforced by food, but only 60% or 75% of correct hint trials were reinforced. We analyzed whether the frequency of choosing a hint-available icon differed between the highly familiar and novel lists. The results showed that two of the four pigeons chose the hint trial significantly more often before receiving a novel list than the familiar list. Moreover, one bird did this reliably even in the earliest test sessions. Our procedure allowed us to reject the possibility of response competition or length of delay until the task, because the pigeons decided to seek a hint or not before the list items appeared; this eliminates the possibility that pigeons used public cues, as proposed by Hampton (2009). Also, the pigeons were not explicitly trained to go for a "hint trial" after a cue of novel lists; in our training phase, metacognitive judgment icons never followed self-start icons that could serve as discriminative cues for the type of lists ("familiar" or "novel" lists). Therefore, our study suggested that the pigeons could control their hint-seeking behavior according to their reference memory or knowledge states; this finding illustrates the importance of applying suitable cognitive tasks for the subject species (Iwasaki et al. 2019).

Non-Verbal Metacognitive Procedures in Human Developmental Studies

Non-verbal metacognitive procedures can be advantageous for studying infants and young children who have poor verbal skills. Call and Carpenter (2000) analyzed infants' looking patterns during an information-seeking task and found that they looked into the tubes more often when they had not seen the hiding event than when they had seen it. Experiments using escape tasks or information-seeking tasks have shown that even young infants can monitor their perception, memory, or knowledge state (e.g., Balcomb and Gerken 2008; Goupil and Kouider 2016; Lyons and Ghetti 2011). However, as with some animal studies, "public information" accounts that make no reference to metacognition may be able to explain the results.

Using an explicitly non-verbal paradigm, Iwasaki et al. (2020) asked whether children would prospectively seek information about upcoming events. In the experiment, human children (4- and 5-year-olds) had to find a sticker located under an opaque or transparent cup (see, Fig. 15.2). At the beginning of a trial, an experimenter informed the children about which type of cup would cover a sticker. While the experimenter hid a sticker under a cup in a "hiding room," the child sat waiting on a sofa in a waiting room. An opaque partition separating the two rooms had a small peephole through which the child could peek while the experimenter hid the sticker. A sticker under a transparent cup could be found easily without peeking, but the sticker under an opaque cup could not be seen directly, and so in the opaque

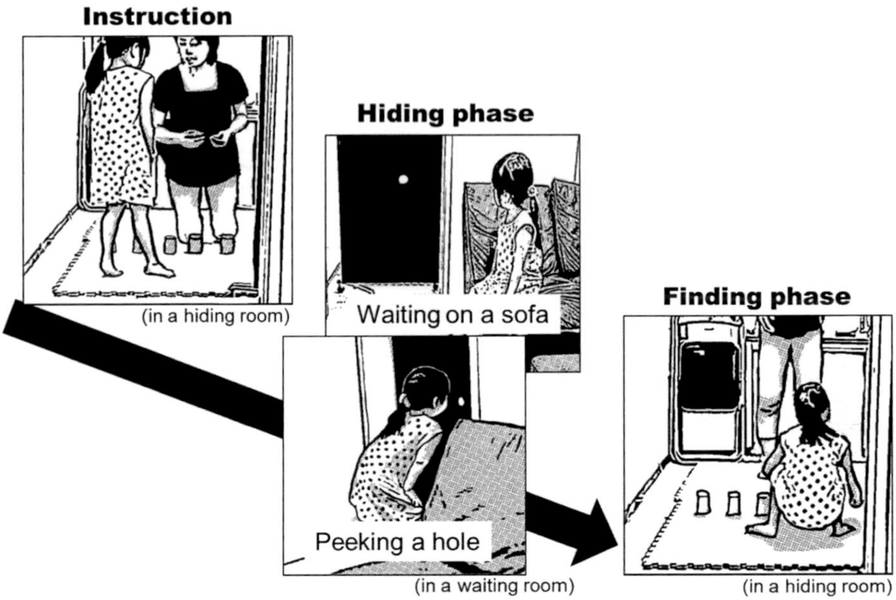

Fig. 15.2 A flow diagram of experiment 1 in Iwasaki et al. (2020)

cups condition it would be advantageous for children to seek information prospectively. The result showed that 5-year-olds peeked for longer in the opaque than the transparent condition, but 4-year-olds did not; this suggested that 5-year-olds know when they will need information.

However, it was still unclear whether 5-year-olds could recognize precisely "what" they would need. So we next examined whether children would know exactly what information they should collect for solving a task. Children (5- and 6-year-olds) had to find stickers in two rooms; in one room an actor hid a sticker under one of the five opaque cups, whereas in the other room another actor placed a sticker under one of the five transparent cups. Children could observe what the actors were doing via a monitor in another room, and they then chose a room in which to search. After the actors finished hiding stickers, children were allowed to enter the two rooms and guess where the sticker was. We analyzed their observing behavior during the hiding phase and found that children in both age groups watched events in the opaque cups room longer than the transparent cups room in the first trial. Together, these results suggest that at least 5-year-olds can prospectively collect appropriate information for an upcoming event.

The procedure used in our second experiment above is similar to that used by Watanabe et al. (2014), in which jays were required to find foods in two compartments: the free and the forced compartments. The free compartment had four potential hiding cups so that peeking was a worthwhile behavior, whereas the forced compartment had only one, open cup, so that the subjects could find rewards without peeking. While an experimenter hid rewards simultaneously in the two compartments, the subjects could look through a peephole to obtain information where the rewards were. After the hiding phase, subjects were allowed to recover the food in both compartments. It was found that the scrub-jays appropriately allocated longer observation time to the free compartment than the forced compartment, a result similar to ours with human children. Despite some difference in training or instructions between the studies in human children and scrub-jays, both focused on the type of situation in which peeking would occur, and for how long it occurred. Non-verbal behavioral tests developed in comparative psychology can be applied to developmental studies in human.

Developmental metacognitive studies may provide some insights into the evolution of metacognition. A developmental psychologist, Perner (2012) proposed that sensitivity to arousal level (e.g., anxiety) precedes the emergence of explicit metacognition. Fletcher and Carruthers (2012) pointed out that metacognition about one's own reasoning or decision-making is a late-developing skill in humans, and there may be variability in the ability. These views from a development perspective appear to support the possibility of a variety of metacognitive processes in non-human animals; sophisticated metacognitive abilities might have evolved from simpler precursors.

Capuchin Monkeys and Information-Seeking in the Context of Potential Forgetting

Until recently, most studies of metacognition in non-human animals focused on current knowledge or memory (e.g., Iwasaki et al. 2013; Brown et al. 2019). However, internal states often change as time progresses. For example, longer passages of time lead to memories being forgotten (Blough 1959; Hunter 1913) and access to some long-term memories becomes more difficult (Godden and Baddeley 1975). Because human adults are aware of this aspect of their memories, they often use reminder devices or notes prospectively to guard against the potential loss of important information. Although such preparatory behavior would also be advantageous for other species, it is unknown whether non-human animals are aware that they might forget and adjust their behavior accordingly.

Previous studies found that non-human animals controlled their mnemonic strategy based on cognitive load required for a task (e.g., Kishimoto et al. 2020; Wasserman et al. 1982). To examine preparatory behavior for potential forgetting, we designed a non-verbal information-seeking procedure (Kishimoto et al, 2020). We modified a procedure that requires subjects to monitor their current memory states (Roberts et al. 2009; Beran et al. 2011). In those studies, subjects performed a matching-to-sample (MTS) task and were given the option to see a sample or not. The sample was sometimes omitted and then subjects took a matching test without a sample. If they recognized their lack of knowledge (their current knowledge state), they should request to see the sample when it was omitted. Rhesus monkeys were found to make such a request, but neither capuchin monkeys nor pigeons did so. For our investigation of preparatory information-seeking against potential forgetting, we applied a DMTS and gave subjects the option to review a sample or not at beginning of the delay (in other words, immediately after the sample). In this task, subjects were required to make a choice to seek the review or not before their memory trace became weaker. The option was presented simultaneously with a "delay cue" which announced a length of upcoming delay. In trials of an announced long delay, subjects could remember a sample if they decided to review it, but if they waited until after the delay they were most likely to forget it. If subjects prospectively seek review to counter potential forgetting, they should request to review more frequently in trials that signaled long delay than short delay.

By using this procedure, we assessed whether capuchin monkeys would opt to re-encode a sample when their memory was likely to be forgotten. Capuchin monkeys have shown mixed results in metacognition studies. As stated above, in earlier work capuchin monkeys did not show optimal information-seeking behavior (Beran et al. 2011; Paukner et al. 2006), despite other evidence for metacognition in this species (Beran et al. 2016; Vining and Marsh 2015). Fujita (2009) suggested that one capuchin monkey showed metacognition during a DMTS task, but Takagi and Fujita (2018) reported failure to monitor detailed memory contents in the monkeys. Takagi and Fujita (2018) attributed the failure to cognitive resource depletion due to the complex experimental design rather than to a lack of metacognition; this species should be tested in more varied settings to clarify their metacognitive abilities. It is worth applying the new information-seeking paradigm to capuchin monkeys.

Before testing, we trained three capuchin monkeys on DMTS tasks with delay cues of 2 s or 16 s. Either a *Review* or a *Non-Review icon* was also presented immediately after a sample, which the monkey responded to within 2 s. When monkeys responded to the *Review icon*, a review sample (re-presenting the original sample) was presented for 1 s or 1.5 s (depending on subjects) immediately before the matching test. In contrast, if they responded to the *Non-Review icon*, they proceeded to the test without a re-presented sample; i.e., they saw a 1-s blank. After this training, subjects took the following test (see, Fig. 15.3): At the start of a test trial, each monkey was required to touch on a sample five times. After this, two icons (*Review* and *Non-Review icon*) and a delay cue were presented simultaneously. Monkeys had to choose the *Review* or the *Non-Review icon* within 2 s. In the matching test, soon after the monkey touched any stimulus, all stimuli disappeared. All correct trials where the *Non-Review icon* had been selected resulted in the monkey receiving a high-value reward, but incorrect responses resulted in a 5-s timeout with a 0.5-s buzzer sound. The *Review* meant that the monkey could perform correctly in the matching test even if it forgot its pre-encoded memory; therefore, correct responses following *Review* icon selections were rewarded at the reduced rate of 60%. If monkeys were able to seek information prospectively to counter potential forgetting, they should seek *Review* more often in trials announcing a longer delay (16 s) than a shorter delay (2 s).

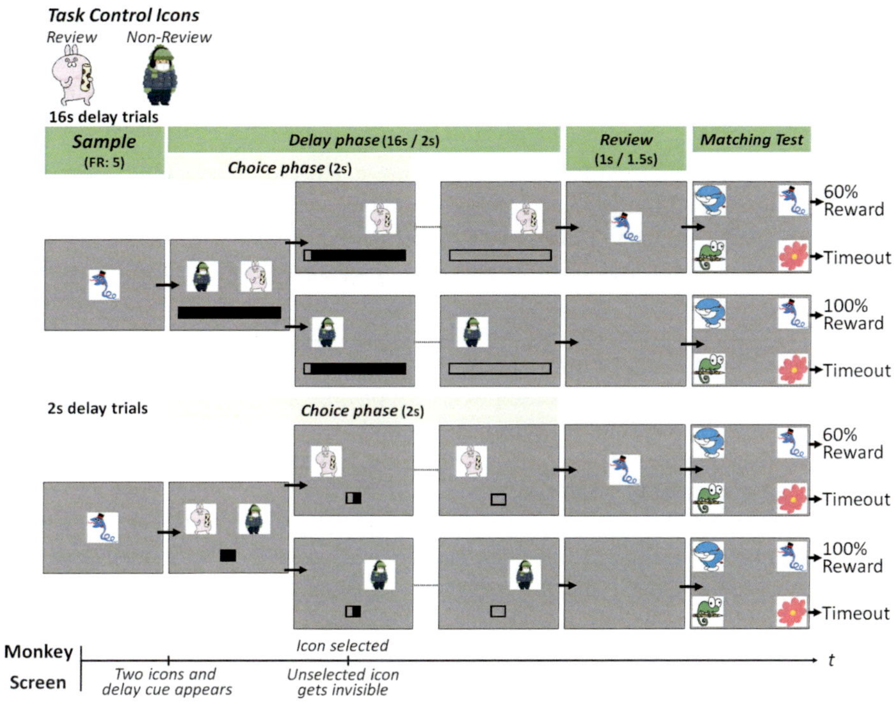

Fig. 15.3 Schematic representation of the experimental sequence. The timeline is not accurate to scale. The length of the delay phase changed across trials and accordingly, and the length of the delay signal was adjusted. To control the response bias for images used as icons, the function of each image was reversed across monkeys. From Kishimoto et al. (2020) with some modifications

We compared the frequencies of seeking *Review* in trials announcing the longer and shorter delays. Results showed that two of the three monkeys more frequently sought *Review* when the longer delay was signaled. This trend might indicate prospective information-seeking to counter forgetting in capuchin monkeys. However, the behavior could be explained as emerging from "public" or external cues (i.e., simply discriminating the length of time bar) (Hampton 2009; Jozefowiez et al. 2009; Le Pelley 2012), rather than anticipation of potential forgetting.

To further explore whether the monkeys' *Review*-seeking depended on "private" information or "public" cues, we tested the monkeys with a sample-omission procedure previously used by Hampton (2001). In a sample-omission trial, the monkey had no knowledge of any sample because none was presented. If monkeys sought *Review* based on private information or monitoring their memory state, they should seek *Review* in sample-omission trials even when a shorter delay was announced. Conversely, if monkeys sought *Review* simply by discriminating delay cue length or some other external cue, they should seek *Review* especially in trials announcing a longer delay, regardless of whether a sample was presented or not. Our results showed that no monkey sought *Review* more frequently in sample-omission trials than in sample-presentation trials, and that one monkey sought *Review* significantly more often in longer-delayed trials regardless of sample presentation. The latter monkey's *Review*-seeking behavior might therefore reflect simple discrimination of delay cue length.

In subsequent tests, samples were always presented but "distraction" responses were sometimes required before the choice of *Review* decision. These distraction responses impaired monkeys' memory trace. If monkeys sought *Review* depending on their memory state, they should seek *Review* in distraction trials even when a shorter delay was announced. Again, no monkey showed more *Review*-seeking in distraction trials. In summary, our experiments failed to show that capuchin monkeys engaged in preparatory information-seeking via metacognitive cues.

Although we found no compelling evidence of preparatory information-seeking to counter potential forgetting in capuchin monkeys, we call for caution before concluding that they lack the requisite cognitive abilities. We need further studies that use tasks with reduced cognitive load (Iwasaki et al. 2019) and carefully controlled manipulation of task difficulty and reward rate (Beran et al. 2016). Moreover, studies similar to ours described above should be extended to great apes and rhesus monkeys, species for which stronger evidence of metacognition exists. The field is a challenging one, but it can help shed new light on metacognition in non-human animals.

Future Perspectives

We hope that our laboratory studies of prospective information-seeking might lead to stronger studies of metacognitive processes in the future. In comparative psychology it is important that cognitive tasks are suitable for the subject species; otherwise, we run the risk reporting false negatives. Furthermore, non-verbal procedures used for exploring metacognition in animals can also be adapted for studies of human children. Developmental studies can also feed into the endeavor to elucidate the evolutionary processes underlying metacognition. Finally, further efforts should be made to develop new procedures to explore various forms of metacognition. Animal studies have been largely focused on monitoring of "current" knowledge or memory states, but metacognition can also play a role in planning ahead (see Miyata, this volume). Kishimoto et al. (2020) designed a procedure to investigate preparatory information-seeking in relation to future-oriented cognition. Additionally, human adults are aware not only of our own knowledge or memory state, but also our emotional state. Do other species monitor their own emotional states? To investigate this question we need to develop new methods (Fujita 2010), taking the study of animal metacognition into new areas.

References

Balcomb FK, Gerken L (2008) Three-year-old children can access their own memory to guide responses on a visual matching task. Dev Sci 11:750–760. https://doi.org/10.1111/j.1467-7687.2008.00725.x

Basile BM, Hampton RR, Suomi SJ, Murray EA (2009) An assessment of memory awareness in tufted capuchin monkeys (*Cebus apella*). Anim Cogn 12:169–180. https://doi.org/10.1007/s10071-008-0180-1

Belger J, Bräuer J (2018) Metacognition in dogs: Do dogs know they could be wrong? Learn Behav 46:398–413. https://doi.org/10.3758/s13420-018-0367-5

Beran MJ, Perdue BM, Church BA, Smith JD (2016) Capuchin monkeys (*Cebus apella*) modulate their use of an uncertainty response depending on risk. J Exp Psychol Anim Learn Cogn 42:32–43. https://doi.org/10.1037/xan0000080

Beran MJ, Smith JD (2011) Information seeking by rhesus monkeys (*Macaca mulatta*) and capuchin monkeys (*Cebus apella*). Cognition 120:90–105. https://doi.org/10.1016/j.cognition.2011.02.016

Blough DS (1959) Delayed matching in the pigeon. J Exp Anal Behav 2:151–160. https://doi.org/10.1901/jeab.1959.2-151

Brown EK, Basile BM, Templer VL, Hampton RR (2019) Dissociation of memory signals for metamemory in rhesus monkeys (*Macaca mulatta*). Anim Cogn 22:331–341. https://doi.org/10.1007/s10071-019-01246-5

Call J (2012) Seeking Information in non-human animals: weaving a metacognitiove web. In: Beran MJ, Brandl JL, Perner J, Proust J (eds) *Foundations of metacognition*. Oxford University Press, Oxford, pp 62–75

Call J, Carpenter M (2000) Do apes and children know what they have seen? Anim Cogn 3:207–220. https://doi.org/10.1007/s100710100078

Castro L, Wasserman EA (2012) Information-seeking behavior: exploring metacognitive control in pigeons. Anim Cogn 16:241–254. https://doi.org/10.1007/s10071-012-0569-8

Fletcher L, Carruthers P (2012) Metacognition and reasoning. Phil Trans R Soc B 367:1366–1378. https://doi.org/10.1098/rstb.2011.0413

Foote AL, Crystal JD (2007) Metacognition in the rat. Curr Biol 17:551–555. https://doi.org/10.1016/j.cub.2007.01.061

Foote AL, Crystal JD (2012) "Play it again": a new method for testing metacognition in animals. Anim Cogn 15:187–199. https://doi.org/10.1007/s10071-011-0445-y

Fujita K (2009) Metamemory in tufted capuchin monkeys (*Cebus apella*). Anim Cogn 12:575–585. https://doi.org/10.1007/s10071-009-0217-0

Fujita K (2010) Comparative studies of metacognition. Jpn Psychol Rev 53:270–294. https://doi.org/10.24602/sjpr.53.3_270

Godden DR, Baddeley AD (1975) Context-dependent memory in two natural environments: On land and underwater. Br J Psychol 66:325–331. https://doi.org/10.1111/j.2044-8295.1975.tb01468.x

Goto K, Watanabe S (2012) Large-billed crows (*Corvus macrorhynchos*) have retrospective but not prospective metamemory. Anim Cogn 15:27–35. https://doi.org/10.1007/s10071-011-0428-z

Goupil L, Kouider S (2016) Behavioral and neural indices of metacognitive sensitivity in preverbal infants. Curr Biol 26:3038–3045. https://doi.org/10.1016/j.cub.2016.09.004

Hampton RR (2001) Rhesus monkeys know when they remember. Proc Natl Acad Sci USA 98:5359–5362. https://doi.org/10.1073/pnas.071600998

Hampton RR (2009) Multiple demonstrations of metacognition in nonhumans: converging evidence or multiple mechanisms? Comp Cogn Behav Rev 4:17–28. https://doi.org/10.3819/ccbr.2009.40002

Hampton RR (2019) Monkey metacognition could generate more insight. Anim Behav Cogn 6:230–235. https://doi.org/10.26451/abc.06.04.02.2019

Hampton RR, Zivin A, Murray EA (2004) Rhesus monkeys (*Macaca mulatta*) discriminate between knowing and not knowing and collect information as needed before acting. Anim Cogn 7:239–246. https://doi.org/10.1007/s10071-004-0215-1

Hunter WS (1913) The delayed reaction in animals and children. Behav Monogr 2(1): Serial No. 6

Inman A, Shettleworth SJ (1999) Detecting metamemory in nonverbal subjects: a test with pigeons. J Exp Psychol Anim Behav Process 25:389–395. https://doi.org/10.1037/0097-7403.25.3.389

Iwasaki S, Kuroshima H, Arahori M, Fujita K (2020) Prospective information-seeking in human children (*Homo sapiens*): when to seek, and what to seek. J Comp Psychol 134:341–348. https://doi.org/10.1037/com0000217

Iwasaki S, Kuroshima H, Fujita K (2019) Pigeons show metamemory by requesting reduced working memory loads. Anim Behav Cogn 6:247–253. https://doi.org/10.26451/abc.06.04.04.2019

Iwasaki S, Watanabe S, Fujita K (2013) Do pigeons (*Columba livia*) seek information when they have insufficient knowledge? Anim Cogn 16:211–221. https://doi.org/10.1007/s10071-012-0566-y

Iwasaki S, Watanabe S, Fujita K (2018) Pigeons (*Columba livia*) know when they will need hints: prospective metacognition for reference memory? Anim Cogn 21:207–217. https://doi.org/10.1007/s10071-017-1153-z

Jozefowiez J, Staddon JER, Cerutti DT (2009) Metacognition in animals: how do we know that they know? Comp Cogn Behav Rev 4:29–39. https://doi.org/10.3819/ccbr.2009.40003

Kishimoto R, Iwasaki S, Fujita K (2020) Do capuchins (*Sapajus apella*) know how well they will remember? Analysis of delay length-dependency with memory strategies. J Comp Psychol 133:340–350. https://doi.org/10.1037/com0000164

Kishimoto R, Iwasaki S, Fujita K (2020) Capuchin monkeys (*Sapajus apella*) failed to seek information for their potential forgetting in a computerized task. Primates 61:623–632. https://doi.org/10.1007/s10329-020-00804-7

Kornell N, Son L, Terrace H (2007) Transfer of metacognitive skills and hint seeking in monkeys. Psychol Sci 18:64–71. https://doi.org/10.1111/j.1467-9280.2007.01850.x

Lambert ML, Osvath M (2020) Investigating information seeking in ravens (*Corvus corax*). Anim Cogn 23:671–680. https://doi.org/10.1007/s10071-020-01372-5

Le Pelley ME (2012) Metacognitive monkeys or associative animals? Simple reinforcement learning explains uncertainty in nonhuman animals. J Exp Psychol Learn Mem Cogn 38:686–708. https://doi.org/10.1037/a0026478

Lyons KE, Ghetti S (2011) The development of uncertainty monitoring in early childhood. Child Dev 82:1778–1787. https://doi.org/10.1111/j.1467-8624.2011.01649.x

Marsh HL (2014) Metacognitive-like information seeking in lion-tailed macaques: a generalized search response after all? Anim Cogn 17:1313–1328. https://doi.org/10.1007/s10071-014-0767-7

Nakamura N, Watanabe S, Betsuyaku T, Fujita K (2011) Do birds (pigeons and bantams) know how confident they are of their perceptual decisions? Anim Cogn 14:83–93. https://doi.org/10.1007/s10071-010-0345-6

Nelson TO (1996) Consciousness and metacognition. Am Psychol 51:102–116. https://doi.org/10.1037/0003-066X.51.2.102

Paukner A, Anderson JR, Fujita K (2006) Redundant food searches by capuchin monkeys (*Cebus apella*): a failure of metacognition? Anim Cogn 9:110–117. https://doi.org/10.1007/s10071-005-0007-2

Perner J (2012) MiniMeta: in search of minimal criteria for metacognition. In Beran MJ, Brandl J, Perner J, Proust J (eds) Foundations of metacognition. Oxford University Press, pp 94–116. https://dx.doi.org/10.1093/acprof:oso/9780199646739.003.0007

Roberts WA, Feeney MC, McMillan N, MacPherson K, Musolino E, Petter M (2009) Do pigeons (*Columba livia*) study for a test? J Exp Psychol 35:129–142. https://doi.org/10.1037/a0013722

Smith JD, Beran MJ, Couchman JJ, Coutinho MVC (2008) The comparative study of metacognition: Sharper paradigms, safer inferences. Psychon Bull Rev 15:679–691. https://doi.org/10.3758/PBR.15.4.679

Smith JD, Coutinho MVC, Church BA, Beran MJ (2013) Executive-attentional uncertainty responses by rhesus macaques (*Macaca mulatta*). J Exp Psychol Gen 142:458–475. https://doi.org/10.1037/a0029601

Smith JD, Schull J, Strote J, McGee K, Egnor R, Erb L (1995) The uncertain response in the bottlenosed dolphin (*Tursiops truncatus*). J Exp Psychol Gen 124:391–408. https://doi.org/10.1037/0096-3445.124.4.391

Smith JD, Shields WE, Allendoerfer KR, Washburn DA (1998) Memory monitoring by animals and humans. J Exp Psychol Gen 127:227–250. https://doi.org/10.1037/0096-3445.127.3.227

Smith JD, Shields WE, Schull J, Washburn DA (1997) The uncertain response in humans and animals. Cognition 62:75–97. https://doi.org/10.1016/S0010-0277(96)00726-3

Suda-King C (2008) Do orangutans (*Pongo pygmaeus*) know when they do not remember? Anim Cogn 11:21–42. https://doi.org/10.1007/s10071-007-0082-7

Sutton JE, Shettleworth SJ (2008) Memory without awareness: pigeons do not show metamemory in delayed matching to sample. J Exp Psychol Anim Behav Process 34:266–282. https://doi.org/10.1037/0097-7403.34.2.266

Takagi S, Fujita K (2018) Do capuchin monkeys (*Sapajus apella*) know the contents of memory traces? : a study of metamemory for compound stimuli. J Comp Psychol 132:88–96. https://doi.org/10.1037/com0000087

Templer VL, Lee KA, Preston AJ (2017) Rats know when they remember: transfer of metacognitive responding across odor-based delayed match-to-sample tests. Anim Cogn 20:891–906. https://doi.org/10.1007/s10071-017-1109-3

Vining AQ, Marsh HL (2015) Information seeking in capuchins (*Cebus apella*): a rudimentary form of metacognition? Anim Cogn 18:667–681. https://doi.org/10.1007/s10071-015-0835-7

Wasserman EA, Grosch J, Nevin JA (1982) Effects of signaled retention intervals on pigeon short-term memory. Anim Learn Behav 10:330–338. https://doi.org/10.3758/BF03213719

Watanabe A, Clayton NS (2016) Hint-seeking behaviour of western scrub-jays in a metacognition task. Anim Cogn 19:53–64. https://doi.org/10.1007/s10071-015-0912-y

Watanabe A, Grodzinski U, Clayton NS (2014) Western scrub-jays allocate longer observation time to more valuable information. Anim Cogn 17:859–867. https://doi.org/10.1007/s10071-013-0719-7

Yuki S, Okanoya K (2017) Rats show adaptive choice in a metacognitive task with high uncertainty. J Exp Psychol Anim Learn Cogn 43:109–118. https://doi.org/10.1037/xan0000130

Chapter 16
Worth the Wait: Evidence for Self-Control in Nonhuman Primates

Michael J. Beran

Abstract Self-control is defined as the ability or capacity to obtain a subjectively more valuable outcome rather than a subjectively less valuable outcome through choosing and then tolerating a longer delay or a greater effort requirement for obtaining that more valuable outcome. In this chapter, approaches to studying self-control (and impulsivity) in nonhuman primates are reviewed. These include tests that present intertemporal choices of smaller-sooner versus larger later-rewards, tests that assess delay of gratification, and other tasks that have been developed to measure self-control across primate species. The general performances of different primate species are described from tasks designed and used by Professor Kazuo Fujita and others for determining what affects whether nonhuman primates wait for better rewards or take more immediate rewards. The relation of these effects to what we know about human self-control also is explored.

Keywords Self-control · Delay of gratification · Nonhuman primates · Impulsivity · Choice behavior

Introduction

When I agreed to write this chapter, I felt that I had all the time I needed and would surely turn in the first draft well ahead of the proposed schedule. Instead, it is now a week late. And, this evening as I finished the last bit of writing, I had every intention of leaving the last of the holiday candy alone so that I might get back to healthy eating, and yet if you were here to see me, you would see candy wrappers all on my desk. So, what happened? Why did my best of intentions come to be failures? For the simple reason that I lacked the *self-control* needed to avoid immediate distractions (and stick to my writing) and avoid the more immediately available and appealing

M. J. Beran (✉)
Language Research Center, Georgia State University, Atlanta, GA, USA
e-mail: mberan1@gsu.edu

J. R. Anderson, H. Kuroshima (eds.), *Comparative Cognition*,
https://doi.org/10.1007/978-981-16-2028-7_16

treats so that I could have the more distant, healthier self. What is self-control? There are different definitions, but I have defined it *as the ability or capacity to obtain a subjectively more valuable outcome rather than a subjectively less valuable outcome through choosing and then tolerating a longer delay or a greater effort requirement for obtaining that more valuable outcome* (Beran 2018).

We are all faced with these kinds of choices, often many times per day. Choices between two or more outcomes, some of which are more immediately rewarding, but also which are not in our longer-term interests. And, we know that we also struggle to choose the option that leads to a more delayed, but objectively better outcome. These are sometimes called *intertemporal choices*. When we choose to wait for the more delayed, and better outcome, this is called a *self-control* choice. When we choose the sooner item this is called an *impulsive* response (Logue 1988; Madden and Bickel 2010; Stevens 2017). For scientists, educators, employers, and policy makers, these are some of the most important choices to be understood. Choosing impulsively can lead to devastating consequences such as problematic gambling, alcoholism, aggression, academic failures, risky sexual behavior, and even chronic tardiness and fraudulent behavior (e.g., Rachlin 2000). For this reason, comparative research into self-control has a long history, and has encompassed a wide variety of species. In this chapter, I focus on work with nonhuman primates for two reasons: (1) because they have been tested in most of the experimental paradigms that have been developed for studying self-control and (2) because this volume celebrates the sustained, high impact research career of Professor Kazuo Fujita, whose seminal work on primate behavior and cognition includes important studies on primate self-control. For those interested in a longer and broader discussion of self-control in animals (and people), please see Beran (2018).

Intertemporal Choice Tasks

Intertemporal choice tasks have been used in psychological research with humans for decades (e.g., Green and Myerson 1993, 2004; Koriat and Nisan 1978; Mazur 1987). This work has demonstrated the phenomenon of *delay discounting*, which is the process of devaluing future outcomes relative to more immediate outcomes (Madden and Bickel 2010). Intertemporal choice tasks have been used even with young children, using appetitive rewards or non-food rewards (e.g., Imuta et al. 2014; Lemmon and Moore 2007; Schwarz et al. 1983). And, intertemporal choice tasks have been used with nonhuman species. Pigeons often are reported to choose impulsively during intertemporal choice tasks (e.g., Logue and Peña-Correal 1984, 1985) although the delays to both rewards and the difference in preference levels between the sooner option and the later option can generate some self-controlled choices (e.g., Ainslie and Herrnstein 1981; King and Logue 1990). Rats also show variability in their choices, depending on the nature of the comparison offered. In some cases, they may show more self-control than pigeons (e.g., van Haaren et al. 1988), but in other cases they are largely impulsive (e.g., Tobin et al. 1993).

Importantly, it seems that for rats there is substantial stability in their choice behavior across different tests and across time (e.g., Peterson et al. 2015; Marshall et al. 2014) much like is true for humans (e.g., Jimura et al. 2011).

Primates also can be tested for intertemporal choice behavior. For example, Tobin et al. (1996) tested two macaque monkeys and found that they chose the larger-later option on nearly all trials. Subsequent work has presented this type of task to lemurs (Stevens and Mühlhoff 2012), cotton-top tamarins and common marmosets (Stevens et al. 2005a, b), brown capuchins (Addessi et al. 2011; Amici et al. 2008), black-handed spider monkeys (Amici et al. 2008), long-tailed macaques (Amici et al. 2008), lowland gorillas (Amici et al. 2008), orangutans (Amici et al. 2008), and bonobos and chimpanzees (Rosati et al. 2007). The best way to summarize this work is that there appear to be some major differences across species, with apes and some monkeys showing more self-control than others. There are also highly variable performances within some of these species, which is a reminder that individual differences are an important source of information for researchers. Numerous factors have been proposed to account for the species differences, including metabolic rate, diet, body size, life span, group social behavior, and reward maximization theory (see Amici et al. 2008; Stevens and Mühlhoff 2012; Stevens et al. 2005a, b).

Self-control has been studied using other tasks, and this also can be done across species. For the remainder of this chapter, I will focus on tasks that were designed to be presented to nonhuman primates. In some cases, Professor Fujita provided some of the most compelling and important data using those tests. Some of the work with nonhuman primates that went beyond traditional intertemporal choice tasks was based on adapting tasks used with humans. The most famous of these, the marshmallow test (Mischel 2014), was adapted in different ways, for example. Another test, also adapted for use with nonhuman primates (and then other animals), was the accumulation task (Beran 2002; Toner 1981). In other cases, new assessments of self-control were designed for use first with nonhuman primates, and in some cases those tasks then were given to humans or other nonprimate species. Those creative tests were generated to match some of the natural tendencies of the species under study, or in some cases were the fortuitous outcome of efforts to ask questions unrelated to self-control, at least initially. The most well-known of these latter cases is the reverse-reward contingency task.

The Reverse-Reward Contingency Task

The task appears to be a simple one. Point to what *you do not want*. It goes away, and you get what remains. That is how Sally Boysen originally designed the task for presentation to chimpanzees, and in that case the task involved pointing to candies (Boysen and Berntson 1995). What she found was that, over and over, the chimpanzees pointed to the larger amount, which went away, and they got the smaller amount. This was the first test of the *reverse-reward contingency task*. Even when rocks were used instead of candies (with each rock in the unchosen set leading to one candy reward), the chimpanzees struggled. It was only with the substitution of Arabic numerals that the numeral-experienced chimpanzees learned to point to the smaller number to get the larger amount (Boysen et al. 1996, 1999; also see Addessi and Rossi 2010 for a similar finding using tokens in capuchin monkeys). And, it was not only chimpanzees that struggled with this task (see Shifferman 2009, for a review). Silberberg and Fujita (1996) were the first to try this task with Japanese macaques. These monkeys repeatedly failed to point to one raisin when it was offered against four raisins. But then Silberberg and Fujita tried something new. If the monkeys pointed to four raisins, they got nothing, whereas pointing to one raisin led to them getting four. With that modification, the monkeys learned to point to one raisin to get the big reward. And, this result suggested that the problem for subjects in this test was not about being unable to point away from more food when it was offered. Silberberg and Fujita argued that something else besides (or in addition to) an inability to inhibit pointing to the bigger amount of food might be at work when animals failed at the task. If that was true, the monkeys should have continued to struggle to point at one raisin no matter the outcome because the other option still held four raisins. Subsequent work sometimes showed similar effects of varying the contingency from getting what was not chosen to getting nothing for choosing more. In the former case, poor performance usually occurred, but in the latter case better performance emerged (e.g., Kralik 2005, 2012; also see Murray et al. 2005). However, in other experiments failure to point to the smaller set was the outcome no matter the contingency (e.g., Kralik et al. 2002).

Professor Fujita conducted other experiments on the reverse-reward task with James Anderson, using squirrel monkeys. At first, those monkeys also failed the test even when no food was given for points to the larger set. However, repetition of the same trials over many instances led the monkeys finally to point to the smaller set (Anderson et al. 2000). This pattern of required massive practice to produce some success was also found with lemurs (Genty et al. 2004, 2011). Additional studies with great apes showed that many failed in the way that Boysen's chimpanzees had initially failed although some showed a greater ability to learn to point to the smaller set even under the reversed-reward contingency (e.g., Uher and Call 2008; Vlamings et al. 2006).

Another of Professor Fujita's contributions to this area came from a modification of the task, involving use of qualitatively different foods. Anderson et al. (2008) found that capuchin monkeys would learn to point to a low preference food to get a high preference food, but they could not do this when food type was constant and food quantity varied, again with the need to point to the smaller amount to get the larger amount. This research team also was interested in how well monkeys would remember the rules of the reverse-reward contingency task. When the squirrel monkeys were given an 8-month delay between the end of their first exposure to the task and being re-exposed, they all continued to point to the smaller amount and thereby receive the larger one (Anderson et al. 2004). However, when tested 5 years later, they showed a mixed result: none of the monkeys pointed more often to the smaller amount, but they also did not point more often to the larger amount, which contrasted with the very strong choose-larger bias they had exhibited in their first exposure to the task (Adachi et al. 2011).

These papers and others inspired my lab to return to this method for testing chimpanzees. I had (on more than one occasion!) replicated Sally Boysen's results with the chimpanzees at the Language Research Center, but in our last attempt we tried some new variations of the task to see what component behaviors were attainable by chimpanzees to allow them to point away from larger piles of food in order to gain those. In that study (Beran et al. 2016a), we presented the chimpanzees with a single food item, and an empty bowl. As you would expect, they pointed to the food item. In this task, that meant that the item was transferred to the bowl. It was still out of reach, and another item took the place of the first item. If it was selected again, it went into the bowl, and so on until the bowl was selected, at which point the content was given to the chimpanzee and the trial ended. The idea was that, after the second or third choice, the chimpanzees would have to point to one item rather than an increasingly larger pile of items in the bowl at each choice point if they wanted to get the most food. And, quite unexpectedly, they learned to do this almost immediately. Importantly, when it was clear that they had mastered this task, we gave them the typical reverse-reward task again. And, they failed miserably, always pointing to the larger amount. We alternated the tasks and again found success when they added one item at a time to an accumulation by pointing at one item instead of the accumulation, but they failed when they immediately got whatever they *did not point toward*. Again, this was initially a surprising result from the perspective of both tasks requiring the same response—point to smaller amounts. But the results make more sense if you consider the timing. The reverse-reward task is about getting something now—whatever you do not point at. But our modified accumulation version was about getting something small now, or making a bigger pile of something for later. Another way to consider this is to imagine the typical reverse-reward task as being about losing what one points toward whereas our new variation was about keeping what one points to (even if it is delayed in its delivery). The first case is difficult for chimpanzees, but the second is easier. Whether this would be true of other species is unknown at present, since no one has tried to replicate this result. What is clear is that chimpanzees seem to tolerate accumulating rewards without much problem, a result that converged with other studies that have been done using the so-called accumulation task.

Delayed Gratification and the Accumulation Task

The accumulation task was originally designed and used with children. Although less well known than Mischel's marshmallow task (Mischel 2014), it was equally effective at demonstrating some of the key environmental factors that impact delay of gratification. The key difference is that the marshmallow test was about waiting for something better versus taking whatever was available now. That available item never changed in value, it simply remained available. In the accumulation task, the presently available, and tempting, reward continued to grow the longer a child waited before taking it. Toner and Smith (1977) first used the task to show that specific instructions to the children could have major impacts on how long they would wait. When children talked about the appealing properties of the items that were accumulating, delay was short. When they talked about non-consummatory features (such as shape or color), delay was longer. Subsequent experiments showed other complementary influences on delay of gratification that Mischel had shown with the marshmallow task (e.g., Toner 1981; Toner et al. 1979). Most importantly for comparative researchers, the task required almost no training, and was easy to implement in almost any setting.

I used the accumulation task first with chimpanzees and an orangutan (Beran 2002). In that test, I moved food items to a bowl that the apes could take at any time and eat from. But, when they did, the trial ended, and any remaining foods not yet transferred would be forfeited. On the first trial, they took the first item, losing the rest. But, after that, they mostly waited while items were transferred within reach. The orangutan was the least proficient, but later work with other orangutans would show them to be quite capable in the task as well (Parrish et al. 2014). Over the years, we modified the task. In an effort to remove the experimenter from the test we used an automated dispenser that delivered the items to a receptacle within reach. Most chimpanzees still were quite good at waiting for all items to accumulate (Beran and Evans 2006), with delays of more than 10 min to get all of the food that they could. Seeing food accumulating was surely tempting, and with children it was clear that certain strategies were more effective for delaying gratification (e.g., Miller and Karniol 1976; Mischel and Ebbesen 1970; Mischel and Moore 1973; Mischel et al. 1972). For example, self-distraction is a highly effective technique to remove attention from the appealing, immediate reward (e.g., Yates and Mischel 1979). We wanted to know if chimpanzees also could engage in self-distraction.

To do this, we presented them with three kinds of trials (Evans and Beran 2007a). On some trials, food accumulated within reach of the chimpanzees using the automated apparatus, and that was all that was present in the test setting. On other trials, food accumulated within reach, but the chimpanzees also were given some items they could manipulate or play with during the trial. As one might expect, they did play with those items, and they did wait longer when they had them. This confirmed that having distractions improved *delay maintenance* (i.e., the continued delay of one's gratification). But, it could not tell us anything about whether the chimpanzees were aware of the relation between self-distraction and delay maintenance. That comparison only told us that they played with objects when they had them, and that in so doing they waited longer. To know about strategic self-distraction required a third condition. Again, toys were present, but now the accumulating rewards *were not in reach*. Instead, the number of items to accumulate was matched to a previous trial where the chimpanzees determined the trial's end. So, the delay was the same, the foods were visible and collecting, but there was no temptation, and no need to engage in self-control because the food was out of reach. If chimpanzees could strategically self-distract, they should have used the toys more when the food was within reach, because it was then when they were at greatest risk of giving in and taking those rewards. And, this was true for three of four chimpanzees. The chimpanzees looked very much like high-performing human children in the marshmallow task—they engaged in self-distraction when needed to sustain delay maintenance.

The accumulation task, in different forms, has now been presented to other primates. Bonobos successfully waited for rewards to accumulate (Stevens et al. 2011), suggesting that great apes, in general, are good at delaying gratification. Monkey species have shown more limited abilities, but with the right contextual cues can also allow food to accumulate. With rhesus monkeys, we initially found little evidence that they would accumulate food items of an identical type. However, if we embedded a higher preference item in a sequence of other identical foods, they would wait for those to accumulate so that they could reach the point at which the highest preference item was added to the accumulation (Evans and Beran 2007b). In our laboratory, capuchin monkeys showed highly variable performances in an accumulation test (Beran et al. 2016b), but they also showed improvements over time and with more experience in the task (also see Evans et al. 2012). Anderson et al. (2010) gave squirrel monkeys and capuchin monkeys a series of tests using the accumulation method. When transferring identical food items, one at a time, none of the capuchin monkeys waited at all, and only one squirrel monkey did so. However, when each additional item was larger than the previous one, this increased delay maintenance in the task for some capuchin monkeys, much as the similar manipulation had worked for Evans and Beran (2007b). And, when the monkeys tested by Anderson et al. (2010) were tested again with identical items, performance remained stable, illustrating that some monkeys could maintain delay of gratification in that form of the accumulation task.

Another outcome of the development and use of the accumulation task and the results from extensive use of the reverse-reward task were a reconsideration of some kinds of intertemporal choice tests used with primates. In some of these tests, animals chose between a smaller-sooner option or a larger-later option, and they did so by pointing at or reaching toward a choice option (e.g., Addessi et al. 2011; Amici et al. 2008; Rosati et al. 2006, 2007; Stevens et al. 2005a). The difficulty with this is that responses to a larger quantity over a smaller quantity are taken to reflect self-control, because those responses also lead to longer delay to food. But, as I described earlier, we know that primates struggle to point to anything other than the larger set of food when making these comparisons, and so there is an interpretive difficulty with these data.

Genty et al. (2012) highlighted this interpretive problem. They manipulated the visibility of the food items when presenting them to monkeys, and they found that this affected choice behavior specifically in a way that challenged the validity of previous experiments where pointing to larger sets was scored as a self-control response. In their task, monkeys made more so-called self-control responses when food was visible than when it was not visible. This should not have happened if visibility of food items *increases* the attraction to more immediate reward. Rather, these results suggested that the self-control responses were more likely to be impulsive, prepotent tendencies to reach toward more food, at least in tests with monkeys (see also Addessi et al. 2014).

We became interested in this problem also, and from this interest came a new test, called the hybrid delay task. This task combines the choice phase of the intertemporal choice tasks with an accumulation component. The procedure consists of first presenting a small set and a large set of items. If the small set is selected, it is given to the animal immediately. But, if the large set is selected, then that set becomes the possible reward that is delivered through the one-at-a-time accumulation method. The idea behind the task was to have a way to discern whether points to larger sets were matched with the self-control needed to allow those sets to accumulate. And, for the first time, researchers could objectively define errors in the intertemporal choice test. Normally, this is not possible. One can define self-control choices or impulsive choices, but those are still choices for which the underlying motivation of the chooser is unknown. The hybrid delay task, instead, could show that an animal might point to a larger set, but then fail to wait long enough to justify that choice. The reason is that a smaller set (e.g., four items delivered immediately) is a better outcome than a larger set (e.g., 12 items) that one does not wait for, and instead takes as soon as the first item is delivered. When that happens, the animals actually *wait longer to get less food*. This is objectively an error in choice behavior.

In the first test of the hybrid delay task, with capuchin monkeys, our concerns were confirmed (Paglieri et al. 2013). Monkeys often chose the larger set, but then much less often waited long enough to make that choice more rewarding than just choosing the smaller-sooner set from the start. In fact, only one of 18 monkeys tested clearly justified choices of the larger set by waiting long enough to get more than four items from that set. And yet, if we had only measured the choices of the larger set over the smaller set, we would have concluded that the majority of monkeys showed good self-control. These results led us to then test chimpanzees, to see whether the pattern was the same. It was not. For chimpanzees at the Language Research Center, and also for a larger group of less task-savvy chimpanzees from the Yerkes National Primate Research Center, choices of the larger set were frequently matched with delay maintenance that justified those choices (Beran et al. 2014). Chimpanzees showed variability in how often they chose to wait, but more importantly they were able to wait when they chose to do so.

In a final experiment with the hybrid delay task, Bill Hopkins and I examined in a sample of 40 chimpanzees the relation between performance in the hybrid delay task and performance on a battery of cognitive measures used to provide a general intelligence score for chimpanzees (Beran and Hopkins 2018). The results were very interesting. General intelligence scores were closely related to a specific measure one can generate from the hybrid delay task that we called *efficiency*. This was a score that combined how well chimpanzees waited to accumulate items when they did choose the larger set, combined with how often they instead chose to take the smaller, sooner set. More efficient chimpanzees in the hybrid delay task had higher general intelligence scores. Importantly, there was only a limited relation between intelligence and how often chimpanzees chose the larger set. So, it was not that smarter chimpanzees tried to wait more often, but rather that smarter chimpanzees were better at waiting in those trials where they chose to try to wait. This result aligns with research on children and human adults showing a similar relation between delay of gratification and various measures of intelligence and cognitive functioning (e.g., Mischel et al. 1988; Shoda et al. 1990; Shamosh and Gray 2008).

Delayed Gratification and Exchange Tasks

Another widely used task to assess delay of gratification is the exchange task. The idea here is also a simple one, making it easy to demonstrate contingencies to animals without need for much training. Items are presented, which can then be returned for subsequent items. In this way, a participant can exchange what they have (the immediate reward) for something they will get in return (the delayed reward). Drapier et al. (2005) was one of the first studies to use this method. Capuchin monkeys were given low-value items, and if they returned those, they received higher value rewards. Or, in some cases, they could return smaller-sized pieces of a given type of food for a larger-sized piece of the same food type. In both cases, when the delayed reward was a more substantial upgrade in preference or size, the probability of an exchange increased. Drapier et al. (2005) also included a particularly interesting test where more than one exchange could occur. Now, the monkeys would have to give back a stone to get a pellet, but then give back the pellet to get a more preferred piece of carrot, and finally give back the carrot to get the more preferred piece of apple. And, as part of this, they had to hold each food item for a few seconds, which ensured that there was some temptation to eat it. Five of six monkeys did this perfectly, demonstrating that sequential exchanges could be made to obtain the highest preference item.

We also have studied sequential exchanges in chimpanzees (Beran et al. 2016c). We designed a test in which chimpanzees could see a sequence of food items out of reach on a bench. Those items were presented in their spatial delivery order, as long as each delivered item was given back to an experimenter. In our test, on each trial the chimpanzees could see everything that would be given on that trial. This allowed us to present a variety of trial types. Sometimes each exchange led to a progressively more preferred reward. But, in other cases some exchanges early in the sequence required giving back a more preferred item for a less preferred item, that then could be exchanged for the most preferred item. In this way, we ensured that the chimpanzees were looking at the whole sequence that was possible, and then deciding whether the exchange or eat items as they were presented. This experiment produced some of the widest ranges of responding we had seen in our chimpanzees. The chimpanzee Lana did not exchange very well in any of the trials, whereas chimpanzees Mercury and Sherman did. In another variation of the task, we used different-sized items, rather than qualitatively different food types. Again, two chimpanzees were very good at exchanging smaller pieces of banana up to the point where they got the largest possible piece on a trial, but a third chimpanzee failed to exchange the smallest piece of food on any trial, and thus received very little reward.

In a variation of the exchange task that made use of "symbolic" stimuli, Judge and Essler (2013) gave capuchin monkeys choices between real food items, or different kinds of tokens that could later be exchanged for specific foods. Monkeys first received a low-value token. If they exchanged it, they could choose between a real low-value food item, or another token, which could then be exchanged for a higher value food item. Some monkeys used the token value to guide their choices although others did not. In other words, successful monkeys exchanged edible stimuli for inedible ones when such exchanges ultimately led to better reward (for similar findings in capuchins see also De Petrillo et al. (2019)).

Other exchange experiments made use of much longer delay periods. For example, Ramseyer et al. (2006) tested capuchin monkeys with delay times ranging between 10 s and 10 min from receiving an item to being able to exchange it. The delay period had a major impact on exchange rates, but even so some monkeys were able to hold and not consume a food item for several minutes before exchanging it. This was more often the case when exchanges led to qualitative changes in reward type, rather than quantitative increases in reward size. In the latter case, monkeys struggled to return small pieces of highly preferred food, even when doing so could eventually result in as much as a 40-fold increase in item size. Dufour et al. (2007) gave five chimpanzees a similar series of tests where they could avoid eating and then exchange smaller pieces of cookie for larger pieces, and they found that their performance was markedly different from that of capuchins! When the delayed food was two times, four times, or eight times bigger, most of the chimpanzees gave back the smaller piece to get the larger piece even with delays of 4 min. Delays could be increased to 8 min if the larger item was 40 times bigger, matching some of the longest delays seen in the accumulation tests with apes and many of the various "marshmallow tasks" given to children.

Before one concludes that apes always outperform monkeys in self-control tests, consider one final exchange experiment. Pelé et al. (2010) tested long-tailed macaques with the two, four, and eight-fold increases in item size, and these monkeys performed much better than capuchin monkeys had. In fact, they performed more like chimpanzees. Furthermore, the same interesting pattern emerged for macaques and chimpanzees when the investigators looked at exactly when an item was eaten versus kept for later exchange. When chimpanzees or macaques ate the first item (i.e., they never exchanged), they did so earlier in the delay interval than would be expected if they were not taking into account the delay time. In other words, as with the more intelligent chimpanzees in Beran and Hopkins (2018) who were more efficient at the hybrid delay task, the macaques and chimpanzees that were engaged in the exchange task seemingly made a strategic decision early in the trial to take what they had or to wait to exchange later.

In a nice convergence of different self-control tests, Pelé et al. (2010) also gave their macaques an accumulation test to examine consistency between performances on the two tasks. The macaques' delay times were roughly comparable to those reported in earlier tests with other macaques (Evans and Beran 2007b), and the maximum tolerated delay times were comparable for the same monkeys who were given the exchange and accumulation trials (also see Pelé et al. 2011). These data suggest that exchange tasks and accumulation tasks tap into similar mechanisms that support maintenance of delayed gratification, at least in the species that have been tested to date.

Summary

There are additional tasks that I do not have space to describe in detail, but that provide other interesting data for understanding when primates show self-control and when they do not. One example is the rotating tray task we developed, in which food items pass by subjects as long as they do not take any. In this way, passive behavior in the face of presented rewards is the necessary response to gain something better (Bramlett et al. 2012). Capuchin monkeys learn to do this, even when they must remember food items that are occluded (Perdue et al. 2015). Other tests make use of ecological factors and natural foraging behaviors such as movement through space. For example, Stevens et al. 2005b) gave cotton-top tamarins and marmosets the choice to approach one of two foods, each at a different distance from the participant. The better food was further away, and so the monkeys had to pass the less preferred food to get to the more preferred food. In this test, tamarins showed greater self-control than marmosets. Stevens et al. 2005b) suggested that this was because traveling through space is more energetically costly than staying in one location, and marmosets have less need to travel widely than do tamarins in natural feeding contexts. Marmosets rely more on food sources that are spatially fixed (such as gum and sap that exude from trees), whereas tamarins forage on ephemeral sources such as insects that require the tamarins to move through space to locate new feedings opportunities. Thus, natural travel tendencies of these two species in support of foraging needs could be reflected in behavior in the self-control task. This result was in contrast to another comparison of marmosets and tamarins where the latter were the more impulsive responders, and marmosets showed more self-control. That test was a more traditional intertemporal choice test Stevens et al. (2005a). Again, this difference could be accounted for by foraging ecology. Marmosets rely more on food such as tree sap that can take longer to obtain as it accumulates, whereas tamarins often grab at food that appears suddenly, for example, insects. These are important outcomes because they remind us that what might evoke self-control in one context for one individual or species may not work in a different context. Ecological factors, experiential factors, genetic factors, and other factors help explain individual variation within and across species, and they remain important variables for future studies of self-control in primates.

As I consider the research that has been done to this point, I have five points of summary regarding self-control in primates (see Beran 2018, for extended discussion of these). The first is that nonhuman primates are not always impulsive creatures. In fact, they are more likely than not to show some degree of self-control in most tests. So, when humans consider themselves capable of overcoming their "animal impulses," they should also consider that other animals, too, can overcome impulses. The second point is that using only one task with a species runs the risk of misestimating the self-control of that species. Humans sometimes demonstrate relations among degrees of self-control across tasks (e.g., someone who saves for retirement also may avoid spontaneous shopping sprees). But in other areas, self-control in one context may not predict similar behavior in another (e.g., someone who avoids unhealthy food, alcohol, and cigarettes might yet engage in risky sexual behavior). So, one test does not reflect the general self-control of any organism, and I have outlined some examples where there was consistency or inconsistency when the same individuals or species were given multiple tasks. The third point is that tests such as the hybrid delay task highlight that some methods run the risk of not assessing self-control but instead may reflect impulsivity or inhibitory control. Intertemporal choice tests with direct points to food rewards probably have this risk inherent in their design and should be used cautiously. The fourth point is that individual differences are a rich source of information about self-control. In primate research, it is often difficult to find large enough sample sizes to adequately assess individual differences. However, new possibilities are emerging where multiple labs, field sites, zoos, and sanctuaries could provide data from larger numbers of primate participants. The recent *ManyPrimates* consortium is one example of this (ManyPrimates et al. 2019). Even so, we should remember that reporting what a *group* did often fails to reflect how different the performances of *individuals* might have been. Finally, we have reached a point where our knowledge of self-control in primates could be leveraged to aid those working on potential interventions for improving self-control where it needs to be improved (e.g., reducing gambling, smoking, drug use, impulsive spending, aggression, etc.). Those who work with rats and pigeons have provided animal models for such interventions, and this could be true also for primate researchers. For me, this is specifically about behavioral interventions, given that most of the research with primates focuses on the relation between environmental features of tasks and choice behavior. For example, I described how in some tests simply having greater experience has led to gains in self-control responses. Whether this could translate to educational settings for children or adolescents with impulsivity issues or other areas of human choice behavior is unknown, but it presents a compelling future avenue for research.

References

Adachi I, Anderson JR, Fujita K (2011) Reverse-reward learning in squirrel monkeys (*Saimiri sciureus*): retesting after 5 years, and assessment on qualitative transfer. J Comp Psychol 125:84–90

Addessi E, Bellagamba F, Delfino A, De Petrillo F, Focaroli V, Macchitella L et al (2014) Waiting by mistake: symbolic representation of rewards modulates intertemporal choice in capuchin monkeys, preschool children and adult humans. Cognition 130:428–441

Addessi E, Paglieri F, Focaroli V (2011) The ecological rationality of delay tolerance: insights from capuchin monkeys. Cognition 119:142–147

Addessi E, Rossi S (2010) Tokens improve capuchin performance in the reverse–reward contingency task. Proc Roy Soc Lond 278:1707. https://doi.org/10.1098/rspb20101602

Ainslie G, Herrnstein RJ (1981) Preference reversal and delayed reinforcement. Anim Learn Behav 9:476–482

Amici F, Aureli F, Call J (2008) Fission–fusion dynamics, behavioral flexibility, and inhibitory control in primates. Curr Biol 18:1415–1419

Anderson JR, Awazu S, Fujita K (2000) Can squirrel monkeys (*Saimiri sciureus*) learn self-control: a study using food array selection tests and reverse-reward contingency. J Exp Psychol Anim Behav Process 26:87–97

Anderson JR, Awazu S, Fujita K (2004) Squirrel monkeys (*Saimiri sciureus*) choose smaller food arrays: long-term retention, choice with nonpreferred food, and transposition. J Comp Psychol 118:58–64

Anderson JR, Hattori Y, Fujita K (2008) Quality before quantity: rapid learning of reverse-reward contingency by capuchin monkeys (*Cebus apella*). J Comp Psychol 122:445–448

Anderson JR, Kuroshima H, Fujita K (2010) Delay of gratification in capuchin monkeys (*Cebus apella*) and squirrel monkeys (*Saimiri sciureus*). J Comp Psychol 124:205–210

Beran MJ (2002) Maintenance of self-imposed delay of gratification by four chimpanzees (*Pan troglodytes*) and an orangutan (*Pongo pygmaeus*). J Gen Psychol 129:49–66

Beran MJ (2018) Self-control in animals and people. Academic Press, London

Beran MJ, Evans TA (2006) Maintenance of delay of gratification by four chimpanzees (*Pan troglodytes*): the effects of delayed reward visibility, experimenter presence, and extended delay intervals. Behav Processes 73:315–324

Beran MJ, Evans TA, Paglieri F, McIntyre JM, Addessi E, Hopkins WD (2014) Chimpanzees (*Pan troglodytes*) can wait, when they choose to: a study with the hybrid delay task. Anim Cogn 17:197–205

Beran MJ, Hopkins WD (2018) Self-control in chimpanzees relates to general intelligence. Curr Biol 28:574–589

Beran MJ, James BT, Whitham W, Parrish AE (2016a) Chimpanzees can point to smaller amounts of food to accumulate larger amounts but they still fail the reverse-reward contingency task. J Exp Psychol Anim Learn Cogn 42:347–358

Beran MJ, Perdue BM, Rossettie MS, James BT, Whitham W, Walker B, Futch SE, Parrish AE (2016b) Self-control assessments of capuchin monkeys with the rotating tray task and the accumulation task. Behav Processes 129:68–79

Beran MJ, Rossettie MS, Parrish AE (2016c) Trading up: chimpanzees (*Pan troglodytes*) show self-control through their exchange behavior. Anim Cogn 19:109–121

Boysen ST, Berntson GG (1995) Responses to quantity: perceptual versus cognitive mechanisms in chimpanzees (*Pan troglodytes*). J Exp Psychol Anim Behav Process 21:82–86

Boysen ST, Berntson GG, Hannan MB, Cacioppo JT (1996) Quantity-based interference and symbolic representations in chimpanzees (*Pan troglodytes*). J Exp Psychol Anim Behav Process 22:76–86

Boysen ST, Mukobi KL, Berntson GG (1999) Overcoming response bias using symbolic representations of number by chimpanzees (*Pan troglodytes*). Anim Learn Behav 27:229–235

Bramlett JL, Perdue BM, Evans TA, Beran MJ (2012) Capuchin monkeys (*Cebus apella*) let lesser rewards pass them by to get better rewards. Anim Cogn 15:963–969

De Petrillo F, Caroli M, Gori E, Micucci A, Gastaldi S, Bourgeois-Gironde S, Addessi E (2019) Evolutionary origins of money categorization and exchange: an experimental investigation in tufted capuchin monkeys (*Sapajus spp.*). Anim Cogn 22:169–186

Drapier M, Chauvin C, Dufour V, Uhlrich P, Thierry B (2005) Food-exchange with humans in brown capuchin monkeys. Primates 46:241–248

Dufour V, Pelé M, Sterck EHM, Thierry B (2007) Chimpanzee (*Pan troglodytes*) anticipation of food return: coping with waiting time in an exchange task. J Comp Psychol 121:145–155

Evans TA, Beran MJ (2007a) Chimpanzees use self-distraction to cope with impulsivity. Biol Lett 3:599–602
Evans TA, Beran MJ (2007b) Delay of gratification and delay maintenance by rhesus macaques (*Macaca mulatta*). J Gen Psychol 134:199–216
Evans TA, Beran MJ, Paglieri F, Addessi E (2012) Delaying gratification for food and tokens in capuchin monkeys (*Cebus apella*) and chimpanzees (*Pan troglodytes*): when quantity is salient, symbolic stimuli do not improve performance. Anim Cogn 15:539–548
Genty E, Chung PC, Roeder JJ (2011) Testing brown lemurs (*Eulemur fulvus*) on the reverse-reward contingency task without a modified procedure. Behav Processes 86:133–137
Genty E, Karpel H, Silberberg A (2012) Time preferences in long-tailed macaques (*Macaca fascicularis*) and humans (*Homo sapiens*). Anim Cogn 15:1161–1172
Genty E, Palmier C, Roeder JJ (2004) Learning to suppress responses to the larger of two rewards in two species of lemurs, *Eulemur fulvus* and *E. macaco*. Anim Behav 67:925–932
Green L, Myerson J (1993) Alternative frameworks for the analysis of self-control. Behav Philos 21:37–47
Green L, Myerson J (2004) A discounting framework for choice with delayed and probabilistic rewards. Psychol Bull 130:769–792
Imuta K, Hayne H, Scarf D (2014) I want it all and I want it now: delay of gratification in preschool children. Dev Psychobiol 56:1541–1552
Jimura K, Myerson J, Hilgard J, Keighley J, Braver TS, Green L (2011) Domain independence and stability in young and older adults' discounting of delayed rewards. Behav Processes 87:253–259
Judge PG, Essler JL (2013) Capuchin monkeys exercise self-control by choosing token exchange over an immediate reward. Int J Comp Psychol 26:256–266
King GR, Logue AW (1990) Choice in a self-control paradigm: effects of reinforcer quality. Behav Processes 22:89–99
Koriat A, Nisan M (1978) Delay of gratification as a function of exchange values and appetitive values of the rewards. Motiv Emot 2:375–390
Kralik JD (2005) Inhibitory control and response selection in problem solving: how cotton-top tamarins (*Saguinus oedipus*) overcome a bias for selecting the larger quantity of food. J Comp Psychol 119:78–89
Kralik JD (2012) Rhesus macaques (*Macaca mulatta*) spontaneously generalize to novel quantities in a reverse-reward contingency task. J Comp Psychol 126:255–262
Kralik JD, Hauser MD, Zimlicki R (2002) The relationship between problem solving and inhibitory control: cotton-top tamarins (*Saguinus oedipus*) performance on a reversed contingency task. J Comp Psychol 116:39–50
Lemmon K, Moore C (2007) The development of prudence in the face of varying future rewards. Dev Sci 10:502–511
Logue AW (1988) Research on self-control: an integrating framework. Behav Brain Sci 11:665–709
Logue AW, Peña-Correal TE (1984) Responding during reinforcement delay in a self-control paradigm. J Exp Anal Behav 41:267–277
Logue AW, Peña-Correal TE (1985) The effect of food deprivation on self-control. Behav Processes 10:355–368
Madden GJ, Bickel WK (eds) (2010) Impulsivity: the behavioral and neurological science of discounting. American Psychological Association, Washington, DC
Many Primates, Altschul DM, Beran MJ, Bohn M, Call J, DeTroy S, Duguid SJ et al (2019) Establishing an infrastructure for collaboration in primate cognition research. PLoS One 14: e0223675
Marshall AT, Smith AP, Kirkpatrick K (2014) Mechanisms of impulsive choice: I. Individual differences in interval timing and reward processing. J Exp Anal Behav *102*:86–101

Mazur JE (1987) An adjusting procedure for studying delayed reinforcement. In: Commons ML, Mazur JE, Nevin JA, Rachlin H (eds) Quantitative analysis of behavior: vol. 5. The effect of delay and of intervening events of reinforcement value. Erlbaum, Hillsdale, NJ, pp 55–73
Miller DT, Karniol R (1976) The role of rewards in externally and self-imposed delay of gratification. J Pers Soc Psychol 34:310–316
Mischel W (2014) The marshmallow test: mastering self-control. Little, Brown and Co, Boston
Mischel W, Ebbesen EB (1970) Attention in delay of gratification. J Pers Soc Psychol 16:329–337
Mischel W, Ebbesen EB, Zeiss A (1972) Cognitive and attentional mechanisms in delay of gratification. J Pers Soc Psychol 21:204–218
Mischel W, Moore B (1973) Effects of attention to symbolically presented rewards on self-control. J Pers Soc Psychol 28:172–179
Mischel W, Shoda Y, Peake PK (1988) The nature of adolescent competencies predicted by preschool delay of gratification. J Pers Soc Psychol 54:687–696
Murray EA, Kralik JD, Wise SP (2005) Learning to inhibit prepotent responses: successful performance by rhesus macaques, *Macaca mulatta*, on the reversed-contingency task. Anim Behav 69:991–998
Paglieri F, Focaroli V, Bramlett J, Tierno V, McIntyre J, Addessi E, Evans TA, Beran MJ (2013) The hybrid delay task: can capuchin monkeys (*Cebus apella*) sustain a delay after an initial choice to do so? Behav Processes 94:45–54
Parrish AE, Perdue BM, Stromberg EE, Bania AE, Evans TA, Beran MJ (2014) Delay of gratification by orangutans (*Pongo pygmaeus*) in the accumulation task. J Comp Psychol 128:209–214
Pelé M, Dufour V, Micheletta J, Thierry B (2010) Long-tailed macaques display unexpected waiting abilities in exchange tasks. Anim Cogn 13:263–271
Pelé M, Micheletta J, Uhlrich P, Thierry B, Dufour V (2011) Delay maintenance in Tonkean macaques (*Macaca tonkeana*) and brown capuchin monkeys (*Cebus apella*). Int J Primatol 32:149–166
Perdue BM, Bramlett JL, Evans TA, Beran MJ (2015) Waiting for what comes later: capuchin monkeys show self-control even for nonvisible delayed rewards. Anim Cogn 18:1105–1112
Peterson JR, Hill CC, Kirkpatrick K (2015) Measurement of impulsive choice in rats: same-and alternate-form test–retest reliability and temporal tracking. J Exp Anal Behav 103:166–179
Rachlin H (2000) The science of self-control. Harvard University Press, Cambridge, MA
Ramseyer A, Pelé M, Dufour V, Chauvin C, Thierry B (2006) Accepting loss: the temporal limits of reciprocity in brown capuchin monkeys. Proc Roy Soc Lond 273:179–184
Rosati AG, Stevens JR, Hare B, Hauser MD (2007) The evolutionary origins of human patience: temporal preferences in chimpanzees, bonobos, and human adults. Curr Biol 17:1663–1668
Rosati AG, Stevens JR, Hauser MD (2006) The effect of handling time on temporal discounting in two new world primates. Anim Behav 71:1379–1387
Schwarz JC, Schrager JB, Lyons AE (1983) Delay of gratification by preschoolers: evidence for the validity of the choice paradigm. Child Dev 54:620–625
Shamosh NA, Gray JR (2008) Delay discounting and intelligence: a meta-analysis. Intelligence 36:289–305
Shifferman EM (2009) Its own reward: lessons to be drawn from the reversed-reward contingency paradigm. Anim Cogn 12:547–558
Shoda Y, Mischel W, Peake PK (1990) Predicting adolescent cognitive and self-regulatory competencies from preschool delay of gratification: identifying diagnostic conditions. Dev Psychol 26:978–986
Silberberg A, Fujita K (1996) Pointing at smaller food amounts in an analogue of Boysen and Berntson's (1995) procedure. J Exp Anal Behav 66:143–147
Stevens JR (2017) Impulsivity: how time and risk influence decision making. Springer, Cham
Stevens JR, Hallinan EV, Hauser MD (2005a) The ecology and evolution of patience in two new world primates. Biol Lett 1:223–226
Stevens JR, Mühlhoff N (2012) Intertemporal choice in lemurs. Behav Processes 89:121–127

Stevens JR, Rosati AG, Heilbronner SR, Mühlhoff N (2011) Waiting for grapes: expectancy and delayed gratification in bonobos. Int J Comp Psychol 24:99–111
Stevens JR, Rosati AG, Ross KR, Hauser MD (2005b) Will travel for food: spatial discounting in two new world monkeys. Curr Biol 15:1855–1860
Tobin H, Chelonis JJ, Logue AW (1993) Choice in self-control paradigms using rats. Psychol Rec 43:441–454
Tobin H, Logue AW, Chelonis JJ, Ackerman KT, May JG (1996) Self-control in the monkey *Macaca fascicularis*. Anim Learn Behav 24:168–174
Toner IJ (1981) Role involvement and delay maintenance behavior in preschool children. J Genet Psychol 138:245–251
Toner IJ, Lewis BC, Gribble CM (1979) Evaluative verbalization and delay maintenance behavior in children. J Exp Child Psychol 28:205–210
Toner IJ, Smith RA (1977) Age and overt verbalization in delay-maintenance behavior in children. J Exp Child Psychol 24:123–128
Uher J, Call J (2008) How the great apes (*Pan troglodytes*, *Pongo pygmaeus*, *Pan paniscus*, *Gorilla gorilla*) perform on the reversed reward contingency task II: transfer to new quantities, long-term retention, and the impact of quantity ratios. J Comp Psychol 122:204–212
van Haaren F, van Hest A, Poll NE (1988) Self-control in male and female rats. J Exp Anal Behav 49:201–211
Vlamings PHJM, Uher J, Call J (2006) How the great apes (*Pan troglodytes*, *Pongo pygmaeus*, *Pan paniscus*, and *Gorilla gorilla*) perform on a reversed contingency task: the effects of food quantity and food visibility. J Exp Psychol Anim Behav Process 32:60–70
Yates BT, Mischel W (1979) Young children's preferred attentional strategies for delaying gratification. J Pers Soc Psychol 37:286–300

Part VI
Companion Animals

Chapter 17
Developments in Research on Cat Cognition and Personality

Saho Takagi

Abstract Cats, along with dogs, are one of the most popular companion animals for humans. Across the world, increasing numbers of cats are being kept as pets. Despite their familiarity, cats' cognition has long been shrouded in mystery, mainly because cats were considered largely unsuitable for psychological studies in laboratory settings. The "Cats Team" in Kazuo Fujita's lab has developed several innovative and useful methods for studying cat cognition. In this chapter, I review findings from some of the team's studies of cat cognition, including physical inference, use human social cues, incidental memory, cross-modal integration, jealousy, and third-party social evaluation. I also briefly describe some ongoing work on the relation between genes and personality, and suggest directions in which behavioral and cognitive studies of cats might go.

Keywords *Felis catus* · Cats · Cognition · Personality · Evolution

Introduction

Cats, along with dogs, are one of the most popular companion animals for humans. In fact, in Japan the total number of cats kept by humans has surpassed that of dogs (Pet Food Association of Japan 2019): 9.8 million cats and 8.8 million dogs. The decreasing number of dogs kept by humans in recent years and the slight increase for cats may be linked to how the breeding style of cats matches our typical modern lifestyle. As cats usually do not need to walk around outside for exercise every day, they are easier than dogs to keep for a typical, double-income, nuclear family living in a city. Given the huge number of cats living with humans; it is important to get to know more about their cognition, to support better cat–human relationships.

S. Takagi (✉)
Japan Society for the Promotion of Science, Tokyo, Japan

Azabu University, Kanagawa, Japan

J. R. Anderson, H. Kuroshima (eds.), *Comparative Cognition*,
https://doi.org/10.1007/978-981-16-2028-7_17

Research on dog cognition received a boost from a study in the early 2000s that showed their ability to interpret human pointing cues (Hare et al. 2002). Indeed, some studies suggest that dogs might be better than great apes at using human pointing cues, despite the latter being our nearest evolutionary neighbors (i.e., genetically closest) to us. As many scientists assumed that the great apes were most like humans in cognitive abilities, the impact of that study on dogs was strong, leading to an explosion of new studies. Hare et al. (2002) also showed that dogs outperformed wolves raised by humans, and even dog puppies only a few weeks old followed human pointing cues (but see Udell et al. (2008), (2010), for example, of human-raised wolves outperforming dogs at following human social cues). The effects of domestication on cognition started to get more attention from these studies. In the last two decades it has been shown that dogs have highly developed social cognitive skills in many ways comparable to human skills; they are highly sensitive to human attentional states (e.g., Call et al. 2003), they can recognize human emotional expressions (e.g., Albuquerque et al. 2016), imitate human gestures (Fugazza and Miklósi 2015), understand human utterances (Andics et al. 2016; Kaminski et al. 2004; Pilley and Reid 2011), cooperate with human on a loose string-pulling task (Range et al. 2019), and evaluate human character from third-party perspective (Chijiiwa et al. 2015).

In an early study that compared dogs and cats, Miklósi et al. (2005) showed that cats were also able to use human pointing cues to find hidden food; however, since then the study of cat cognition has lagged behind that of dogs. Their wariness of novelty is one factor that might have contributed to the relatively underdeveloped state of cognitive research on cats. Many cats do not behave normally in unfamiliar surroundings such as an experimental room, regardless of coaxing by their owner, whereas dogs are generally more relaxed and behave normally especially if their owner is also present. Probably because their ancestor, African wildcats (*Felis lybica*) are a territorial species, it seems possible that cats inherited a fear of unfamiliar places, and so they become afraid when taken out of their familiar surroundings. Furthermore, many cats show fear of unfamiliar humans (e.g., experimenters) as well as new places, further complicating research efforts. Proximate factors would also explain their behavioral traits of wariness of novelty; they receive much less familiarization to unfamiliar locations and people compared to dogs. Additionally, cats are notoriously individualistic when it comes to food, for example; compared to dogs many cats are not so highly motivated to perform for food rewards. These factors combine to make experimental studies of cognition more difficult for feline subjects, in terms of how to control their behavior and develop methods that do not require long-term training with food rewards each day.

Members of the cat research team in Kazuo Fujita's lab addressed these issues, and by trial and error finally managed to obtain usable behavioral data from "capricious" cats. First, we conducted experiments in a place that was familiar to the cats, namely their living space. This meant planning experiments involving portable apparatus and that could be completed in one or two visits, instead of experiments requiring long-term training. We also utilized "cat cafés," which have grown to become highly popular and widespread in Japan; this allowed us to test adequate numbers of subjects in one visit. There are many types of cat cafés in Japan, in which visitors can interact freely with cats. Some cafés are a kind of shelter that allows people to adopt cats if conditions and the cats' personality are suitable (e.g., someone who spends a lot of time at home might be matched with a cat who loves human company). Other cafés specialize in specific breeds and are run by a breeder (e.g., a cat café for Bengal cats). We carefully explained our enthusiasm for, and the importance of cognitive research on cats, and gradually the number of cat cafés willing to participate in our projects increased. In the following section, I introduce our studies of cat cognition over the past decade.

Physical Inference from Sounds

Making inferences about the world from different sensory cues is advantageous for survival. In an influential study of this kind of ability, Call (2004) gave captive great apes a "cup task" in which the goal was to find a reward in one of two opaque cups by inference, using visual or auditory information. In the auditory information condition, apes heard the noise of the reward moving around in the baited cup when the experimenter horizontally shook the cup. Although some readers might expect that such a basic inference would be easy for great apes, many subjects failed on the task that presented only auditory information, whereas they passed a comparison, visual inference version of the task. Furthermore, human children did not pass this auditory inference task until they were 3–4 years old (Hill et al. 2012). These results suggested that primates, including humans, would less good at inferences based on auditory information compared to visual information. Some researchers pointed out the possible ecological relevance of modality-specific information in a given species (Maille and Roeder 2012; Plotnik et al. 2014; Takagi et al. 2016). Assuming that cats' evolutionary history of hunting prioritize using auditory information more than visual information (Grastyán, and Vereczkei 1974), we predicted that they would show better auditory than visual inferential ability.

We tested whether cats would infer the presence of an invisible object from noises. We did not use the precise cup task procedure of Call (2004) because of the wide intra- and inter-individual variability in cats' appetite; for example, a cat might participate and eat a reward until the third trial, but then lose interest on the fourth trial. Instead, therefore, we presented cats with a sort of "magic show," to attract them using an expectancy violation method. An experimenter horizontally shook a container with a ball inside in front of the cat for 5 s (shaking phase) and then turned it upside down (turning-over phase). There were four conditions, of which half were congruent with the laws of physics and half were incongruent. In congruent conditions, shaking that produced noise was followed by the ball dropping out and silent shaking was followed by no ball when the container was turned over. In incongruent conditions, silent shaking was followed by a ball dropping out, and noisy shaking was followed by no ball dropping out (Fig. 17.1). We predicted that if cats inferred

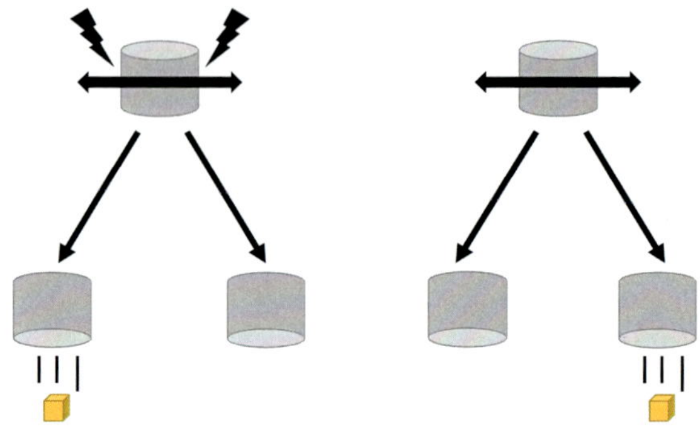

Fig. 17.1 Diagram of physical inference conditions (Takagi et al. 2016). An experimenter shook the container, which produced sounds or no sounds, then turn it upside down so that a ball dropped out or did not. Half of the trials were congruent with physical laws and half were incongruent

the existence of an object inside the container from the noise, they would look longer at the container in the incongruent conditions.

The results supported our prediction: cats looked at the container longer in both incongruent conditions. We also confirmed no difference in looking between congruent and incongruent conditions in the shaking phase. We concluded that cats inferred the existence of the object from noise of it rattling inside the container and expected it to drop out when the container was turned upside down, in accordance with physical laws.

Cats' causal understanding has been regarded as low-level (Bradshaw et al. 2012); for example, cats did not pass a string-pulling task which required causal understanding between string and rewards (Whitt et al. 2009). However, we demonstrated that cats made inferences from auditory information, something that human children do not do until 3–4 years of age. This feline ability can be understood in terms of the ecological features of cats' natural hunting style. They are ambush-style predators, often hunting prey in a low visibility (Turner and Meister 1988) where the ability to infer unseen objects (prey) from noises is presumably advantageous.

Prioritizing Social Cues over Inferential Reasoning

It is advantageous to be able to adjust one's own foraging behavior based on the amount of food that is available at a given time. Individuals in many socially foraging species follow their companions differently depending on the availability of food at a feeding site (e.g., chimpanzees: Hirata and Matsuzawa 2001; dogs: Cooper et al. 2003; capuchin monkeys: Takahashi et al. 2015). If food is abundant, then following others could be advantageous, but if the amount of food appears restricted, a better strategy would be to visit another food site instead. In one comparative study in captivity capuchin monkeys, but not tree shrews, rats, or hamsters showed this kind of inference (Takahashi et al. 2015). The monkeys followed a companion when the food site had plenty of food, whereas they chose an alternative food site when the companion re-visited a site at which all the food had already been eaten by the companion. Note that the subjects did not actually see the companion eating the food, because their view was blocked by a screen; instead, they simply watched the companion visiting one of the food sites. Takahashi et al. (2015) concluded that this monkey inferred the absence of food and adjusted their behavior based on behavioral cues picked up from the companion.

Dogs, who appear to be adept at reading social cues from humans, can sometimes misjudge a situation because of their sensitivity to such cues. For example, Szetei et al. (2003) reported that dogs spontaneously chose a container that contained food when only olfactory cues were available. However, if an experimenter pointed towards an empty container, they chose that one instead, ignoring the olfactory cues from the baited container. This outcome probably reflects a side-effect of domestication during which dogs came to attend carefully and respond to social cues given by humans. Do other domesticated animals show the same tendency?

Chijiiwa et al. (2020) compared dogs and cats in a social inference task. Subjects were presented with two containers with one piece of food in each, which the subject saw in advance. There two main experimental conditions: "eating" and "showing." In the eating condition, the subject observed an experimenter pretending to eat the food from one of the containers. The subject was then allowed to approach one of the containers. In the showing condition the experimenter simply went to one container picked it up, showed it to subject, then replaced it in its original position. It was predicted that if subjects inferred the absence of food from the experimenter's behavior, they would visit the non-visited container in the eating condition and the visited container in the showing condition.

The results showed that dogs and cats both approached the visited container more frequently than chance in both conditions, suggesting that both species had a tendency to prioritize and follow human social cues. However, when we compared between conditions we found a significant difference: subjects followed the experimenter more in the showing condition than the eating condition. Furthermore, we confirmed that both dogs and cats both spontaneously chose a baited container that was presented alongside an empty one. In sum, the animals' choices were heavily affected by human action: they preferred the container human visited, despite presumably knowing it was now empty. Note that they did not merely follow human "automatically," because they chose the human-visited container less often in the eating condition than the showing condition. Perhaps they inferred the absence of food from the human action, but their behavior was still heavily affected by the actions they observed. So, despite their quite different domestication histories, cats and dogs responded similarly to human actions. Future studies should include other domesticated species and help to clarify whether the results are attributable more to the effect of domestication or the consequence of living with humans.

Incidental Memory

Incidental memory is a memory which is not a result of active encoding at the time when the event occurred (Zentall et al. 2001). For example, we can recite what we ate for breakfast in response to suddenly being asked by a friend: "What did you eat this morning?" It was thought to be difficult to examine incidental memory in animals because it was necessary to test just once, because repeated testing would leave open the possibility of subjects somehow preparing to answer on an upcoming trial.

Fujita et al. (2012) introduced a simple memory task to examine whether dogs retrieve and utilize incidentally an encoded memory from a single previous experience. In an "exposure phase" dogs were allowed to eat food from two of four baited containers, but their owners did not allow them to eat from the other two. After 10 min delay, dogs were unexpectedly returned the experimental room and allowed to freely explore the containers (test phase). If dogs recalled the incidentally encoded memory from the exposure phase, they were expected to preferentially approach the baited-uneaten containers. Simple associative learning theory, by contrast, would predict that dogs would choose the baited-eaten containers, in an example of simple one-trial leaning. Results showed that most dogs chose the baited-uneaten containers, which suggested retrieval of the incidentally encoded memory from a single event, and that the dogs retrieved a "where" memory from past experience.

We applied the same procedure as Fujita et al. (2012) to cats (Takagi et al. 2017). In Experiment 1, cats were presented with four baited containers. They were allowed to eat food from two of them but were not allowed to eat from the other two, as in the dog study. In the test phase, after a 15-min delay, cats tended to choose the baited-uneaten container, similar to dogs. Although the result was only marginally significant, it suggested that cats resembled dogs in retrieving the incidentally encoded memory of "where the food was" from a single past experience. Our next question was about what aspects of memory they could retrieve other than "where" information.

In Experiment 2, to examine whether cats retrieve information of "what was inside the container" as well as "where the food was," we changed the container contents, in the following procedure: Cats were presented with four containers; two were baited, one contained a neutral, non-edible object, and one was empty. If cats retrieved only "where" information, they should visit the baited-uneaten container and the container-with-object randomly. However, if they retrieved and utilized "what + where" information in an integrated fashion, they should visit the baited-uneaten container more often than the other container.

Results showed that cats visited the baited-uneaten container more often than the other container and spent more time to explore the container. We concluded that cats can integrate "what + where" memory in an integrated fashion and from a single event, similar to dogs. The performances of dogs and cats in these studies were remarkably similar. The incidental encoding of memories might be shared among carnivores, or more generally mammals, in view of results reported for other mammalians reported for other species (dolphins (*Tursiops truncatus*): Mercado et al. 1998; rats (*Rattus norvegicus*): Zhou et al. 2012). Incidental memory is thought to be related to the other cognitive abilities such as future planning, imagination, and even creativity (Szpunar et al. 2013), because episodic memory including incidental memory would be beneficial when mentally simulating future events. Whether cats engage in future planning is a topic for further study.

Cross-Modal Integration

Humans routinely integrate visual and auditory information. For example, we can recall our mother's face when hearing mother calling our name because we can form an integrated, cross-modal concept of "mother" (Adachi et al. 2007). This helps us to identify people easily in different environments. Giving that even 4- to 6-month-old infants can perceive face–voice relatedness in unfamiliar adults (Bahrick et al. 2005), this cross-modal integration may be a fundamental ability for life in a complex social society.

Many non-human animals also have cross-modal concepts of others. Some have cross-modal representations of conspecifics (e.g., rhesus macaques (*Macaca mulatta*): Adachi and Hampton 2011; lions (*Panthera leo*): Gilfillan et al. 2016; crows (*Corvus macrorhynchos*): Kondo et al. 2012), some animals looked after by humans can also form cross-modal integration of this different species, (squirrel monkeys (*Saimiri boliviensis*): Adachi and Fujita 2007; dogs (*Canis familiaris*): Adachi et al. 2007). The latter study used a cross-modal expectancy violation method, and found that dogs looked at a monitor longer in incongruent conditions where a stranger's face appeared in the monitor after the owner's voice calling the subject's name had been broadcast, and vice versa. These results showed that dogs recalled their owner's face upon hearing their voice; in other words, dogs have a cross-modal concept of their owner.

We examined whether cats also have a cross-modally integrated concept of their owner (Takagi et al. 2019). Although cats are able discriminate their owner's voice from a stranger's (Saito and Shinozuka 2013), it was not clear whether they recall their owner's face upon hearing his or her voice. Given that responses to social stimuli differed between house cats and café cats (Saito et al. 2019), we compared the two types. The procedure was almost the same as in Adachi et al. (2007). Cats were presented with a monitor with a speaker attached, and either the owner's or a stranger's voice was played back through speaker. Immediately after the voice playback, either the owner's or a stranger's face appeared on the monitor. If cats have cross-modal concept of their owner, they should look at the monitor for longer in the incongruent condition, as predicted by the expectancy violation approach.

Results showed that café cats looked at the monitor for longer in both incongruent conditions—when voices and face were mismatched—suggesting that they integrated auditory-visual information and had a cross-modal concept of owner (Fig. 17.2). Unlike the café cats, house cats appeared to attend to the monitor

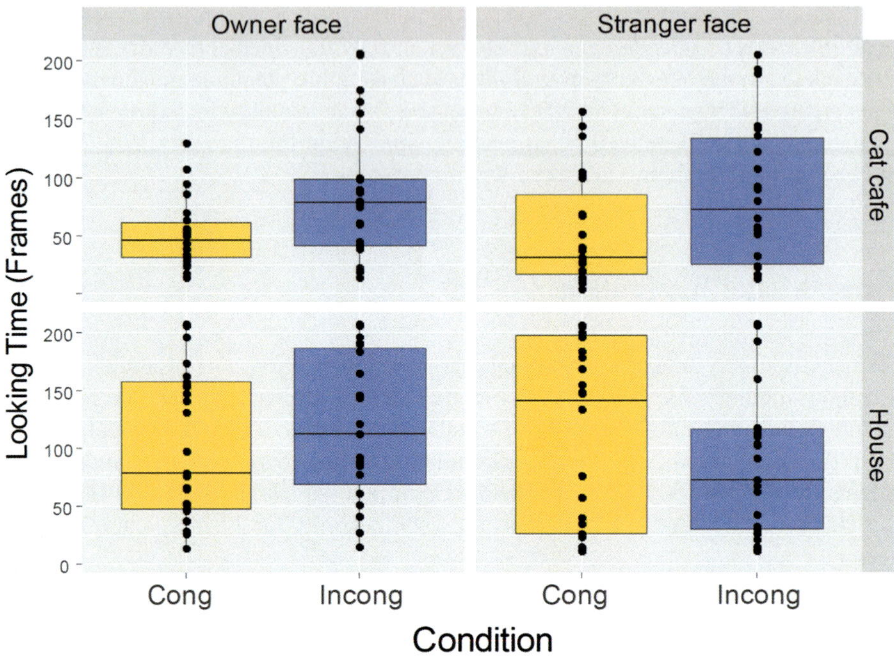

Fig. 17.2 Time spent looking at the monitor in café cats (top) and house cats (bottom). Cong: Congruent; Incong: Incongruent

randomly. In particular, house cats looked at the monitor for longer when a stranger's face appeared after a stranger's voice than when a stranger's face appeared after the owner's voice; contradicting our prediction. Could these different responses be explained in terms of the two groups of cats' rearing environments? One factor might be the cats' previous experiences with strangers. Café cats interact with strangers almost every day, whereas for most house cats only comparable experience might be when a visitor comes to their house. Possibly, house cats' responses to a stranger override any expectancy violation effect. This idea is consistent with house cats' higher attention scores in the stranger voice-stranger face condition. This is an issue for further studies, for which is might be useful to use stimuli such as "mother" vs "father," or to include information about frequency of visitors.

Jealous of Another Cat?

It has been suggested that humans and animals have similar primary emotions such as happiness, fear, and surprise, that emerged through a brain-affective network supposedly homologous in all mammals (Panksepp 2010). These emotions help an individual to quick decide how to behave in a particular situation for survival. In addition, we humans have secondary emotions such as jealousy, envy, or relief that are proposed to arise from higher cognitive processes (Becker-Asano and Wachsmuth 2010). Jealousy has been described as a secondary emotion experienced when an important relationship is perceived to be under the threat from a potential rival (Dillon 2013), and its main function is to protect and maintain the valued relationship. A primary form of jealousy (one that does not require complex cognitive abilities) might be present in human infants (Hart and Carrington 2002; Hart et al. 2004). For example, infants as young as 6 months displayed more jealousy-related behaviors (including negative affections and seeking proximity to the mother) after witnessing their mother interacting with a lifelike baby doll (potential rival) than when she interacted with a non-social object (Hart and Carrington 2002). Similar studies have been conducted with dogs and their owners (Harris and Prouvost 2014; Prato-Previde et al. 2018; Abdai et al. 2018). Abdai et al. (2018) reported that dogs responded with owner-directed behaviors more often when their owner interacted with another dog than with a non-social object, suggesting a primary form of jealousy in this species (but see Prato-Previde et al. 2018).

It has been reported that cats form an emotional attachment to their owner, comparable to the attachment of most human infants to their mothers, and dogs to their owners (Vitale et al. 2019). Cats can clearly distinguish between their owner and a stranger (Saito and Shinozuka 2013; Takagi et al. 2019). However, it is still unclear whether cats possess a secondary emotion such as jealousy. Bucher et al. (2020) assessed whether cats have a primary form of jealousy by using a procedure similar to those used with human infants and dogs. The cat's owner and a stranger (unfamiliar experimenter) sat side by side in front of the subject, which was gently restrained by a second experimenter. The owner and stranger alternately petted a "social" object represented by a realistic-looking soft-toy cat (potential rival), and a non-social object represented by a furry cushion. After each petting action, the subject was released free to explore the testing area. If cats possess a primitive form of jealousy, they should show jealousy-related behaviors more often when their owner (important relationship) petted the soft-toy cat (potential rival). Bucher et al. (2020) analyzed the responses of house cats and café cats separately, as house cats are known to form stronger relationships with their owners than café cats (Saito et al. 2019).

The data showed that house cats in particular reacted more intensely toward the soft-toy cat previously petted by their owner; they looked at and interacted with this "social" longer. Although this result might be taken as support for a jealousy-like type of response in cats, unlike infants and dogs, however, cats did not express other behaviors indicative of jealousy, such as trying to get the owner's attention, or aggression toward the "social" object. Thus, it would be premature to draw firm conclusions about the existence of a primary form of jealousy in cats. One plausible interpretation of Bucher et al. (2020) results was that after a brief interaction with the soft-toy cat, subjects quickly realized that it was not a real cat, and therefore not a potential threat. A similar issue has been raised in dog studies, and some researchers have used real dogs as potential rivals (Abdai et al. 2018). Although it could be difficult to use unknown conspecifics with cats, such an approach should be considered to shed further light on this phenomenon.

Third-Party Evaluation

We often evaluate other people based on their interactions with third parties, even when such interaction is of no direct relevance to us. For example, if we observe someone being cruel to a turtle, we evaluate the person as "bad," even if he or she does not cause us any harm. This cognitive ability to socially evaluate others in third-party contexts plays a role in the highly cooperative societies that characterize humans (Nowak 2006). Researchers have examined whether other species share this cognitive ability.

A study of dogs showed that they did evaluate a person who declined their owner's request for help as negative (Chijiiwa et al. 2015). In the experiment, dogs observed a series of events in which owner's request was refused by an experimenter (Non-helper condition). Subsequently, dogs avoided taking food from the non-helper, choosing to take food instead from another, "neutral" person, who had not interacted with the owner. Interestingly, in a "Helper" condition in which the experimenter responded positively to the owner's request, dogs did not preferentially choose this "good" person. In other words, the dogs showed a negativity bias, which possibly evolved earlier than a positivity bias due to greater survival value (Abdai and Miklósi 2016). Similar results have been reported for young human infants (Hamlin et al. 2010) and capuchin monkeys (Anderson et al. 2013).

How do cats react to people in a third-party context? Chijiiwa et al. (2020) replicated the dog experiment with cats, although using a procedure similar to Chijiiwa et al. (2015). Cats observed an interaction involving their owner and two experimenters. The owner sat in the center, with one experimenter (actor) to one side of the owner and the other experimenter (neutral person) on the other side (Fig. 17.3a). The owner tried to open a transparent container to take out an object from inside the container (Fig. 17.3a). Then the owner asked the actor for help to open it (Fig. 17.3b). In the helper condition, the actor responded by helping her to open it, by holding the bottom of the container (Fig. 17.3c). The owner successfully took the object out and showed it to the subject (Fig. 17.3d). In the non-helper condition, the actor refused the owner's request by simply turning their head away from the owner (Fig. 17.3e), resulting in the owner failing to open the container, and giving up (Fig. 17.3f). After this exposure phase, the actor and the neutral person simultaneously offered a piece of food to the cat (Fig. 17.3 g). If cats evaluated the actor as a "bad" or a "good" person, they should avoid the unhelpful actor, but possibly choose the helpful one.

However, the results showed that cats' behavior was not differentially affected by the two conditions; they neither avoided a non-helper nor preferred a helper. It is therefore possible that cats do not evaluate people from third-party perspective.

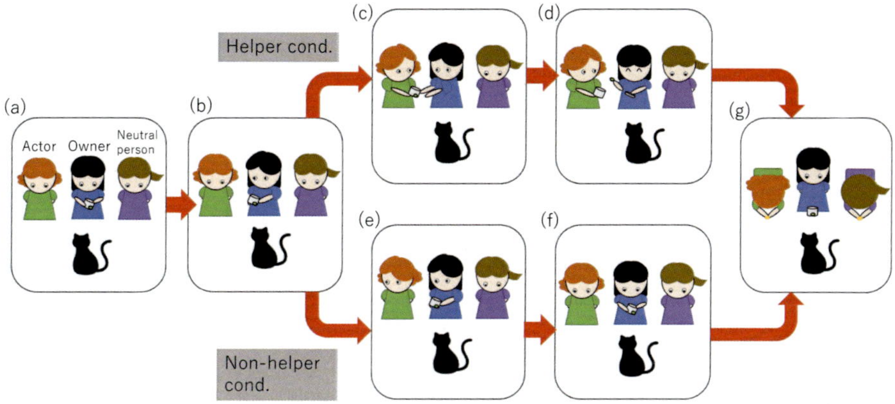

Fig. 17.3 Diagram of third-party evaluation experiment in cats. The owner sat between one person to her left ("actor") and another to her right ("neutral"). (**a**) The owner tries to open the container. (**b**) The owner requests help from the actor. (**c**) The actor helps to open the container. (**d**) The owner opens the container and take the object out. (**e**) The actor refuses the request by turning her head away. (**f**) The owner resumes trying to open the container, but fails. (**g**) The actor and the neutral person simultaneously offer a piece of food to cats

Similar results were recently reported by Leete et al. (2020), who tested cats' discrimination between two unfamiliar experimenters who were either "friendly" or "aggressive" to another cat. To understand the lack of social evaluation in cats, we need to consider that they evolved from African wildcats (*Felis lybica*), a solitary species. By contrast, dogs are descended from group-living, cooperatively hunting ancestors. Furthermore, cats have not been artificially selected for cooperative work with humans (Driscoll et al. 2009a). This background might have contributed to cats' apparent lack of third-party social evaluation tendencies.

Another possibility is that in Chijiiwa et al. (2020) cats did not understand the interaction between the owner and the actor. The extent to which cats can perceive people's goals remains to be studied. We suggest that it is too early to draw any conclusions from these studies (Leete et al. 2020; Chijiiiwa et al. 2020), and that further research is required, including other third-party situations. As mentioned, the differences in responses of dogs and cats in these experiments might reflect ecological aspects of their ancestors or differences in domestication history, and further research is required to clarify their respective contributions.

Evolutionary Changes in Cat Personality

The domestication history of cats is unique compared to other domesticated animals, in that human did not get involved in their breeding until relatively recently (Driscoll et al. 2009b). In general, when humans start to domesticate an animal, they control its breeding for some specific purposes (Price 2002). But cats were not bred for specific reasons at the start of their domestication; they were wanted just for their natural hunting of prey, in particular small rodents that ate stored grains (Driscoll et al. 2009b). Therefore it has been proposed that cats are "semi-domesticated" animals (Serpell 2000). Although the morphology of domestic cats shows some changes from that of from the ancestral African wildcats (*Felis lybica*) as a result of domestication (Wilkins et al. 2014), the changes appear minor compared to the differences between domestic dogs and wolves.

Minori Arahori, a member of the cat research team in Fujita's lab, has studied genetic polymorphisms, rather than morphological changes in cats. It would be difficult for a wary, wild animal to live with another species, much larger than itself, without any changes to its personality. It seems likely that cats who tamed more easily, and who were friendlier and more sociable with humans survived, resulting in continued genetic representations in future generations. Among many polymorphisms, Arahori looked at oxytocin receptor gene (*OXTR*) polymorphisms, which in humans and other primates are thought to be associated with social traits such as empathy and cooperation (Wu et al. 2012; Haas et al. 2013; Staes et al. 2016). As there were no studies linking genetic variations to social behavior or personality in cats, Arahori et al. (2016) firstly examined whether oxytocin receptor gene polymorphisms occurred in cats, and whether the variation was associated with cat personalities as rated by the owners. They found that oxytocin receptor gene polymorphisms in exon but not change amino acid (synonymous) existed in cats. Moreover, cats with the *A* allele in the single nucleotide polymorphism G738A received significantly higher "roughness" scores than cats without the *A* allele from the owner questionnaire data. These results suggested that genetic variation was associated with cat personality traits.

To assess possible evolutionary importance of these genetic variations, they were compared between mixed breed cats, and Tsushima leopard cats (*Prionailurus bengalensis euptilurus*) as a wild felid relative, by focusing on the microsatellites (short tandem repeats) adjacent to *OXTR* (Arahori et al. 2017; Arahori 2019). Results revealed significant differences in allele frequencies between mixed breed cats and Tsushima leopard cats, suggesting genetic differences. Arahrori et al. (2017) also examined the association between genotype and personality scores of mixed breed cats, using owner-completed personality questionnaires. Not only did we find that cats with longer alleles in some regions were scored higher on "Friendliness," but also that mixed breed cats had longer alleles than Tsushima leopard cats, suggesting that the former had more genotypes that were associated with "Friendliness" trait. Taken together, these results suggest that mixed breed cats have evolved toward being of a friendlier disposition than Tsushima leopard cats. It would be interesting to examine African wildcats, to further clarify the trajectory of domestication process from wild cats to house cats.

From these experiments, it appears that friendliness has become more prominent in domestic cat personality through domestication, whereas physical appearance has changed less. Conceivably, more sociable cats had more opportunities to hunt prey around humans, and so their genes continued into subsequent generations even though humans did not artificially select cats for personality traits. The process, called "self-domestication," is described in the next section.

Future Directions

I have reviewed recent findings about cat cognition and personality, several of the relevant studies having been done in Fujita's lab. We are still at the stage of examining what cats do and do not know. Studies comparing their cognition with that of their ancestor, African wildcats, would be valuable for better understanding the influence of domestication on their cognition. Importantly, unlike dogs who have undergone extensive artificial selection by humans (Wobber et al. 2009), cats have only recently been subjected to similar treatment (Driscoll et al. 2009a). As already mentioned, some researchers refer to cats as "semi-domesticated" (Serpell 2000), with the self-domestication process occurring during cats' history of living alongside humans (Driscoll et al. 2009a; Saito 2018). That is, individuals with high tolerance of novel environment including humans are those who were more likely to survive and reproduce successfully. That way, a new ecological niche opened. Humans also developed genetic, anatomical, behavioral, and cognitive characteristics as a result of a kind of domestication process (e.g., Leach 2003; Theofanopoulou et al. 2018; Hare and Tomasello 2005; Hare 2017). For these reasons, studies on the psychological effects of domestication increasingly gaining attention.

In addition to the effects of domestication, it is important to consider individual differences and the living environment. For example, recent research has revealed that emotional contagion between dogs and owners is affected by the amount of time spent together (Katayama et al. 2019). Cats also respond to differently to social stimuli depending on their living environment (Bucher et al. 2020; Saito et al. 2019; Takagi et al. 2019). Compared to dogs and other domestic animals, cats live in a variety of environments: they may be house cats, café cats, island cats living on high cat-density islands where they are fed by people, and feral cats, living freely outside but close to human settlements. These different environments probably have quite different influences as cats go through developmental stages. Cats are good candidates for studying how environment affects cognition and personality.

Another strength of research on companion animals is that aspects of "bilingualism" can be studied. Increasingly fewer companion animals live with conspecifics; indeed, many spend much more time with another species (usually humans) than with their own. To what extent do such animals shift their communicatory repertoires to adjust to that of humans? Cats may be a suitable animal for examining social and cognitive flexibility and plasticity.

Studies of cognition and personality in cats have just begun. Examination of their "mysterious" minds must include effects of domestication, of living environment, and behavioral flexibility including in sociality.

References

Abdai J, Miklósi Á (2016) The origin of social evaluation, social eavesdropping, reputation formation, image scoring or what you will. Front Psychol 7:1772

Abdai J, Terencio CB, Fraga PP, Miklósi Á (2018) Investigating jealous behaviour in dogs. Sci Rep 8:1–8

Adachi I, Fujita K (2007) Cross-modal representation of human caretakers in squirrel monkeys. Behav Process 74:27–32

Adachi I, Hampton RR (2011) Rhesus monkeys see who they hear: spontaneous cross-modal memory for familiar conspecifics. PLoS One 6:e23345

Adachi I, Kuwahata H, Fujita K (2007) Dogs recall their owner's face upon hearing the owner's voice. Anim Cogn 10:17–21

Albuquerque N, Guo K, Wilkinson A, Savalli C, Otta E, Mills D (2016) Dogs recognize dog and human emotions. Biol Lett 12:20150883

Anderson JR, Kuroshima H, Takimoto A, Fujita K (2013) Third-party social evaluation of humans by monkeys. Nat Commun 4:1–5

Andics A, Gábor A, Gácsi M, Faragó T, Szabó D, Miklósi A (2016) Neural mechanisms for lexical processing in dogs. Science 353:1030–1032

Arahori M. (2019) 家畜化がネコ-ヒト間の愛着関係に及ぼした影響の検討. Doctoral thesis, Kyoto University. (in Japanese)

Arahori M, Chijiiwa H, Takagi S, Bucher B, Abe H, Inoue-Murayama M, Fujita K (2017) Microsatellite polymorphisms adjacent to the oxytocin receptor gene in domestic cats: association with personality? Front Psychol 8:2165

Arahori M, Hori Y, Saito A, Chijiiwa H, Takagi S, Ito Y, Watanabe A, Inoue-Murayama M, Fujita K (2016) The oxytocin receptor gene (*OXTR*) polymorphism in cats (*Felis catus*) is associated with "roughness" assessed by owners. J Vet Behav 11:109–112

Bahrick LE, Hernandez-Reif M, Flom R (2005) The development of infant learning about specific face-voice relations. Dev Psychol 41:541–552

Becker-Asano C, Wachsmuth I (2010) Affective computing with primary and secondary emotions in a virtual human. Auton Agent Multi-Agent Syst 20:32–49

Bradshaw JWS, Brown SL, Casey RA (2012) The behaviour of the domestic cat, 2nd edn. CAB International, Wallingford, UK

Bucher B, Arahori M, Chijiwa H, Takagi S, Fujita K (2020) Domestic cats' reactions to their owner and an unknown individual petting a potential rival. Pet Behav Sci 9:16–33

Call J (2004) Inferences about the location of food in the great apes (*Pan paniscus*, *Pan troglodytes*, *Gorilla gorilla*, and *Pongo pygmaeus*). J Comp Psychol 118:232–241

Call J, Bräuer J, Kaminski J, Tomasello M (2003) Domestic dogs (*Canis familiaris*) are sensitive to the attentional state of humans. J Comp Psychol 117:257–263

Chijiiwa H, Kuroshima H, Hori Y, Anderson JR, Fujita K (2015) Dogs avoid people who behave negatively to their owner: third-party affective evaluation. Anim Behav 106:123–127

Chijiiwa H, Takagi S, Arahori M, Hori Y, Saito A, Kuroshima H, Fujita K (2020) Dogs and cats prioritize human action: choosing a now-empty instead of a still-baited container. Anim Cogn 24:65–73

Cooper JJ, Ashton C, Bishop S, West R, Mills DS, Young RJ (2003) Clever hounds: social cognition in the domestic dog (*Canis familiaris*). Appl Anim Behav Sci 81:229–244

Dillon L (2013) Functional aspects of jealousy across the lifespan. Hum Ethol Bull 28:13–26

Driscoll CA, Clutton-Brock J, Kitchener AC, O'Brien SJ (2009a) The taming of the cat. Genetic and archaeological findings hint that wildcats became housecats earlier--and in a different place--than previously thought. Sci Am 300:68–75

Driscoll CA, Macdonald DW, O'Brien SJ (2009b) From wild animals to domestic pets, an evolutionary view of domestication. Proc Natl Acad Sci 106:9971–9978

Fugazza C, Miklósi Á (2015) Social learning in dog training: the effectiveness of the do as I do method compared to shaping/clicker training. Appl Anim Behav Sci 171:146–151

Fujita K, Morisaki A, Takaoka A, Maeda T, Hori Y (2012) Incidental memory in dogs (*Canis familiaris*): adaptive behavioral solution at an unexpected memory test. Anim Cogn 15:1055–1063
Gilfillan G, Vitale J, McNutt JW, McComb K (2016) Cross-modal individual recognition in wild African lions. Biol Lett 12:20160323
Grastyán E, Vereczkei L (1974) Effects of spatial separation of the conditioned signal from the reinforcement: a demonstration of the conditioned character of the orienting response or the orientational character of conditioning. Behav Biol 10:121–146
Haas BW, Anderson IW, Smith JM (2013) Navigating the complex path between the oxytocin receptor gene (*OXTR*) and cooperation: an endophenotype approach. Front Hum Neurosci 7:801
Hamlin JK, Wynn K, Bloom P (2010) Three-month-olds show a negativity bias in their social evaluations. Dev Sci 13:923–929
Hare B (2017) Survival of the friendliest: Homo sapiens evolved via selection for prosociality. Annu Rev Psychol 68:155–186
Hare B, Brown M, Williamson C, Tomasello M (2002) The domestication of social cognition in dogs. Science 298:1634–1636
Hare B, Tomasello M (2005) Human-like social skills in dogs? Trends Cogn Sci 9:439–444
Harris CR, Prouvost C (2014) Jealousy in dogs. PLoS One 9:e94597
Hart S, Carrington H (2002) Jealousy in 6-month-old infants. Infancy 3:395–402
Hart SL, Carrington HA, Tronick EZ, Carroll SR (2004) When infants lose exclusive maternal attention: is it jealousy? Infancy 6:57–78
Hill A, Collier-Baker E, Suddendorf T (2012) Inferential reasoning by exclusion in children (*Homo sapiens*). J Comp Psychol 126:243–254
Hirata S, Matsuzawa T (2001) Tactics to obtain a hidden food item in chimpanzee pairs (*Pan troglodytes*). Anim Cogn 4:285–295
Kaminski J, Call J, Fischer J (2004) Word learning in a domestic dog: evidence for "fast mapping". Science 304:1682–1683
Katayama M, Kubo T, Yamakawa T, Fujiwara K, Nomoto K, Ikeda K, Mogi K, Nagasawa M, Kikusui T (2019) Emotional contagion from humans to dogs is facilitated by duration of ownership. Front Psychol 10:1678
Kondo N, Izawa EI, Watanabe S (2012) Crows cross-modally recognize group members but not non-group members. Proc R Soc B Biol Sci 279:1937–1942
Leach H (2003) Human domestication reconsidered. Curr Anthropol 44:349–368
Leete J, Vonk J, Oriani S, Eaton T, Lieb J (2020) Do domestic cats (*Felis silvestris catus*) infer reputation in humans after direct and indirect experience? Hum Anim Interact Bull 8:35–53
Maille A, Roeder JJ (2012) Inferences about the location of food in lemurs (*Eulemur macaco* and *Eulemur fulvus*): a comparison with apes and monkeys. Anim Cogn 15:1075–1083
Mercado E, Murray SO, Uyeyama RK, Pack AA, Herman LM (1998) Memory for recent actions in the bottlenosed dolphin (*Tursiops truncatus*): repetition of arbitrary behaviors using an abstract rule. Anim Learn Behav 26:210–218
Miklósi Á, Pongrácz P, Lakatos G, Topál J, Csányi V (2005) A comparative study of the use of visual communicative signals in interactions between dogs (*Canis familiaris*) and humans and cats (*Felis catus*) and humans. J Comp Psychol 119:179–186
Nowak MA (2006) Five rules for the evolution of cooperation. Science 314:1560–1563
Panksepp J (2010) Affective consciousness in animals: perspectives on dimensional and primary process emotion approaches. Proc R Soc B Biol Sci 277:2905–2907
Pet Food Association of Japan (2019) National Survey on Dog and Cat Ownership. Retrieved from https://petfood.or.jp/topics/img/191223.pdf (Access date 8th October)
Pilley JW, Reid AK (2011) Border collie comprehends object names as verbal referents. Behav Process 86:184–195
Plotnik JM, Shaw RC, Brubaker DL, Tiller LN, Clayton NS (2014) Thinking with their trunks: elephants use smell but not sound to locate food and exclude nonrewarding alternatives. Anim Behav 88:91–98

Prato-Previde E, Nicotra V, Poli SF, Pelosi A, Valsecchi P (2018) Do dogs exhibit jealous behaviors when their owner attends to their companion dog? Anim Cogn 21:703–713
Price EO (2002) Animal domestication and behavior. CABI, Wallingford, UK
Range F, Kassis A, Taborsky M, Boada M, Marshall-Pescini S (2019) Wolves and dogs recruit human partners in the cooperative string-pulling task. Sci Rep 9:1–10
Saito A (2018) Why are cats (*Felis catus*) companion animals for humans? A consideration of cat domestication from the perspective of comparative cognitive science. Jpn J Anim Psychol 68:77–88
Saito A, Shinozuka K (2013) Vocal recognition of owners by domestic cats (*Felis catus*). Anim Cogn 16:685–690
Saito A, Shinozuka K, Ito Y, Hasegawa T (2019) Domestic cats (*Felis catus*) discriminate their names from other words. Sci Rep 9:1–8
Serpell JA (2000) Domestication and history of the cat. In: Turner DC, Bateson P (eds) The domestic cat: the biology of its behavior, 2nd edn. Cambridge University Press, Cambridge, UK
Staes N, Weiss A, Helsen P, Korody M, Eens M, Stevens JM (2016) Bonobo personality traits are heritable and associated with vasopressin receptor gene 1a variation. Sci Rep 6:1–8
Szetei V, Miklósi Á, Topál J, Csányi V (2003) When dogs seem to lose their nose: an investigation on the use of visual and olfactory cues in communicative context between dog and owner. Appl Anim Behav Sci 83:141–152
Szpunar KK, Addis DR, McLelland VC, Schacter DL (2013) Memories of the future: new insights into the adaptive value of episodic memory. Front Behav Neurosci 7:47
Takagi S, Arahori M, Chijiiwa H, Saito A, Kuroshima H, Fujita K (2019) Cats match voice and face: cross-modal representation of humans in cats (*Felis catus*). Anim Cogn 22:901–906
Takagi S, Arahori M, Chijiiwa H, Tsuzuki M, Hataji Y, Fujita K (2016) There's no ball without noise: cats' prediction of an object from noise. Anim Cogn 19:1043–1047
Takagi S, Tsuzuki M, Chijiiwa H, Arahori M, Watanabe A, Saito A, Fujita K (2017) Use of incidentally encoded memory from a single experience in cats. Behav Process 141:267–272
Takahashi M, Ueno Y, Fujita K (2015) Inference in a social context: a comparative study of capuchin monkeys (*Cebus apella*), tree shrews (*Tupaia belangeri*), hamsters (*Mesocricetus auratus*), and rats (*Rattus norvegicus*). J Comp Psychol 129:402–411
Theofanopoulou C, Gastaldon S, O'Rourke T, Samuels BD, Martins PT, Delogu F, Alamri S, Boeckx C (2018) Self-domestication in *Homo sapiens*: Insights from comparative genomics. Plos One 12(10):e0196700
Turner DC, Meister O (1988) Hunting behaviour of the domestic cat. In: Turner DC, Bateson P (eds) The domestic cat: the biology of its behaviour. Cambridge University Press, Cambridge, UK
Udell MAR, Dorey NR, Wynne CDL (2008) Wolves outperform dogs in following human social cues. Anim Behav 76:1767–1773
Udell MAR, Dorey NR, Wynne CDL (2010) What did domestication do to dogs? A new account of dogs' sensitivity to human actions. Biol Rev 85:327–345
Vitale KR, Behnke AC, Udell MA (2019) Attachment bonds between domestic cats and humans. Curr Biol 29:R864–R865
Whitt E, Douglas M, Osthaus B, Hocking I (2009) Domestic cats (*Felis catus*) do not show causal understanding in a string-pulling task. Anim Cogn 12:739–743
Wilkins AS, Wrangham RW, Fitch WT (2014) The "domestication syndrome" in mammals: a unified explanation based on neural crest cell behavior and genetics. Genetics 197:795–808

Wobber V, Hare B, Koler-Matznick J, Wrangham R, Tomasello M (2009) Breed differences in domestic dogs' (*Canis familiaris*) comprehension of human communicative signals. Interact Stud 10:206–224

Wu N, Li Z, Su Y (2012) The association between oxytocin receptor gene polymorphism (*OXTR*) and trait empathy. J Affect Disord 138:468–472

Zentall TR, Clement TS, Bhatt RS, Allen J (2001) Episodic-like memory in pigeons. Psychon Bull Rev 8:685–690

Zhou W, Hohmann AG, Crystal JD (2012) Rats answer an unexpected question after incidental encoding. Curr Biol 22:1149–1153

Chapter 18
Dog–Human Attachment as an Aspect of Social Cognition: Evaluating the Secure Base Test

Monique A. R. Udell, Lauren Brubaker, Lauren E. Thielke, Shelby S. H. Wanser, Giovanna Rosenlicht, and Kristyn R. Vitale

Abstract The capacity for dogs to form attachment bonds to humans has been recognized by scientists for over two decades. However, evaluations of dog-human attachment styles, including to what extent dogs experience attachment security with their human caregivers, are relatively new. In humans, the development of secure attachments is considered a predictor of social wellbeing and positive cognitive outcomes including future relationship success, persistence, mental wellbeing and executive functioning. A better understanding of dog-human attachment relationships could have important scientific and applied implications. Here we provide an overview of attachment research as it relates to the dog-human bond, and take a closer look at one experimental approach, the Secure Base Test (SBT), currently used to evaluate dog-human attachment styles.

Keywords Dog cognition · Social cognition · Attachment · Human-animal interactions

Introduction

Among the many aspects of dog cognition that have captivated scientists in recent decades, social behavior and cognition have perhaps drawn the greatest amount of attention. While there are many reasons to be interested in the social behavior of dogs, understanding how and why humans share such a close relationship with these animals, and what allows dogs to thrive within human environments have been important areas of inquiry (Hori et al. 2013; Miklosi et al. 2007; Udell et al. 2010; vonHoldt et al. 2017), not least because many other canids struggle to adapt to anthropogenic environments. Some scientists have suggested that human caregivers have familial-like bonds with their dogs, even regarding them as family members

---

M. A. R. Udell (✉) · L. Brubaker · L. E. Thielke · S. S. H. Wanser · G. Rosenlicht · K. R. Vitale
Department of Animal and Rangeland Sciences, Oregon State University, Corvallis, OR, USA

J. R. Anderson, H. Kuroshima (eds.), *Comparative Cognition*,
https://doi.org/10.1007/978-981-16-2028-7_18

(Arahori et al. 2017). Dogs may learn to respond to their human caretakers in a similar manner (Adachi et al. 2007; Archer 1997; Coppinger et al. 2016). Evidence that pet dogs appear to hold cross-modal representations of familiar humans (Adachi et al. 2007) and show preferences for unfamiliar individuals who are kind to their human caregivers (Chijiiwa et al. 2015) provides further evidence of the importance of these ties. In recent years there has been an increased focus on the attachment bonds formed between dogs and humans. New research has asked to what extent the bonds between dogs and human caregivers might be similar to within-species offspring–caregiver relationships, and what implications the kind and quality of these bonds may have for social development, cognition, and welfare across a range of basic and applied contexts.

What Is Attachment?

From an ethological perspective, attachment can be defined as an affiliative tie between two individuals that promotes proximity seeking and contact between an individual and a caregiver/attachment figure (Bowlby 1958). When separated from their attachment figure, the dependent individual displays characteristic distress vocalizations, such as crying, which help draw the attachment figure near the bonded young (Bowlby 1982; Suomi et al. 1973). Attachment behavior may serve several distinct functions in the lives of animals that require provisioning and care after birth. One of the most obvious functions is helping offspring meet their immediate needs, such as obtaining food, shelter, protection, and other resources from caregivers (Bowlby 1958). In both humans and other species, the formation of attachment bonds has also been identified as important to the development of healthy socio-cognitive skills; lack of attachment bonds early in life can result in a wide range of lasting negative impacts to development, welfare, and social behavior (Cross and Harlow 1965; Rutter 1998; Suomi et al. 1973).

Although most studies of attachment behavior have focused on young animals or human infants and their same-species adult caregivers, cross-species attachments have also been well documented (Lorenz 1935, 1952). For some animals—including companion animals—the primary caregiver may be a human and dependency on that caregiver is often life-long. In such cases, classic manifestations of attachment behavior can persist into adulthood (Gácsi et al. 2001; Topál 1998). The fact that an animal displays attachment behavior towards a bonded individual is also only one piece of the puzzle. We can also evaluate the *quality* of attachment formed between an individual and their caregiver, a factor known to have important implications for welfare, social developmental, executive functioning, and learning outcomes (Udell and Brubaker 2016). Quality of attachment is typically evaluated based on the presence or absence of the Secure Base Effect (SBE), first described by ethologist John Bowlby (Bowlby 1958). The SBE is observed when an individual displays a contact-exploration balance in the presence of their attachment figure. In other words, in addition to seeking caregiver proximity, individuals exhibiting the SBE are also more likely to investigate novel environments and unfamiliar situations while periodically "checking in" with the attachment figure (Bowlby 1958). In this context, the attachment figure serves as a source of security and stress reduction that promotes individual growth and learning about the environment.

Origins of Attachment Style Research

Modern attachment theory has origins in both evolutionary biology and ethology. Although the public popularity of human attachment research sometimes creates the impression that attachment theory first developed within human psychology, in fact its origins stem from research on non-human animals (Harlow 1958; van der Horst et al. 2008). Beyond the early field-based studies of imprinting conducted by Lorenz and others, one early experimental approach to studying attachment and security (the SBE) comes from research conducted on infant macaque monkeys by Harry Harlow (1958). In this research, newborn monkeys were separated from their biological mothers, and some were reared only with inanimate, cloth-covered, wire mesh "surrogate mothers." Periodically the infants were assessed in an open field test, where they were introduced alone into an unfamiliar room containing a variety of toys and other objects. This strange situation was designed to serve as a mild stressor that would elevate the infant's attachment behavior towards its cloth mother surrogate upon reunion. The monkey was left in the room for 3-min sessions. If their cloth surrogate mother was absent, the monkeys emitted distress cries and showed little exploratory behavior, instead they froze, remaining immobile. If the cloth surrogate mother was present, the infant monkeys with secure attachments more readily explored their environment and periodically returned to the surrogate, before continuing to explore the novel room and stimuli, thus achieving contact-exploration balance.

John Bowlby, one of the founders of modern human attachment theory, became interested in Harlow's studies (van der Horst et al. 2008) and along with Mary Ainsworth, experimentally investigated attachment bonds in human infants under similar conditions to the open field test conducted by Harlow. What became known as the human Strange Situation Test (SST) consisted of seven episodes, each lasting 3 min, and examined how a human infant behaved in an unfamiliar room in the presence of their caregiver, a stranger, both individuals, and when alone. As in Harlow's research, the most critical episode for attachment style classification was the reunion phase with their caregiver after an absence (Ainsworth and Bell 1970; Bowlby 1982). As in the earlier research, the SBE was used as evidence of secure attachments, with secure individuals displaying open and positive greeting during the reunion and a contact-exploration balance. Children classified as insecurely attached showed evidence of stress in response to being left alone, as well as reduced exploration of the room even upon their caretaker's return. Several distinct categories of insecure attachment were also identified. Children were classified as Insecure Ambivalent if during the caregiver return phase they showed excessive proximity seeking and clinging towards their caretaker while remaining inconsolable. Others, while initially distressed by the absence of their caretaker, excessively avoided their caretaker during the return phase. These children were classified as Insecure Avoidant. Children showing inconsistent patterns of behavior, and those displaying stereotypies or approach-avoidance conflict were classified as having an Insecure-Disorganized attachment (Bretherton 1992).

Over the last several decades, interest in the dog–human relationship has grown exponentially. After at least 14,000 years of human-dog cohabitation (Nobis 1979), it seems fitting that researchers have at last begun to evaluate the nature of the bonds formed between dogs and humans. One of the first behavioral studies specifically evaluating dog–human attachment (Topál et al. 1998) was conducted using a modified version of the Strange Situation Test (Ainsworth and Bell 1970). The researchers looked at the amount of time dogs spent playing and exploring in an unfamiliar environment when their human caregiver (or a stranger) was present, time in contact with the door, and time in contact with the human. Three main factors were found to account for dog behavior during the SST: anxiety (the amount of distress displayed in the unfamiliar environment), attachment (the dog's response to being separated from their caregiver), and acceptance (how the dog reacted to the stranger). Additionally, attachment styles of the dogs were evaluated based on the individual behavioral measures using a cluster analysis. Three main groups (and multiple sub-groups) were identified among the dog participants: Group 1 dogs were less anxious, but varied in their degree of preference for owner vs stranger, Group 2 dogs displayed high levels of anxiety during the stressful situation and also high levels of contact seeking upon caretaker return (but also upon the return of the stranger), Group 3 dogs displayed medium levels of anxiety and acceptance, and varied in behaviors associated with owner attachment, with one subgroup showing few attachment behaviors and the other subgroup engaging in more proximity seeking. Although the attachment style classification methods and categories were not the same as those used in previous research, this initial investigation revealed that the dog–human bond shared many qualities of the attachment relationships previously observed in human parent–child relationships, and that different patterns of attachment behavior exist in dog–human relationships, paving the way for future investigations into dog–human attachment (Topál et al. 1998).

Dog–Human Attachment Methodology

Following the study by Topál et al. (1998) methodological and interpretive concerns were raised about the application of the SST to studies of the dog–human bond. For example, several studies demonstrated that order effects, inherent to the full version of the SST methodology, can alter the way dogs respond the owner or stranger depending on the order in which they entered after the dog had been left alone (Palmer and Custance 2008; Rehn et al. 2013). In addition, the traditional version of the SST adapted for dogs did not include a condition where their caregiver returned directly after an absence; instead, the phase in which dogs remained alone for 3 min was immediately followed by the stranger entering the room and interacting with the dog before the caretaker returned (Prato-Previde et al. 2003). Furthermore, length of testing time (> 21 min for the full SST) and differences in how dog attachments were categorized compared to the human literature were raised as additional concerns (Rehn et al. 2013; Thielke et al. 2017; Wanser and Udell 2019).

Some researchers have since returned to the original attachment test (the foundation of the SST) adapted from Harlow's open field test (Harlow 1958). This Secure Base Test (SBT) also takes place in an unfamiliar room with toys present, and consists of three 2-min phases: a baseline phase in which the dogs primary human caregiver may interact with the dog when it engages in proximity seeking, an alone phase in which the dog remains alone in the room, and a return phase in which the human returns and may again interact with the dog. As in prior studies, the return phase is considered the most relevant for categorizing attachment style, as it allows assessment of dogs' responses to their human caregiver after exposure to a stressful event—being left alone in an unfamiliar room (Rehn et al. 2013; Waters 1978). The SBT presents several advantages for evaluation of attachment relationships between dogs and caretakers. Total testing time is 6 min instead of 21 min, making it more practical in applied settings, order effects are eliminated (as stranger conditions are absent), and dog-caretaker reunion occurs immediately following the alone phase. Although behavior in this test can be analyzed in many ways, it is often used to classify dogs into attachment style categories using an ethogram and methods similar to those used with other species, including human infants (Thielke et al. 2017; Thielke and Udell 2019, 2020).

Another important consideration is how to analyze and categorize the behavior exhibited by dogs in tests of dog–human attachment. Among many possibilities, two primary methods have emerged in the literature. One method has involved coding individual behaviors (duration and/or frequency) and then using statistics to determine if behavioral patterns are consistent with attachment predictions or statistical models to classify groups of dogs displaying the same patterns of behavior. For instance, in one study, researchers used a modified version of the SST and recorded the proportion of time spent in proximity to the owner, time spent in proximity to a stranger, time spent in locomotor activity, and the frequency of vocalizations during each phase of the test. The relative proportions of time or frequency of these behaviors were then compared across phases to evaluate the relative influence of the owner or stranger's presence over the course of the SST (Parthasarathy and Crowell-Davis 2006). Other researchers have used this method in conjunction with a factor analysis or cluster analysis, identifying clusters of behaviors associated with attachment style categories (Topál et al. 1998). Although such studies have yielded important results, one challenge is that differences between the resulting attachment style categories (as well as different methods and criteria used for classification) have made direct cross-species comparisons difficult. Classification methods that rely on large sample sizes and statistical models also pose challenges for use in applied contexts. For example, such methods would be of little value for practitioners or animal behaviorists who would benefit from a test for identifying attachment style or security of a single dog to their caretaker or handler. Therefore, while the canine SST and statistical modeling methods may provide valuable insights into attachment behavior, it is important to also consider more holistic approaches to classification.

A Closer Look at the Secure Base Test (SBT)

Because the abbreviated structure and user-friendly analysis of the SBT make it immediately suitable for applied contexts, it is especially critical to ensure that the results it produces are reliable, stable over time and that the test is measuring what it is intended to measure. Using previously collected videos of dog–human attachment tests (including Thielke et al. 2017; Wanser and Udell 2019 and unpublished data) we were able to evaluate a larger sample of data than has been previously evaluated in the literature, including repeated testing of the same dog–human pairs. We also looked at attachment videos for two different variations of the SBT, one that was more restrictive on human caregiver behavior and one where the human could behave more freely, to determine if this influenced attachment classifications. Our basic aim here was to further probe the robustness of the SBT as a measure of attachment style in dog–human relationships, to assess its efficacy, and identify possible limitations in terms of future applied contexts.

For this analysis we looked at 207 videos of dogs completing the SBT with their primary adult caregiver. Of these videos, 128 were from the first, or in some cases the only, SBT the dog-owner pair had experienced. The dogs included 71 females and 57 males of various breeds and mixes, ranging in age from 6 months to 15 years (mean age 5.68 years) at the time of their first or only visit. The remainder of the videos were from subsequent visits by some of the same dog-human pairs; 65 dogs had complete second visit data and these were used to evaluate test–retest reliability.

Each testing room was unfamiliar to the dog at the time of testing and was barren except for a chair with a 1-meter radius circle taped on the floor around it (to facilitate coding for proximity seeking) and three dog toys. The SBT consisted of three 2-min phases. In baseline, the caretaker sat on the chair and could freely interact with the dog when it entered the circle with at least two paws (including playing, petting, talking, etc.). The human caregiver was instructed to remain neutral when the dog was not in the circle. In the alone phase, the human left the room, leaving the dog alone. In the final phase, the human caregiver returned, sat on the chair, and instructions were identical to baseline. Some dogs were tested using a restricted petting SBT methodology, allowing us to evaluate this aspect of caregiver behavior as a possible variable. In the restricted methods, the human caregiver was instructed that they could only pet their dog twice if it entered the taped circle, but then they had to stop petting unless the dog exited and then re-entered the circle.

Two forms of behavior coding/classification were used. *Holistic Coding* was based on the behavioral pattern of the dog over either the full SBT (6-min) or return phase only (2-min) version of the test under evaluation. See Table 18.1 for classification categories. Two researchers coded each video independently. After

Table 18.1 Holistic coding: dog–human attachment style categories

Secure: Little or no resistance to contact or interaction. Greeting behavior is active, open, and positive. Seeks proximity and is comforted upon reunion, returning to exploration or play
Insecure Avoidant: May show little/no distress on departure. Little/no visible response to return, ignores/turns away but may not resist interaction altogether (e.g., rests or stands without bodily contact, out of reach or at a distance)
Insecure Ambivalent: Shows exaggerated proximity seeking and clinging behavior, but may struggle if held by owner. Mixed persistent distress with efforts to maintain physical contact and/or physically intrusive behavior directed toward the owner. (Dogs who the judges agreed seemed essentially secure but with ambivalent tendencies were included in the secure group)
Insecure Disorganized: Evidence of strong approach-avoidance conflict or fear on reunion, for example, circling owner, hiding from sight, rapidly dashing away on reunion, "aimless" wandering around the room. May show stereotypies on return (e.g., freezing or compulsive grooming). Lack of coherent strategy shown by contradictory behavior. "Dissociation" may be observed, that is, staring into space without apparent cause; still or frozen posture for at least 20 seconds (in the non-resting, non-sleeping dog)
Unclassifiable: Classifiers unable to reach consensus on final placement for dogs from this classification category

Table 18.2 Duration coding ethogram

Proximity seeking: Laying, sitting, or standing inside of the blue circle taped around the owner's chair
Exploring: Moving around the room or walking in a non-repetitive manner (i.e., not pacing)
Object play: Picking up/making contact with toys
Avoiding: Sitting, standing, or laying out of reach outside circle.

independent inter-rater reliability was calculated, the two coders met to review any videos with different classifications together to determine the final attachment style classification. *Duration Coding* of individual behaviors was conducted using the behavior analysis coding application *Countee*. See Table 18.2 for Ethogram. Coders were different individuals from those who completed the holistic coding and were blind to the attachment style classifications of the dogs, to ensure that duration coding was completely independent from the holistic coding.

Consistent with prior research (Thielke et al. 2017), all videos (100%) were analyzed by two independent coders for the holistic coding method. 70 out of a total of 207 SBT videos were also analyzed by an independent coder for all duration measures (34%) during the return phase of testing to evaluate inter-rater reliability. Strong inter-rater reliability was present when coding secure base videos for both holistic and duration measures. Holistic Attachment Style Classifications: 79% IRR, Binomial Test, $p < 0.001$, 100% were classified together during the final classification phase of coding. Duration IRR for Proximity Seeking: Pearson correlation $x = 0.81$, $r(68) = 0.90$, $p < 0.001$. Duration IRR for Exploration: Pearson correlation $R2 = 0.63$, $r(68) = 0.79$, $p < 0.001$. Duration IRR for Avoidance: Pearson correlation $R2 = 0.85$, $r(68) = 0.92$, $p < 0.001$. Duration coding for Object Play, Pearson correlation $R2 = 0.96$, $r(68) = 0.98$, $p < 0.0001$.

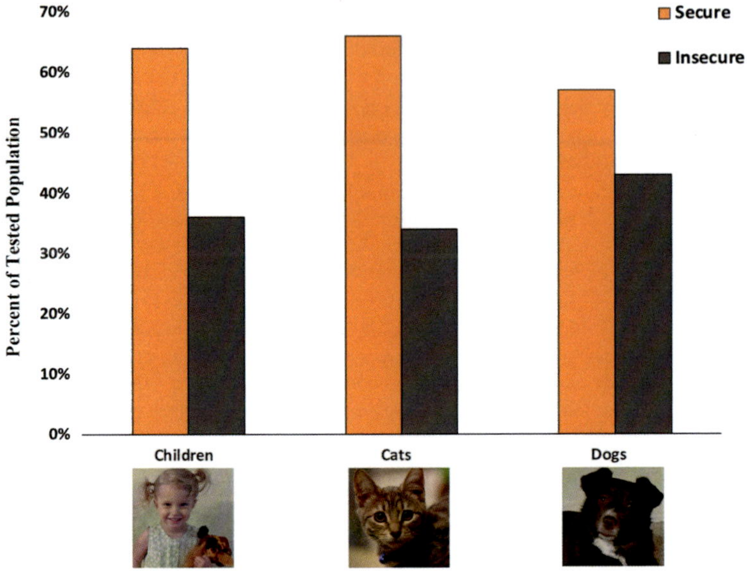

Fig. 18.1 Percentage of individuals classified as securely attached or insecurely attached to their caretakers in attachment research on human infants (Waters 1978), cats (Vitale et al. 2019), and in the current study with pet dogs. Photo credit: Monique Udell and Kristyn Vitale

Analysis of Compiled Data

We had complete data sets for the first visit of 128 dogs. At the time of their first attachment test, 73 (57%) of these dogs were categorized as having a secure attachment to their human caregiver, 42 (33%) were categorized as having an insecure ambivalent attachment, 10 (8%) had an insecure avoidant attachment, and 3 (2%) had an insecure disorganized attachment. Given the small number of dogs in the disorganized category, this classification was not included in the remaining statistical analyses. These relative percentages of attachment classification are similar to those reported in the human infant literature (Waters 1978) and also in a recent study on cat–human attachment style (Vitale et al. 2019) (Fig. 18.1). There were not significant differences in the average age of dogs (H (2,125) = 3.87, p = 0.14) or in the number of male and female dogs within each attachment style category (Fisher's Exact Test, p = 0.66).

Internal Validity

To assess internal validity, we looked at four primary patterns of behavior that would be expected to differ between dogs within each holistic attachment style classification based on attachment theory: contact/exploration balance, proximity seeking, avoidance, and object play. We then analyzed independently coded duration data to determine if dogs classified into different attachment categories displayed significantly different behavioral patterns for these measures. For this analysis, we used only the first visit data (128 dogs) to avoid repeated measures for some dogs.

Attachment theory predicts that secure individuals should have a contact-exploration balance: On average, secure individuals spent roughly 57% of the return phase seeking the proximity of their human caregiver and 54% of the time engaged in play + exploration, striking almost perfect balance between contact and exploratory behavior. Note: total percentage can be over 100% because object play is not mutually exclusive from the other categories.

Insecure ambivalent individuals should engage in prolonged proximity seeking compared with secure and avoidant individuals: Ambivalent individuals spent more time, 94% of trial duration, in close proximity to their caregiver upon return from a brief absence compared with dogs classified as secure or avoidant (Kruskal–Wallis, $H(2,125) = 72.83$, $p < 0.001$). At the group comparison level, secure dogs spent significantly less time in caregiver proximity, 57%, than insecure ambivalent dogs (Mann–Whitney $U = 241$, $z = -7.50$, $p < 0.001$). Insecure avoidant individuals spent the least amount of time, only 9%, in close proximity to their caregiver during the return phase, significantly less than both insecure ambivalent dogs (Mann–Whitney $U = 0$, $z = 4.86421$, $p < 0.001$) and secure dogs (Mann–Whitney $U = 34.5$, $z = 4.6164$, $p < 0.001$).

Attachment theory predicts that avoidant individuals are more likely to actively avoid contact with their caregiver during the return phase (time spent outside of proximity, but not in play or exploration): Dogs categorized as insecure avoidant spent 70% of their time in avoidance behavior on average, compared to only 9% in secure dogs and only 7% in insecure ambivalent dogs (Kruskal–Wallis, H $(2,125) = 27.21$, $p < 0.001$). As predicted, statistically significant differences for this behavioral measure were found between dogs classified as insecure avoidant and secure (Mann–Whitney U = 10.5, z = −4.95214, $p < 0.0001$) and between dogs classified as insecure avoidant and ambivalent (Mann–Whitney $U = 13.5$, $z = -4.55076$, $p < 0.001$), but not between dogs classified as secure and insecure ambivalent (Mann–Whitney $U = 1259.5$, $z = 1.59$, $p = 0.11$).

Dogs with secure attachments should spend more time engaging with objects, including play: In some cases, duration of time spent engaging with unfamiliar objects in the presence of a human caretaker has been used directly as a measure of the Secure Base Effect in dogs (Horn et al. 2013). Here we asked if dogs classified as secure spent significantly more time engaged with play objects in the room. Ninety-two dogs for which we had appropriate data spent on average 35% of the return phase engaged in object play, compared to only 2% in insecure ambivalent dogs and 0% insecure avoidant dogs (Kruskal–Wallis, $H(2,90) = 21.75$; secure Vs ambivalent, Mann–Whitney $U = 332.5$, $z = 4.37$, $p < 0.001$; secure vs avoidant, Mann–Whitney $U = 66.5$, $z = 2.373$, $p = 0.02$; ambivalent vs avoidant, Mann–Whitney $U = 85.5$, $z = -0.04$, $p = 0.97$). Like the other duration measures, the relative duration of object play behavior was consistent with attachment theory, with securely attached dogs showing the most interest in playing with objects after a stressful event.

Test–Retest Reliability

A total of 65 dogs were tested twice with the same human partner, and these videos were used to evaluate the stability of attachment style classification. The second testing session occurred between 5 days and 829 days after the first visit (mean latency 66 days). Attachment style classifications matched exactly (when considering all four attachment classifications: secure, insecure ambivalent, insecure avoidant, insecure disorganized) for 82% of dogs (Binomial test, $p < 0.001$). Attachment style stability was independent of testing latency (matching outcomes group: $M = 70$ days, different outcomes: $M = 47$ days, $t(63) = 0.54$, $p = 0.59$), suggesting changes in attachment style are not merely a product of time. Instead, established attachment styles toward particular individuals appear robust over time, but as in humans, they may change under some conditions, for example, following significant life-altering events (Bretherton 1992).

We also evaluated test–retest reliability of four duration measures and found positive correlations between the behavior of dogs within the SBT during visits 1 and 2 (proximity seeking, Pearson correlation $R2 = 0.33$, $r(63) = 0.57$, $p < 0.001$; exploration, Pearson correlation $R2 = 0.08$, $r(63) = 0.27$, $p = 0.03$; avoidance, Pearson correlation $R2 = 0.43$, $r(63) = 0.65$, $p < 0.001$; play, Pearson correlation $R2 = 0.39$, $r(63) = 0.62$, $p < 0.0001$).

Restricted vs Unrestricted Petting Methods

Of the 128 dogs, 54 were tested with the human caregiver's behavior restricted (which only allowed two petting interactions per dog approach during the return phase) and 74 were tested with no restrictions on owner petting during the return phase on their first visit. Of these, three dogs were classified as disorganized; these dogs were dropped from the statistical analysis; however, all three were tested using the unrestricted method. We used a Fishers Exact test to determine if there were significant differences in the number of dogs with different attachment styles according to restricted vs. unrestricted petting. No such differences were found (restricted petting methods: 36 of 54 dogs were classified as secure, 13 ambivalent and 5 avoidant; unrestricted petting methods: 37 of 71 dogs were classified as secure, 29 ambivalent and 5 avoidant; Fishers Exact Test, $p = 0.14$). We predicted that if restricted petting influenced dog behavior the effect would be seen in the amount of time spent proximity seeking, as continuous petting by the human might prolong the amount of time a dog spent in close proximity. Despite a noticeable a trend in this direction, there was no significant difference between dogs tested with the two slightly different methodologies (unrestricted: 69% of time on average in proximity; restricted: 59% of time; Mann–Whitney $U = 1592.5$, $z = 1.95$, $p = 0.05$).

Coding Considerations

The return phase (the reunion phase after the caretaker's absence) has often been the only phase used to evaluate attachment security in dogs (Thielke et al. 2017; Waters 1978). To determine to what extent this approach was sufficient, we asked if two independent coders—one watching the full 6-min SBT and one watching only the 2-min return phase of the same SBT—would classify dogs into the same category. We also looked at the independent Inter-Rater Reliability scores when two independent coders evaluated the same 6-min SBT videos, and when another two independent coders evaluated just the 2-min Return phase of those videos.

To this aim 79 attachment videos were analyzed twice, once looking at the full 6-min test, and once looking at only the 2-min return phase. In total, four independent coders, blind to the purpose of the study coded the videos to evaluate attachment style (two for the full-test coding, and two for the return-phase-only coding).

Rate of independent inter-observer agreement (holistic) for the full 6-min SBT was calculated for 35 (44%) of the videos (IRR = 80%), this was equivalent to the IRR for the holistic coding of the return only phase (IRR = 79%). When comparing the final classification style of the same dogs using the two coding methods, IRR was 68%, higher than expected by chance (binomial test, $p < 0.001$). This suggests that, consistent with findings in the human literature (Waters 1978), holistic coding of the 2-min return phase is at least as reliable as coding the full attachment test; it might also be preferable given that doing so cuts the video coding time by 1/3. This may be especially useful in applied settings where time available for testing and analysis may be limited or even determine what evaluations are feasible.

Conclusions

Although the dog–human bond has been categorized as an attachment relationship in the scientific literature for roughly two decades (Topál et al. 1998), the methodologies used to evaluate these such relationships have been evolving to address important concerns about order effects, testing time, and analysis complexities associated with early versions of the SST modified for use with dogs (Rehn et al. 2013). The ability to classify dogs into traditional attachment style categories is now possible, with the SBT serving as one method for addressing confounds in earlier studies, while also providing a more user-friendly assessment for applied applications (Thielke et al. 2017, Thielke and Udell 2019, 2020; Wanser and Udell 2019).

Here we conducted further evaluation of several aspects of reliability, internal validity, testing methodology, and coding approach for the SBT test, based on 207 videos. Overall, consistent with prior dog–human attachment studies (Schöberl et al. 2016; Thielke et al. 2017, Thielke and Udell 2019, 2020; Wanser and Udell 2019) dogs could be readily categorized into the primary attachment styles common to the broader established attachment literature (Bretherton 1992). The percentage of dogs classified into each attachment style category was similar to that previously observed in the human infant–mother, dog–human, and cat–human attachment literatures (Waters 1978; Vitale et al. 2019), using similar coding methods and attachment style definitions.

The SBT methodology was found to have high independent inter-rater agreement, high test–retest reliability, and strong internal validity. Reliability measures were just as high whether evaluating the 2-min return phase of the SBT test or the full 6-min test. While the full 6-min test methodology (2 min baseline, alone, return phases) is still necessary to establish mild distress (during the alone phase) associated with inducing the attachment responses in this test, we have shown the feasibility of cutting the behavioral coding time down from 6 to 2 min per dog without loss of data integrity. We also found that minor variations in caregiver behavior (e.g., restricted versus unrestricted petting) in the return phase did not significantly alter attachment behavior or the final attachment classification in this test. However, it is possible that more substantial variations between methodology and owner instructions could lead to meaningful differences, thus such factors should still be considered when interpreting or comparing study outcomes. Furthermore, with these methods basic attachment style classifications are possible with a single dog and without the use of models or statistics. Overall the SBT appears to be a highly reliable and robust measure of dog–human attachment styles and the Secure Base Effect, with methodology well suited for both scientific and applied environments.

Although the SBT lacks a stranger condition, caregiver vs stranger preference can be assessed as a separate test (e.g., a paired attachment test, or a sociability test), or the SBT can be tested on another day with a stranger. These optional tests would still result in a total testing time under 21 min and help reduce concerns about the other potential confounds associated with the full SST. However, it is possible that stranger comparisons, especially for adult dogs, may not achieve what was intended in the original SST. Even young human children tend to display a strong preference for one individual only during a brief period of development (typically between 7 and 12 months of age); the development of multiple attachments by 18 months of age is common (Schaffer and Emerson 1964). In other words, secure individuals would be expected to be able to be comfortable around new people eventually, making stranger conditions generally less relevant after infancy. Furthermore, many domestic dogs are hyper-social, a trait that has been linked to genetic changes associated with domestication that result in exaggerated social responses and less avoidance of strangers (vonHoldt et al. 2017), consequently dogs often behave prosocially to new humans, and in some cases form attachments quickly (Gàcsi et al. 2001). Like humans, dogs in some populations also display *disinhibited attachment*, meaning that despite a stable attachment style toward their human caregiver they might not show more proximity seeking toward this individual versus a stranger, in at least some contexts (Thielke and Udell 2020). Therefore, stranger conditions may not always be necessary or desirable when the primary goal is to evaluate attachment security/the Secure Base Effect (stress reduction and contact-exploration balance) in the presence of a bonded individual.

Future Considerations

Several studies have investigated how dog–human attachment styles might influence dog welfare and success in working environments. For example, one study found that therapy dogs who were insecurely attached to their handler spent more time during a mock therapy session looking back towards their handler than securely attached dogs (Wanser and Udell 2019). Such factors might influence therapy participants' responses to their interactions with dogs (e.g., they may feel a dog that is looking away from them likes them less, or they may show less of an oxytocin response). Another study suggested that search and rescue dogs may have stronger attachments to their caregivers compared to pet dogs, perhaps due to the amount of time these working dogs spend with their human partner (Mariti et al. 2013). The SBT has also been used to evaluate attachment bonds between dogs in shelter and foster environments and their caretakers. Dogs in foster homes display similar attachment styles toward their caretaker as pets toward their owners, whereas greater differences exist between shelter dogs and pets, with fewer shelter dogs categorized as having a secure attachment to their primary caretaker. Understanding to what extent such outcomes are influenced by traits of the dog and the quality of relationships prior to relinquishment, versus a product of their current environmental circumstances will require more research. Attachment style also appears to correspond with measures of social cognition and persistence in shelter and foster settings (Thielke and Udell 2019, 2020). However, more research is needed in pet, working and shelter contexts to help determine any causal relationships and also to clarify how attachment style varies with social, cognitive, behavioral, and welfare measures.

Another area of interest may be pet "parenting styles," as beliefs and behaviors in how parents raise and care for their offspring have been linked to child attachment styles, and consequently a variety of cognitive well-being outcomes (Baumrind 1991). Recently, pet parenting styles have been identified among dog caretakers (van Herwijnen et al. 2018; Brubaker and Udell in prep), and preliminary evidence suggests that there is likely a relationship between pet parenting style and dog–human attachment style. More broadly, positive factors within a human–dog relationship (such as play, positive reinforcement training, and lack of punishment) have been found to profoundly influence the human–dog bond, including correlations between attachment reported by the human caregiver and pro-social behaviors by the dog, dog training success, and reduced problem behaviors in dogs (Hiby et al. 2004; Rehn et al. 2013, 2014, 2017; Rooney and Bradshaw 2002, 2003; Rooney and

Cowan 2011). Therefore, with more research it may be possible to predict likely attachment outcomes for dog-human pairs based on parenting style and human behavior, and perhaps to educate humans on best practices for establishing a secure relationship with their dog, similar to the promotion of positive parenting practices when parenting human children.

Although less research exists on how dog attachment styles specifically influence canine behavior in dimensions such as problem-solving, aggression, and pro-social behavior towards strangers, at least one study has found a similar trend in dogs, with human caregiver "parenting" behavior being connected to dog attachment styles (Brubaker and Udell in prep), social behavior, and problem-solving ability (Brubaker and Udell in prep). An important next step will be to use this methodology to further evaluate the external validity of attachment styles in a wider variety of experimental and naturalistic settings, including working dog success, prediction and modification of problem behaviors, placement success, and mutual welfare outcomes associated with strengthening the dog–human bond.

Acknowledgements We would like to thank all the participant volunteers and research assistants in the Human Interaction Lab, including but not limited to Holly Duvall and Champayne Master for coding work related to this project. Owner consent for dog participation was obtained for all dogs in the original studies that allowed for the collection of these videos. The research protocols and consent processes for these studies were in accordance with the ethical standards of the United States and Oregon State University, approved under Oregon State University ACUP # 4837.

References

Adachi I, Kuwahata H, Fujita K (2007) Dogs recall their owner's face upon hearing the owner's voice. Anim Cogn 10(1):17–21

Ainsworth MDS, Bell SM (1970) Attachment, exploration, and separation: illustrated by the behavior of one-year-olds in a strange situation. Child Dev 41(1):49–67

Arahori M, Kuroshima H, Hori Y, Takagi S, Chijiiwa H, Fujita K (2017) Owners' view of their pets' emotions, intellect, and mutual relationship: cats and dogs compared. Behav Process 141:316–321

Archer J (1997) Why do people love their pets? Evol Hum Behav 18(4):237–259

Baumrind D (1991) Parenting styles and adolescent development. In: Brooks-Gunn J, Lerner RM, Petersen AC (eds) The encyclopedia on adolescence. Garland Publishing, New York, pp 746–758

Bowlby J (1958) The nature of the child's tie to his mother. Int J Psychoanal 39(5):350–373

Bowlby J (1982) Attachment and loss: attachment, vol 1, 2nd edn. Basic Books, New York

Bretherton I (1992) The origins of attachment theory: John Bowlby and Mary Ainsworth. Dev Psychol 28(5):759–775

Brubaker L, Udell MAR (in prep). Pet Parenting and Canine Behavior

Chijiiwa H, Kuroshima H, Hori Y, Anderson JR, Fujita K (2015) Dogs avoid people who behave negatively to their owner: third-party affective evaluation. Anim Behav 106:123–127

Coppinger R, Coppinger L, Beck AM (2016) What is a dog? The University of Chicago Press, Chicago, IL

Cross H, Harlow H (1965) Prolonged and progressive effects of partial isolation on the behavior. J Exp Res Pers 1(1):39–49

Gàcsi M, Topàl J, Miklòsi À, Dòka A, Csànyi V (2001) Attachment behavior of adult dogs (*Canis familiaris*) living at rescue centers: forming new bonds. J Comp Psychol 115(4):423–431
Harlow HF (1958) The nature of love. Am Psychol 13(12):673–685
Hiby EF, Rooney NJ, Bradshaw JWS (2004) Dog training methods: their use, effectiveness and interaction with behaviour and welfare. Anim Welf 13:63–69
Hori Y, Kishi H, Inoue-Murayama M, Fujita K (2013) Dopamine receptor D4 gene (DRD4) is associated with gazing toward humans in domestic dogs (*Canis familiaris*). Open J Anim Sci 3 (1):27422
Horn L, Huber L, Range F (2013) The importance of the secure base effect for domestic dogs—evidence from a manipulative problem-solving task. PLoS One 8(5):e65296
Lorenz K (1935) Der Kumpan in der umwelt des Vogels. Der Artgenosse als auslösendes moment sozialer Verhaltungsweisen [the companion in the bird's world. The fellow-member of the species as releasing factor of social behavior]. J Ornithol 83:137–213
Lorenz K (1952) King Solomon's ring: new light on animal ways. Methuen, London
Mariti C, Ricci E, Carlone B, Moore JL, Sighieri C, Gazzano A (2013) Dog attachment to man: a comparison between pet and working dogs. J Vet Behav Clin Appl Res 8(3):135–145
Miklósi Á, Topál J, Csányi V (2007) Big thoughts in small brains? Dogs as a model for understanding human social cognition. Neuroreport 18(5):467–471
Nobis G (1979) Der älteste Haushunde lebte vor 14000 Jahren. [the oldest domestic dog lived 14,000 years ago]. Umschau 79:610
Palmer R, Custance D (2008) A counterbalanced version of Ainsworth's strange situation procedure reveals secure-base effects in dog–human relationships. Appl Anim Behav Sci 109 (2–4):306–319
Parthasarathy V, Crowell-Davis SL (2006) Relationship between attachment to owners and separation anxiety in pet dogs (*Canis lupus familiaris*). J Vet Behav Clin Appl Res 1(3):109–120
Prato-Previde E, Custance DM, Spiezio C, Sabatini F (2003) Is the dog-human relationship an attachment bond? An observational study using Ainsworth's strange situation. Behaviour 140 (2):225–254
Rehn T, Beetz A, Keeling LJ (2017) Links between an owner's adult attachment style and the support-seeking behavior of their dog. Front Psychol 8:2059
Rehn T, Lindholm U, Keeling L, Forkman B (2014) I like my dog, does my dog like me? Appl Anim Behav Sci 150:65–73
Rehn T, McGowan RTS, Keeling LJ (2013) Evaluating the strange situation procedure (SSP) to assess the bond between dogs and humans. PLoS One 8(2):e56938
Rooney NJ, Bradshaw JW (2002) An experimental study of the effects of play upon the dog–human relationship. Appl Anim Behav Sci 75(2):161–176
Rooney NJ, Bradshaw JW (2003) Links between play and dominance and attachment dimensions of dog-human relationships. J Appl Anim Welf Sci 6(2):67–94
Rooney NJ, Cowan S (2011) Training methods and owner–dog interactions: links with dog behaviour and learning ability. Appl Anim Behav Sci 132(3–4):169–177
Rutter M (1998) Developmental catch-up, and deficit, following adoption after severe global early privation. J Child Psychol Psychiatry 39(4):465
Schaffer HR, Emerson PE (1964) The development of social attachments in infancy. Monogr Soc Res Child Dev 29(3):1–77
Schöberl I, Beetz A, Solomon J, Wedl M, Gee N, Kotrschal K (2016) Social factors influencing cortisol modulation in dogs during a strange situation procedure. J Vet Behav Clin Appl Res 11:77–85
Suomi SJ, Collins ML, Harlow HF (1973) Effects of permanent separation from mother on infant monkeys. Dev Psychol 9(3):376
Thielke LE, Rosenlicht G, Saturn SR, Udell MAR (2017) Nasally-administered oxytocin has limited effects on owner-directed attachment behavior in pet dogs (*Canis lupus familiaris*). Front Psychol 8:1699

Thielke LE, Udell MAR (2019) Evaluating cognitive and behavioral outcomes in conjunction with the secure base effect for dogs in shelter and foster environments. Anim Open Access J MDPI 9 (11):932
Thielke LE, Udell MAR (2020) Characterizing human–dog attachment relationships in foster and shelter environments as a potential mechanism for achieving mutual wellbeing and success. Animals 10(1):67
Topál J, Miklósi Á, Csányi V, Dóka A (1998) Attachment behavior in dogs (*Canis familiaris*): a new application of Ainsworth's (1969) strange situation test. J Comp Psychol 112(3):219–229
Udell MAR, Brubaker L (2016) Are dogs social generalists? Canine social cognition, attachment, and the dog-human bond. Curr Dir Psychol Sci 25(5):327–333
Udell MAR, Dorey NR, Wynne CDL (2010) What did domestication do to dogs? A new account of dogs' sensitivity to human actions. Biol Rev 85(2):327–345
van der Horst FCP, LeRoy HA, van der Veer R (2008) "When strangers meet": John Bowlby and Harry Harlow on attachment behavior. Integr Psychol Behav Sci 42(4):370–388
van Herwijnen I, van der Borg J, Naguib M, Beerda B (2018) The existence of parenting styles in the owner-dog relationship. PLoS One 13(2):e0193471
Vitale KR, Behnke AC, Udell MAR (2019) Attachment bonds between domestic cats and humans. Curr Biol 29(18):R864–R865
vonHoldt BM, Shuldiner E, Koch IJ, Kartzinel RY, Hogan A, Brubaker L, Wanser S, Stahler D, Wynne CDL, Ostrander EA, Sinsheimer JS, Udell MAR (2017) Structural variants in genes associated with human Williams-Beuren syndrome underlie stereotypical hypersociability in domestic dogs. Sci Adv 3(7):e1700398
Wanser SH, Udell MAR (2019) Does attachment security to a human handler influence the behavior of dogs who engage in animal assisted activities? Appl Anim Behav Sci 210:88–94
Waters E (1978) The reliability and stability of individual differences in infant-mother attachment. Child Dev 49(2):483–494

Index

J. R. Anderson, H. Kuroshima (eds.), *Comparative Cognition*,
https://doi.org/10.1007/978-981-16-2028-7

Kolibri 360 GmbH
Ferdinand-Jühlke-Straße 7
99095 Erfurt, Deutschland
produktsicherheit@kolibri360.de